FESTKÖRPERPROBLEME

ADVANCES IN SOLID STATE PHYSICS 29

FESTKÖRPER PROBLEME
ADVANCES IN SOLID STATE PHYSICS 29

Plenary Lectures of the Divisions

Semiconductor Physics
Thin Films
Dynamics and Statistical Physics
Magnetism
Metal Physics
Surface Physics
Low Temperature Physics

of the German Physical Society (DPG),
Münster, April 3 to 7, 1989

Edited by
Ulrich Rössler

With 204 Figures

Springer Fachmedien Wiesbaden GmbH

Editor:

Prof. Dr. Ulrich Rössler
Institut für Theoretische Physik
Universität Regensburg
P.O. Box 397
D-8400 Regensburg, FRG

Originally published by Friedr. Vieweg & Sohn Verlagsgesellschaft mbH, Braunschweig in 1989.

Printed by Lengericher Handelsdruckerei, Lengerich
Bound by Hunke + Schröder, Iserlohn
Cover design: Barbara Seebohm, Braunschweig

ISBN 978-3-662-16068-8   ISBN 978-3-540-75350-6 (eBook)
DOI 10.1007/978-3-540-75350-6

Foreword

The annual spring meeting of the Solid State Physics Division (Arbeitskreis Festkörperphysik) of the German Physical Society took place in 1989 once more in Münster. As in previous years the plenary and invited talks presented at this meeting attracted and hold together the larger part of the attendents every day, which otherwise were dispersed over the numerous parallel sessions. A selection of these plenary and invited talks is the content of this volume of Advances in Solid State Physics/ Festkörperprobleme. They are representative for the widespread activities in condensed matter physics and signalize particular achievements and concentration of research activities on special topics.

The Walter-Schottky Prize 1989 was shared by three scientists, U. Eckern, G. Schön and W. Zwerger, who received the award for their theoretical contributions to the understanding of quantum effects in superconducting tunnel-contacts and granular films. Pattern formation in liquid crystals, local probing of atomic arrangements in condensed matter by EXAFS, surface studies by scanning-tunnel microscopy, anisotropic propagation of heat pulses, lattice dynamics of surfaces including reconstructions, and fast relaxation of excitons and free carriers in semiconductors are topics of individual contributions. Two groups of papers, presented mainly in symposia during the conference, cover the current investigations of deep centers in semiconductors and the fabrication and physics of low dimensional semiconductor structures.

In order to lower the cost of production and likewise the market prize this volume is produced from camera-ready manuscripts. On the other hand, the intention was to conserve the homogeneous appearance as a printed book. Thus, the authors had to carry also the burden of preparing a TeX-nically perfect manuscript. Not all of them succeeded and their manuscripts had to be rewritten, at least partially. This tremendous work has been done with great care by Angela Reisser. The TeX-instructions have been prepared by Dr. Franz Malcher. Their help and effort was essential for the completion of this volume. Finally, I acknowledge the ever smooth cooperation with Björn Gondesen (Vieweg-Verlag).

Regensburg, April 1989 *Ulrich Rössler*

Contents

Charge Transfer between Weakly Coupled Normal Metals and Superconductors at Low Temperatures

Ulrich Eckern

Institut für Theorie der Kondensierten Materie, Universität Karlsruhe, P.O.Box 6980, D-7500 Karlsruhe, Federal Republic of Germany

Gerd Schön

Department of Applied Physics, Delft University of Technology, Lorentzweg 1, NL-2628 CJ Delft, The Netherlands

Summary: We give a review of recent advances in the theoretical description of ultrasmall tunnel junctions, i.e. weakly coupled normal metals and superconductors, at very low temperatures. The description, which is based on the microscopic theory, accounts for the quantum effects associated with Cooper pair tunneling, and for the dissipation due to single electron tunneling (SET) or normal current flow. The former is described by the familiar periodic potential which, in the quantum regime, leads to energy bands, and to coherent Bloch oscillations in response to an external current. The stochastic SET, which may be blocked by the Coulomb interaction, yields voltage oscillations even in normal junctions, and also modifies the Bloch oscillations in Josephson junctions. The theory is extended to networks and granular superconducting films as well.

1 Introduction

During the past ten years, a great deal of theoretical as well as experimental efforts have been devoted to what is now commonly called Quantum Mechanics of Macroscopic Variables [1]. The experimental system which has been studied extensively is a Josephson junction, i.e. a system of two weakly coupled superconductors. It is well known that the relevant variable for the description of the dynamics of a Josephson junction is the difference φ of the phases of the superconducting order parameters in the two electrodes. If the capacitance of the junction is large enough, the dynamics of the phase is determined by the following classical equations:

$$\hbar\dot{\varphi} = 2eV \; ; \quad C\dot{V} + V/R + I_c\sin\varphi = I_x \tag{1}$$

where the first is Josephson's equation, and the second expresses the balance of currents; V, I_x, C, R, and I_c denote the voltage, the external current, the capacitance, the resistance, and the critical current, respectively. Obviously, Eq. (1) describes the damped motion of a particle in a tilted periodic potential $U(\varphi)$, given by

$$U(\varphi) = -E_J\cos\varphi - \hbar I_x \cdot \varphi/2e \tag{2}$$

where the Josephson energy $E_J = \hbar I_c/2e$ and the charging energy $E_C = e^2/2C$ are the important scales of energy. Generally, R^{-1} can be thought of as a sum of two contributions, $R^{-1} = R_t^{-1} + R_s^{-1}$, where R_t is the (strongly temperature dependent) resistance due to quasiparticle tunneling, and R_s denotes an external shunt (if present).

Of course, a description based on Eq. (1) is incomplete in several ways. First of all, it is a general theorem of statistical mechanics that dissipation and fluctuations are intimately related; for the present case, this means that I_z in Eq. (1) has to be replaced by $I_z + \delta \tilde{I}$, where $\delta \tilde{I}$ denotes a random contribution to the current. Considering the tunneling of electrons, it was noted eighty years ago by Schottky [2] that in situations, in which the energy gained by tunneling from one metal to another is much larger than the thermal energy, i.e. $eV >> kT$, the resulting *Schroteffekt* or shot noise can considerably exceed the thermal noise. Shot noise is a direct consequence of the charge being transported in elementary units; thus it proved to be an independent (and at that time [2] quite accurate) possibility to measure the charge of an electron. The classical stochastic description of a normal tunnel junction, and in particular the influence of the Coulomb interaction (Coulomb blockade) on the dynamics, is discussed in the next chapter.

Secondly, the classical description fails in the quantum regime of very low temperatures, where the phase has to be considered an operator-valued quantity, with the charge $\hat{Q}$ being its canonically conjugate variable [3] according to the commutator relation $[\hat{Q}, \hat{\varphi}] = -2e \cdot i$. In the non-dissipative case it is obvious that a discussion of quantum mechanical effects is to be based on the Hamiltonian

$$\mathcal{H} = \hat{Q}^2/2C + U(\hat{\varphi}) \, . \tag{3}$$

On the other hand, the general treatment [4], which starts from microscopic theory, proceeds most conveniently by working with path integral methods in the Lagrangian formulation. In particular, an earlier approach [5] in which the coupling to the environment, i.e. the microscopic degrees of freedom, is modeled in a phenomenological way, can be confirmed and extended [4]. The microscopic theory of weakly coupled superconductors is briefly reviewed in Ch. 3. In the final two chapters, we describe some of the consequences of the quantum description: macroscopic quantum tunneling [6], energy levels [7], macroscopic quantum coherence, Bloch oscillations [8, 9], and further consequences of single electron tunneling (SET), and give a brief discussion of granular superconducting films and networks of Josephson junctions [10]. A detailed review is in preparation [11].

2 Normal Tunnel Junctions

We first review the classical description of stochastic tunneling of electrons in a normal junction, which proceeds analogous to [2]. The tunneling current is composed of statistically independent SET processes, forward and backward across the barrier, at random times $\{t_i^+\}$ and $\{t_j^-\}$. Since the time for tunneling is negligible, the current is given by a sum of delta-functions as follows:

$$I(t) = e \sum_i \delta(t - t_i^+) - e \sum_j \delta(t - t_j^-) \tag{4}$$

The statistics governing these processes is Poissonian, characterized by the rates ν^+ and ν^-. Thus the probability density for n particles to tunnel forward at times $0 < t_1^+, \ldots, t_n^+ < t_f$ is given by

$$P_n^+(t_1^+, \ldots, t_n^+) = e^{-\nu^+ t_f} \cdot (\nu^+)^n / n! \tag{5}$$

and similar for the backward tunneling. Detailed balance determines the ratio $\nu^+/\nu^- = \exp(-\epsilon/kT)$, where ϵ is the energy difference needed for the forward tunneling process. The average current and the current fluctuations are expressed by ν^+ and ν^- as follows:

$$< I(t) >= e(\nu^+ - \nu^-) \tag{6}$$

$$< \delta I(t)\delta I(t') >=< I(t)I(t') > - < I >^2 = e^2 \delta(t - t') \cdot (\nu^+ + \nu^-) \ . \tag{7}$$

Let us assume that for a fixed voltage, which means fixed energy difference $\epsilon^\pm = \pm eV$, the tunneling current voltage characteristic $I_t(V)$ is known (in simple cases it is given by $I_t(V) = V/R_t$). This provides a relation for the average current, $I_t(V) = e(\nu^+ - \nu^-)$, and combined with the detailed balance requirement it fixes the rates to

$$\nu^\pm = \frac{1}{e} I_t(\frac{\epsilon^\pm}{e}) \big[\exp(\frac{\epsilon^\pm}{kT}) - 1\big]^{-1} \ . \tag{8}$$

Thus the fluctuations of the voltage biased junction are given by

$$< \delta I(t)\delta I(0) >= e\delta(t)I_t(V)\coth(\frac{eV}{2kT}) \tag{9}$$

which obviously interpolates between Schottky's result and the standard Nyquist formula for thermal noise ($eV >> kT$ and $eV << kT$, respectively). After generalization to the quantum mechanical form, the power spectrum is found to be given by [12]

$$S_I(\omega) = \frac{e}{4\pi} \sum_\pm I_t(V \pm \hbar\omega/e)\coth(\frac{eV \pm \hbar\omega}{2kT}) \ . \tag{10}$$

In more general situations, the energy is not fixed by the external voltage but depends on the state of the system. In particular in junctions with very small capacitance, the energy is dominated by the charging energy $E(Q) = Q^2/2C$, which changes in the tunneling process since $Q \to Q \pm e$. The detailed analysis [13] shows that the tansition rates are still given by Eq. (8), however, the energy differences depend on the charge according to

$$\epsilon^{\pm}(Q) = E(Q \pm e) - E(Q) \ . \tag{11}$$

At zero temperature the rates reduce to

$$\nu^{\pm}(Q, T = 0) = e^{-1} \cdot I_t(|\epsilon^{\pm}|/e) \quad \text{for } \epsilon^{\pm} < 0 \tag{12}$$

and vanish for $\epsilon^{\pm} > 0$. Thus tunneling occurs only if during the process the energy decreases, i.e. only for charges exceeding $|Q| > e/2$. This phenomenon is called the *Coulomb blockade*. At finite temperature, also transitions occur by thermal excitation which are unfavourable with regard to energy.

We consider now a normal tunnel junction with $E(Q) = Q^2/2C$ which is biased by an external current $I_x(t)$. This current increases the charge on the junction electrodes in a deterministic, continuous fashion. But the charge changes also stochastically, in discrete units of e due to tunneling with rates given by Eq. (8). Thus we may write

$$dQ/dt = I_x(t) + \dot{Q}|_{tunneling} \ . \tag{13}$$

The statistical properties of the last term on the rhs have been discussed above. Because of the discrete nature of the tunneling, the allowed values of the charge Q at each moment can take values only from the discrete set $Q = Q_x(t) + q \cdot e$, where q is an integer, and

$$Q_x(t) = \int_{-\infty}^{t} dt' \, I_x(t') \ . \tag{14}$$

The segmentation of the current in units of e and the related charging and discharging of the junction leads to an oscillatory time dependence of the voltage across the junction. On the average, it oscillates with the fundamental frequency [13]

$$f_{SET} = I_x/e \ . \tag{15}$$

The strict discreteness of the tunneling, combined with the Coulomb interaction which favours transitions to states with small charge, leads also to long-time correlations in $< Q(t)Q(0) >$ which are noticable for small currents and low temperatures [13, 11].

From the time averaged charge we obtain the *dc* current-voltage characteristic which is shown in Fig. 1. In particular, the voltage for small currents is

$$V_0 = (\pi I_x R_t e/2C)^{1/2} \qquad \text{for} \qquad I_x R_t C << e, \ T = 0 \tag{16}$$

whereas at large currents the *I-V*-curve becomes linear but shifted with respect to the Ohmic line:

$$V_0 = I_x R_t + \text{sign} I_x \cdot e/2C \qquad \text{for} \qquad I_x R_t C >> e \ . \tag{17}$$

4

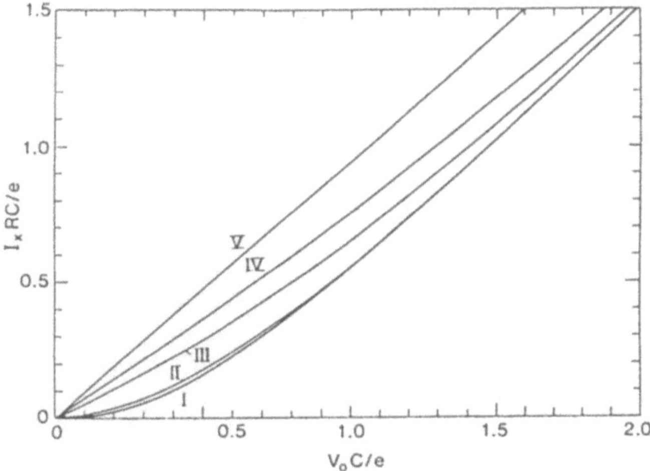

Fig. 1 The current voltage characteristic of a normal junction driven by a *dc* current I_x is shown for different temperatures. (I) to (V): $kT/E_C = 0.01$, 0.1, 0.5, 1, and 5. At low temperature a pronounced deviation from an Ohmic form is found.

It extrapolates to a nonzero *dc* voltage $V_g = e/2C$ at $I_x = 0$ which is called the *Coulomb gap*. The Coulomb gap and other manifestations of the discrete nature of single electron tunneling have been observed by now in several experiments on small capacitance junctions [14] and granular materials.

If the junction in addition is shunted by a parallel Ohmic resistor R_s, the junction also discharges continuously, which can be accounted by the following equation:

$$dQ/dt = I_x(t) + \dot{Q}|_{tunneling} - Q/R_s C + \delta\tilde{I}(t) \ . \tag{18}$$

Here $\delta\tilde{I}(t)$ denotes the Gaussian current noise of the shunt resistor, whose power spectrum is given by

$$S_I(\omega) = (2\pi)^{-1} < \delta\tilde{I}(t)\delta\tilde{I}(t') >_\omega = kT/\pi R_s \ . \tag{19}$$

A weak shunt mostly modifies the low voltage part of the *I-V*-characteristics. The noise associated with it destroys the long-time correlations in the charge correlation function $< Q(t)Q(0) >$.

3 Microscopic Theory of Weakly Coupled Superconductors

The derivation of an effective action for Josephson junctions [4] from microscopic theory has been described in detail in the literature, and will only briefly be summarized here. The starting point is the Hamiltonian

$$\mathcal{H} = \mathcal{H}_L + \mathcal{H}_R + \mathcal{H}_Q + \mathcal{H}_t \tag{20}$$

where $\mathcal{H}_{L,R}$ are appropriate model Hamiltonians describing the two electrodes of the junction, including the attractive interaction which leads to superconductivity; $\mathcal{H}_Q$ describes the capacitive Coulomb interaction between charges accumulating in the vicinity of the oxide barrier, and $\mathcal{H}_t$ is the coupling (tunnel Hamiltonian) given by

$$\mathcal{H}_t = \sum_{\vec{k},\vec{q}} (t\, C_{\vec{k}}^+ C_{\vec{q}} + h.c.) \; . \tag{21}$$

Here $(C_{\vec{k}}^+, C_{\vec{k}})$ and $(C_{\vec{q}}^+, C_{\vec{q}})$ denote the electron creation and annihilition operators of the two electrodes, respectively, and the spin is omitted for brevity. It turns out to be convenient to start from the path integral representation for the partition function

$$Z = \mathrm{tr}\,\exp(-\beta\mathcal{H}), \quad \beta = (kT)^{-1} \tag{22}$$

and handle the interaction terms by introducing auxiliary fields. Of course, this procedure is ambiguous, and we have to resort to physical intuition, which means that the auxiliary fields should be chosen in a sensible way. Considering a single superconductor, it is known that the pairing interaction can be treated by introducing a complex order parameter field, Δ, such that the standard BCS theory is recovered from the saddle point of the resulting path integral. [This is a possibility to define the meaning of "sensible way" used above.] For the present model, we thus arrive at a description involving two order parameter fields, Δ_L and Δ_R, and a voltage field V to represent the Coulomb interaction, such that

$$Z = \int \mathcal{D}\Delta_L \mathcal{D}\Delta_R \mathcal{D}V \; e^{-S_{eff}/\hbar} \tag{23}$$

where S_{eff} depends on $\Delta_L(\vec{r},\tau)$, $\Delta_L(\vec{r},\tau)$, and $V(\vec{r},\tau)$. However, the analysis shows that further simplifications are possible. First of all, except in a very tiny temperature interval close to the critical temperature, fluctuations of the magnitude of the order parameters are negligible. Secondly, the coupling may be assumed to be weak such that an expansion with respect to $\mathcal{H}_t$ is possible. Introduce the dimensionless parameter α by

$$\alpha = R_0/R_N = 2\pi^2 |t|^2 \mathcal{N}_L(0)\mathcal{N}_R(0) \tag{24}$$

where $R_0 = h/4e^2 = 6.45\ k\Omega$, R_N is the normal state resistance, and $\mathcal{N}_{L,R}(0)$ the normal state density of states of the two electrodes at the Fermi surface. Then weak in the above sense means $\alpha << \mathcal{A}\cdot k_F^2$, where $\mathcal{A}$ is the area of the interface and k_F the Fermi wavevector; note that $\mathcal{A}k_F^2 >> 1$. Thirdly, the assumption of local equilibrium in the electrodes leads to Josephson's relation $\hbar\dot{\varphi} = 2eV$. Again, deviations from this relation are suppressed by energies proportional to the volume of the electrodes, and can be ignored. As a final result, we arrive at a description in terms of the relevant variable (namely the phase difference φ) alone, and we obtain the following representation:

$$Z = \int \mathcal{D}\varphi \, e^{-S[\varphi]/\hbar} \; . \tag{25}$$

Here $S = S_0 + S_1 + S_2$, with

$$S_0 = \int_0^{\hbar\beta} d\tau \left[\frac{\hbar^2 C}{8e^2} \dot{\varphi}^2 + U(\varphi) \right] \tag{26}$$

where the potential is given in Eq. (2), and

$$S_1 = \hbar \int_0^{\hbar\beta} d\tau \int_0^{\hbar\beta} d\tau' \alpha(\tau - \tau') \left[1 - \cos\frac{\varphi(\tau) - \varphi(\tau')}{2} \right] \; . \tag{27}$$

The contribution S_2, which is classically related to the "$\cos\varphi$" term in the equation of motion (which has been omitted in Eq. (1)), has (up to now) not led to any important consequences in the quantum behavior, and thus will not be discussed here. The contribution S_1 represents the single electron tunneling, which becomes evident by noting that the Fourier transform, $\alpha(\omega_j)$, where $\omega_j = 2\pi j/\hbar\beta$ denotes the Matsubara frequency, is related to the normal current, $I_n(V)$, as follows:

$$\alpha(\omega_j) = \frac{1}{e} \int \frac{d\nu}{2\pi} \frac{\nu}{\nu^2 + \omega_j^2} I_n\left(\frac{\hbar\nu}{e}\right) \; . \tag{28}$$

Thus, considering identical electrodes for simplicity, we find for ideal junctions in the zero temperature limit (in which case $I_n(V) = 0$ for $V < 2|\Delta|/e$), that $\alpha(\tau)$ decreases exponentially $\sim \exp(-2|\Delta|\tau/\hbar)$ for large times; this clearly reflects the gap in the single particle excitation spectrum. On the other hand, for short times such that $|\Delta|\tau << \hbar$, we obtain $\alpha(\tau) = \alpha/(\pi\tau)^2$, which is the characteristic behavior also found [5] in the case of an Ohmic shunt. Note that Eq. (28) covers both cases and in addition more general ones, provided we consider $I_n(V)$ an experimentally determined input for the theory. We mention that, after the fact, other "microscopic" models can be devised [15,16] which reduce to the effective action given above. As an important point, these models involve coupling $\sin(\varphi/2)$ and $\cos(\varphi/2)$ to a suitably chosen environment, in contrast to φ in the phenomenological treatment [5], the latter nevertheless being adequate in cases where a shunt resistor is present.

We emphasize that the appearance of the trigonometric functions in Eqs. (26) and (27) is intimately related to the discreteness of charge transfer, which is also directly connected with the boundary conditions in the path integral (25), and with the allowed eigenvalues of the charge operator [15]. As an illustration, consider S_0 only, and the simplest choice (i): $\varphi(0) = \varphi(\hbar\beta)$; consequently, Eq. (25) can be interpreted as the partition function of a *particle* in the potential $U(\varphi)$, and the corresponding *momentum* (i.e. the charge) can take values from a continuous set. The important point is that values of φ differing by 2π are considered *distinguishable*, which in fact is realized in the presence of an external current source or a shunt resistor, or in a SQUID where the energy of the enclosed magnetic flux adds a term quadratic in φ.

However, taking $I_x = 0$ we realize that S_0 is invariant under $\varphi \to \varphi + 2\pi$; and, indeed, there are situations where states differing by 2π are *indistinguishable*. This leads to choice (ii): $\varphi(0) = \varphi(\hbar\beta) + 2\pi n$, where n is an integer; summation over n has to be included in Eq. (25) in this case. In usual quantum mechanical language, φ is then a true *phase*, i.e. the wave-functions are 2π periodic, and the conjugate variable corresponds to an *angular momentum*. Thus the eigenvalues of the charge operator are found to be given by $2e \cdot q$, q being an integer, reflecting the charge transfer in units of $2e$ through Cooper pair tunneling. Finally, taking SET described by S_1 into account, we realize that only states differing by 4π are indistinguishable. The consistent choice for this case is (iii): $\varphi(0) = \varphi(\hbar\beta) + 4\pi n$. As a consequence, neighboring minima of the periodic potential are distinct, however, second nearest neighbors may be considered equivalent. We conclude similar to case (ii), if SET is important, that the eigenvalues of $\hat{Q}$ are $e \cdot q$. The above considerations can be generalized to the presence of an external charge Q_x [15], and are decisive for the phenomenon of Bloch oscillations [13].

Finally, we remark that further detailed microscopic investigations, in particular with regard to the external current contribution, and of other weak links, are reviewed in [11].

4 Quantum Phenomena in Josephson Junctions

4.1 Macroscopic Quantum Tunneling (MQT)

Considering a Josephson junction biased by an external current $I_x < I_c$, it is clear that classically the phase is trapped in one of the (metastable) minima of the potential. Due to thermal noise, the system can escape over the barrier, which is visible experimentally as a transition to a nonzero voltage state; the decay rate for this process is given by the standard thermal activation formula. As a result of quantum mechanics, however, the junction can escape from one of the local minima even at zero temperature, by "macroscopic quantum tunneling" through a region which is classically not accessible. It has been pointed out that the rate for the tunneling process decreases with increasing dissipation; the result for $T = 0$ and weak damping is given by [5]

$$\Gamma_q = c\,\omega_0 \exp\left[-a\,\frac{\Delta U}{\hbar\omega_0}\left(1 + \frac{b}{RC\omega_0}\right)\right] \ . \tag{29}$$

Here ω_0 is the oscillation frequency in the local minimum, ΔU is the potential barrier, and a, b, c are numerical coefficients depending on the form of the potential. Typical parameters for which quantum tunneling dominates over thermal activation are $C \leq 10^{-12}F$ and $T \leq 0.1K$; also, for $E_C << E_J$ the current has to be chosen close to I_c in order for quantum tunneling to occur at an observable rate. Eq. (29) has been confirmed quantitatively in experiments [6].

4.2 Energy Levels

Another immediate consequence of quantum mechanics is (for weak dissipation) the appearance of discrete energy levels. Though in the tilted washboard potential of a current biased junction these states are not true eigenstates, they are still well defined since the tunneling rate is small. The levels show up as resonances if the frequency f_z of an external perturbation matches an energy difference, i.e. $hf_z = E_{n+1} - E_n$. On resonance, the probability for transitions out of a metastable minimum increases [7].

4.3 Bloch Oscillations

In an unbiased junction, we can realize a perfectly periodic potential (see Eq. (2)). If, in addition, the capacitance is in the range $C < 10^{-15}F$, i.e. smaller than typically in the MQT and energy level experiments, the wavefunction is not longer sharply localized in a single minimum. Hence, the appropriate boundary conditions (see Ch. 3) become crucial. Formally, the eigenstates of the Hamiltonian (3) (with $I_z = 0$) are Bloch states depending on the parameter Q_x, which we denote as "quasi-charge" in analogy to the quasi-momentum of a Bloch state in a periodic potential:

$$\psi_n(\varphi + 2\pi) = e^{i2\pi Q_x/2e}\psi_n(\varphi) \ . \tag{30}$$

The energy levels thus develop into bands [13], $E_n(Q_x)$, and the equivalent of the first Brillouin zone extends between $-e < Q_x < e$. The detailed form of $E_n(Q_x)$ depends on the ratio E_C/E_J. In the limit $E_J << E_C$, corresponding to the "nearly free electron" limit, the lowest energy band for small Q_x is $E_0(Q_x) = Q_x^2/2C$, and the bandsplitting at $|Q_x| = e$ is given by $E_1(e) - E_0(e) = E_J$.

The bands have a transparent physical interpretation. The quasi-charge is the charge provided in a continuous way by the external circuit, just as the external charge Q_x in a normal junction (see Eq. (14)). For Q_x close to $\pm e$, Cooper pair tunneling mixes the states with $Q_x = +e$ and $Q_x = -e$, leading to a reduction of the energy at the zone boundary. In addition, if Q_x is increased adiabatically beyond e, Cooper pairs tunnel (*Umklapp processes*) such that the actual charge remains small, $|Q_x| < e$, i.e. $E_0(Q_x) < E_C$. Considering the case where the junction is driven by a weak external current $I_x = \dot{Q}_x$, we conclude that the energy and the voltage oscillate with the fundamental frequency [13]

$$f_{\text{Bloch}} = I_x/2e \ . \tag{31}$$

These oscillations are the complete analog of Bloch oscillations of an electron in a periodic potential driven by an electric field. Note that, compared to the classical ac Josephson effect, the role of current and voltage are reversed. For example, if the junction is driven by a dc plus ac current, one expects to find resonance phenomena. In fact, Yosihiro et al. [8] see a sequence of activation steps in a granular superconductor, which they associate with this resonance. We remark that a similar though quantitatively different picture is obtained in the tight binding limit $E_J >> E_C$; for example, the width of the lowest band is found to be given by

$$\Delta_0 = 32\Big(\frac{E_J E_C}{\pi}\Big)^{1/2}\Big(\frac{E_J}{2E_C}\Big)^{1/4}\exp\Big[-(8\frac{E_J}{E_C})^{1/2}\Big] \tag{32}$$

and hence is exponentially small.

4.4 Weak Single Electron Tunneling

As long as the quasiparticle tunneling (as described microscopically by Eq. (27)) is weak, we can take it into account perturbatively, as first discussed by Averin and Likharev [13]. It allows stochastic transitions between states according to the selection rule $Q_x \leftrightarrow Q_x \pm e$ (in an extended zone scheme), with transition probabilities $\nu^{\pm}_{n,m}(Q_x)$ which are straightforward extensions of Eq. (8), in which $\epsilon^{\pm}$ is to be replaced by $\epsilon^{\pm}_{n,m} = E_m(Q_x \pm e) - E_n(Q_x)$. Note also that in general $I_t(V)$ depends on the superconducting gap, $|\Delta|$, though presently we assume for simplicity $I_t(V) = V/R_t$; here R_t is to be identified with the subgap conductance. As a dimensionless parameter, it is convenient to define $\alpha_t = 4R_0/\pi^2 R_t$.

As before, assume that the quasi-charge Q_x is increased adiabatically by an imposed dc current I_x. For small currents, $I_x << e/R_t C$, Q_x rarely reaches the boundary of the Brillouin zone, due to SET processes. Thus the situation is similar to a normal junction, with the result given in Eq. (16). For large currents, $I_x > e/R_t C$, on the other hand, Q_x is frequently driven to the zone boundary, and Bloch oscillations may develop. As a result, the dc voltage is given by

$$V_0 = E_C/(12 I_x R_t C) \qquad \text{for} \qquad I_x R_t C >> e \ . \tag{33}$$

The resulting I-V-characteristic is shown in Fig. 2(a) for $E_J << E_C$; it is apparent that the Coulomb gap of the normal junction is replaced by a nose-like structure. Typical current and voltage scales of this structure are given by $e/R_t C$ and the bandwidth E_C/e, respectively. A similar behavior is found in the limit $E_J >> E_C$, however, the scale is set by Δ_0 in this case. *A nose-like I-V-characteristic, i.e. a negative differential resistance regime, has very recently [9] been seen in experiments on very small capacitance junctions, which is the first direct observation of a consequence of Bloch oscillations.* Furthermore, for large currents, $I_x > e/R_t C$, we expect that resonance phenomena related to SET and Bloch oscillations, i.e. with

10

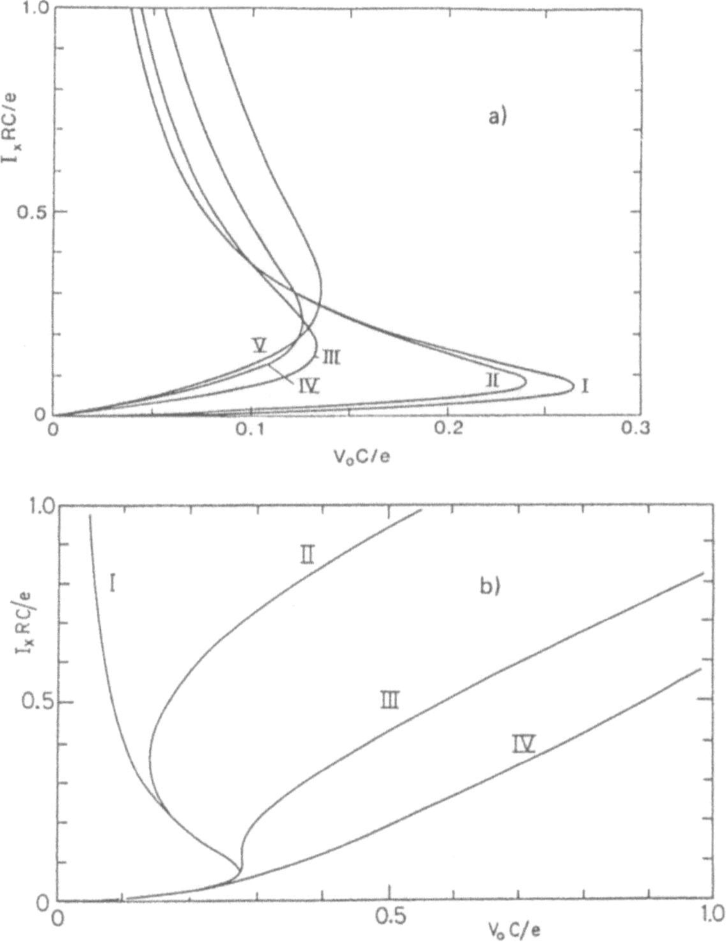

Fig. 2 The *dc* current voltage characteristic of a superconducting tunnel junction for $E_J = 0.2E_C$. (a) $\alpha_t \ll 1$, and different temperatures, (I) to (V): $kT/E_C = 10^{-3}$, 0.1, 0.5, 1, 5. (b) Same as (a), but $kT/E_C = 10^{-3}$ fixed, and (I) to (IV): $\alpha_t = 10^{-3}$, 0.01, 0.05, 0.5. The Zener tunneling increases with increasing α_t, suppressing the nose structure.

characteristic frequencies I_x/e and $I_x/2e$, respectively, show up when the junction is driven by a *dc* plus *ac* current.

Finally, we remark that Zener tunneling between neighboring bands leads to an additional modification, which can be taken into account for $E_J \ll E_C$ by the standard formula for the tunneling probability. As a result, it is found that Zener tunneling is only negligible for very small dissipation, namely $\alpha_t \ll (E_J/E_C)^2$. In the opposite case, one observes a transition from the nose-like *I-V*-characteristic, typical for a system where bandgaps prevent transitions to higher bands, to an I-V-characteristic of the shape similar to a normal junction (see Fig. 2(b)). At intermediate strength of Zener tunneling, the Coulomb gap appears both as a nose as well as a shift of the high voltage asymptote [17].

4.5 Strong Single Electron Tunneling

For strong quasiparticle tunneling, i.e. $\alpha_t \sim 1$, we have fully to take into account the contribution S_1 (see Eq. (27)) to the action, and in particular boundary conditions (iii). As a consequence, the Brillouin zone is reduced by a factor two, which also implies a doubling of the energy bands, now denoted by $E_{n,m}(Q_x)$, where $m = 0, 1$, and $|Q_x| < e/2$. The modified band picture is shown in Fig. 3(a) for $E_J << E_C$.

In order to account for quasiparticle tunneling, we can proceed e.g. in a renormalization procedure, i.e. we integrate out successively high frequencies, or in an instanton approximation. This leads to quantitative — and for large α_t to further qualitative — modifications of the bands [15]. For example, considering the energy difference between the two lowest bands, $\Delta_0(Q_x)$, it is necessary to introduce the corresponding renormalized quantity, $\Delta_r(Q_x)$. In the tight binding limit $E_J >> E_C$, where $\Delta_0(Q_x) = \Delta_0|\cos(\pi Q_x/e)|$, the renormalization procedure yields

$$\Delta_r(Q_x) = \Delta_0(Q_x)\Big[\Delta_0(Q_x)/2\hbar\omega_1\Big]^{\alpha_t/(\alpha_{t,c}-\alpha_t)} \quad \text{for} \quad \alpha_t < \alpha_{t,c} \qquad (34)$$

where $\hbar\omega_1 = (8E_J E_C)^{1/2}$, and $\Delta_r(Q_x) = 0$ for $\alpha_t > \alpha_{t,c}$. A phase transition occurs at the critical value

$$E_J/E_C >> 1: \quad \alpha_{t,c} = 1 + \Delta_0(Q_x)/\hbar\omega_1 \qquad (35)$$

which depends on Q_x, and the values of E_J and E_C. Above the transition, the energy difference between the two lowest bands vanishes. In the opposite limit $E_J << E_C$, the same phase transition is found, with the result that the renormalized band splitting vanishes in a finite region close to $|Q_x| = e/2$, for $\alpha_t > \alpha_{t,c} = 2$. On the other hand, the band splitting for $Q_x \simeq 0$ remains finite until α_t exceeds the value $2/\pi^2 \cdot \ln(AE_C/E_J)$, where $A \sim 1$. The phase diagram is shown in Fig. 3(b) (see [15] for further details).

The drastic modifications of the energy bands described above show up e.g. in the I-V-characteristic. In the tight binding limit $E_J >> E_C$, the nose structure, characteristic for weak dissipation, is found to disappear as the bandwidth decreases with increasing α_t. The phase transition is essentially a transition from strong quantum behavior — a finite dc plus ac voltage — to classical behavior with vanishing dc voltage. On the other hand, in the nearly free electron limit $E_J << E_C$, and for the interesting currents $I_x \sim e/R_t C$, the Zener tunneling gains importance with increasing α_t, before the phase transition at $\alpha_{t,c} > 1$ is reached. As a result, the characteristic changes from a nose shape to the shape as in a normal junction (see Fig.

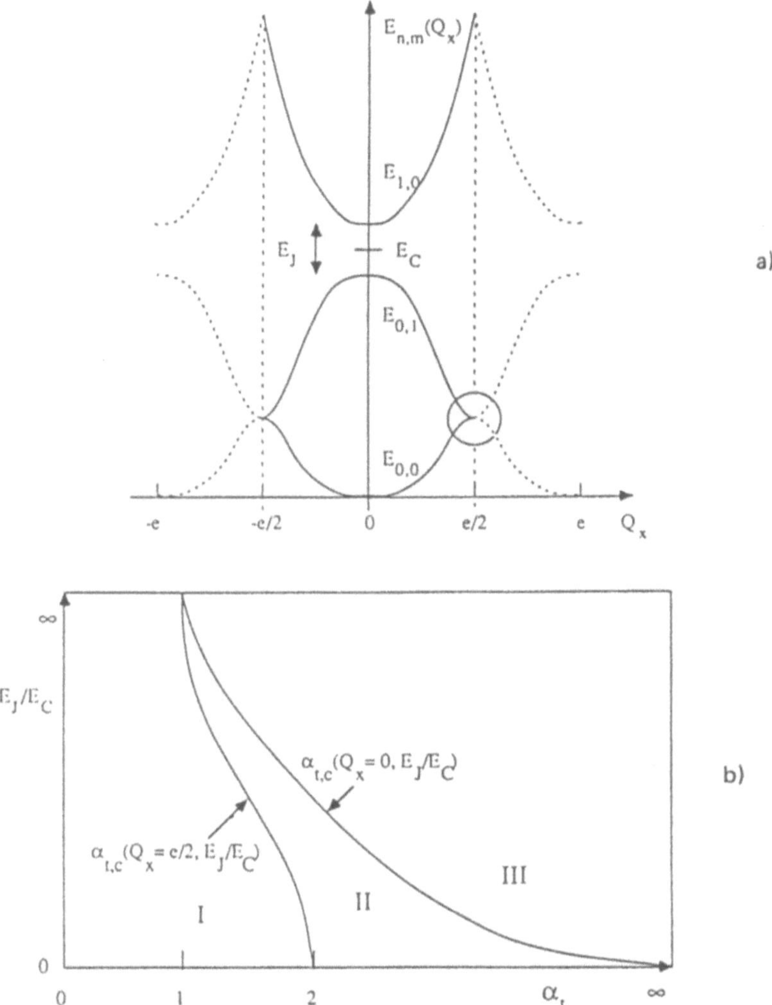

Fig. 3 (a) Overall picture of the energy bands in the "nearly free electron" limit $E_J \ll E_C$. (b) Phase diagram for strong quasiparticle tunneling. In region (I), the two lowest bands are degenerate only at $|Q_x| = e/2$; in (II), they are degenerate in a finite interval near the zone boundary; in (III), the bands are degenerate for all Q_x.

2(b)). Only at larger values, $1 < \alpha_t < \alpha_{t,c}$, the modifications of the energy bands begin to become important. The band gap increases, and hence the probability of Zener tunneling decreases, leading to a reappearance of the nose (but on a smaller current and voltage scale). Above the phase transition, the two lowest bands are degenerate. Eventually, for very large α_t, the classical superconducting response (i.e. zero voltage) is recovered [15, 11].

We emphasize that the described phase transition occurs as a function of the dimensionless subgap conductance α_t; in contrast, in the presence of a shunt resistor (and for $\alpha_t = 0$), a phase transition is found from quantum to classical behavior upon increasing $\alpha_s = R_0/R_s$. The critical value is $\alpha_{s,c} = 1$, independent of E_J/E_C [18].

5 Networks and Granular Superconducting Films

Recent experiments on granular superconducting films [19] and networks of Josephson junctions [9] have revealed a remarkable feature, namely that the low temperature properties of the films depend crucially on the value of the normal state resistance R_N (per square). More precisely, it is found that the films become superconducting only for $R_N < R_0^c$, where $R_0^c \sim R_0 = h/4e^2$, while for $R_N > R_0^c$, the $T = 0$ resistance is finite, and depends exponentially on $R_N - R_0^c$ [19]. Since a phase transition from extended to localized states is found in single *resistively shunted junctions* [18], which would be consistent with a transition from a strong fluctuation (finite resistance) to a classical (superconducting) regime in a film, at precisely R_0 ($\alpha_s = 1$), it is tempting to use this model (appropriately generalized). We find it difficult, however, to imagine how shunt resistors appear e.g. in a granular film, which is believed to consist of grains of size ~ 100 Å, weakly coupled by atomic bridges.

Instead, we wish to elaborate on a consequence of the microscopic theory, namely a strong renormalization of the capacitance [4], and discuss its possible relevance for the experiments mentioned above. Considering zero temperature and an ideal junction (see Ch. 3), an inspection of Eq. (27) shows that for low frequencies $\hbar\omega << |\Delta|$, the quasiparticle part of the effective action leads to a renormalization of the geometrical capacitance in Eq. (26) (henceforth denoted C_0) such that the total capacitance is given by [4]

$$C = C_0 + C_{qp} \, , \; C_{qp} = 3e^2\alpha/16|\Delta| \; . \tag{36}$$

Note that for grains of size ~ 100 Å and $\alpha \sim 1$, we have $C_{qp} \sim 10^2 C_0$. The appropriate generalization of Eq. (3), considering an ordered lattice of identical junctions for simplicity, is

$$\mathcal{H} = \frac{1}{2} \sum_{i,j} \hat{Q}_i (\mathcal{C}^{-1})_{ij} \hat{Q}_j - E_J \sum_{<i,j>} \cos(\hat{\varphi}_i - \hat{\varphi}_j) \tag{37}$$

where $\hat{Q}_i$ and $\hat{\varphi}_i$ denote the charge and the phase operators on the i-th grain, and $\mathcal{C}$ is the capacitance matrix, in two dimensions given by $C_{ii} = c+4C$, $C_{ij} = -C$ for ij nearest neighbors, while the other matrix elements vanish. Here we also include a ground capacitance $c\,(c \sim C_0)$, with corresponding charging energy $E_c = e^2/2c$. For $C = 0$, the phase diagram of the model Eq. (37) has been studied in detail by Monte-Carlo methods [20]. Clearly, for $E_c << E_J$, the familiar Kosterlitz-Thouless-Berezinskii transition is recovered, upon

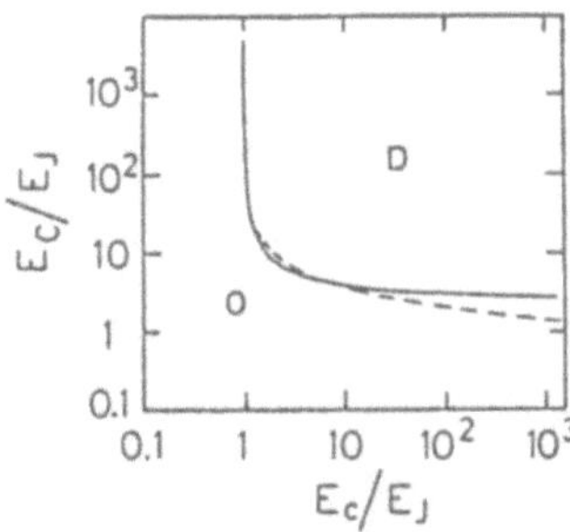

Fig. 4 Zero temperature phase diagram based on the mean-field Hamiltonian Eq. (38). D and O indicate the disordered and the ordered phase, respectively. A deviation from the result of [22] (dashed line) appears only for very small c.

decreasing the temperature, with a critical temperature given (implicitly) by $kT \sim E_J$. On the other hand, for $T = 0$, the model is in the same universality class as the 3D-XY model, and shows the standard second order transition with the well-known exponents, upon decreasing E_c/E_J, at a critical value ~ 1. In order to gain further insight into the phase diagram, e.g. the reentrant behavior found by the numerical investigation [20], it seems appropriate to consider the dual description in terms of vortices and vortex loops [21].

The more interesting case, however, is $C >> c$ as discussed above. Employing path integral methods, it is possible to derive an effective Ginzburg-Landau-Wilson functional [22], confirming that at $T = 0$ the transition is of the 3D-XY type. Generally, it is found that the ordered region of the phase diagram is increased upon increasing the nearest neighbor capacitance C. In fact, a mean-field analysis based on the ansatz [23]

$$\mathcal{H}_{mf} = \frac{1}{2} \sum_{i,j} \hat{Q}_i (C^{-1})_{ij} \hat{Q}_j - \sum_i h_i^{mf} \cos\hat{\varphi}_i \tag{38}$$

leads to the critical value $E_C = 2E_J$ for $c = 0$. Thus, inserting the renormalized capacitance Eq. (36) in the limit $C_0 << C_{qp}$, and the standard relation $E_J = \alpha|\Delta|/2$, one finds a phase transition [22] at $\alpha \sim 1$, consistent with experiment. More precisely, mean-field theory predicts a transition at $\alpha_c = (8/3)^{1/2}$, though the $c = 0$ limit is approached only very slowly. The zero temperature phase diagram [23], based on the optimal choice (in the usual sense) of h_i^{mf} in Eq. (38), is shown in Fig. 4. However, we point out that for ideal junctions, the response of the system necessarily involves frequencies of the order $\hbar\omega \sim 2|\Delta|$, which is beyond the adiabatic limit Hamiltonian Eq. (37); thus it becomes crucial to consider the full effective action, in order to include correctly the process of quasiparticle creation [21]. The origin of dissipation at zero temperature, and especially the strong dependence of the resistance on the normal state resistance, require further investigations.

Acknowledgement

We are grateful to Vinay Ambegaokar, with whom we had the pleasure to collaborate on some of the topics discussed above, and to our mentor Albert Schmid for many stimulating discussions and contributions.

References

[1]   A. J. Leggett, Jap. J. Appl. Phys. 26, Suppl. 26-3, 1986 (1987)

[2]   W. Schottky, Ann. d. Phys. 57, 541 (1918); 68, 157 (1922); W. Schottky and H. Rothe, in: Handbuch der Experimentalphysik XIII/2, ed. by W. Wien and F. Harms (Akademische Verlagsgesellschaft, Leipzig 1928), p. 1

[3]   P. W. Anderson, in: Lectures on the Many Body Problem, ed. by E. R. Caianiello (Academic Press, New York 1964), Vol. 2, p. 113

[4]   V. Ambegaokar, U. Eckern, and G. Schön, Phys. Rev. Lett. 48, 1745 (1982); V. Ambegaokar, in: Percolation, Localization and Superconductivity, NATO Advanced Study Institute, Series B: Physics, Vol. 109, ed. by A. M. Goldman and S. A. Wolf (Plenum, New York 1984), p. 43; A. I. Larkin and Yu. N. Ovchinnikov, Phys. Rev. B28, 6281 (1983); U. Eckern, G. Schön, and V. Ambegaokar, Phys. Rev. B30, 6419 (1984)

[5]   A. O. Caldeira and A. J. Leggett, Phys. Rev. Lett. 46, 211 (1981); Ann. Phys. (N.Y.) 149, 374 (1983)

[6]   R. F. Voss and R. A. Webb, Phys. Rev. Lett. 47, 265 (1981); S. Washburn, R. A. Webb, R. F. Voss, and S. M. Faris, Phys. Rev. Lett. 54, 2712 (1985); D. B. Schwartz, B. Sen, C. N. Archie, and J. E. Lukens, Phys. Rev. Lett. 55, 1547 (1985); M. H. Devoret, J. M. Martinis, and J. Clarke, Phys. Rev Lett. 55, 1908 (1985)

[7]   J. M. Martinis, M. H. Devoret, and J. Clarke, Phys. Rev. Lett. 55, 1543 (1985)

[8]   K. Yoshihiro, J. Kinoshita, K. Inagaki, C. Yamanouchi, S. Kobayashi, and T. Karasawa, Jap. J. Appl. Phys. 26, 1379 (1987); K. Yoshihiro, Physica B152, 207 (1988)

[9]   L. J. Geerligs, M. Peters, L. E. M. de Groot, A. Verbruggen, and J. E. Mooij, preprint

[10] See also W. Zwerger, in: Festkörperprobleme/ Advances in Solid State Physics, Vol. 29, ed. by U. Rößler (Vieweg, Braunschweig 1989), following paper

[11] K. K. Likharev, G. Schön, and A. D. Zaikin, to be published

[12] A. J. Dahm, A. Denenstein, D. N. Langenberg, W. H. Parker, D. Rogovin, and D. J. Scalapino, Phys. Rev. Lett. 22, 1461 (1969)

[13] K. K. Likharev and A. B. Zorin, J. Low Temp. Phys. 59, 347 (1985); D. V. Averin and K. K. Likharev, J. Low Temp. Phys. 62, 345 (1986); Zh. Eksp. Teor. Fiz. 90, 733 (1986) [Sov. Phys. JETP 63, 427 (1986)]

[14] See e. g. L. J. Geerligs and J. E. Mooij, Physica B152, 212 (1988)

[15] F. Guinea and G. Schön, Europhys. Lett. 1, 585 (1986); G. Schön, in: SQUID '85 - Superconducting Quantum Interference Devices and their Applications, ed. by H. D. Hahlbohm and H. Lübbig (de Gruyter, Berlin, New York 1985), p. 251; F. Guinea and G. Schön, J. Low Temp. Phys. 69, 219 (1987)

[16] *V. Ambegaokar* and *U. Eckern*, Z. Phys. **B69**, 399 (1987)

[17] *U. Geigenmüller* and *G. Schön*, Physica **B152**, 186 (1988)

[18] *A. Schmid*, Phys. Rev. Lett. **51**, 1506 (1983); *M. P. A. Fisher* and *W. Zwerger*, Phys. Rev. **B32**, 6190 (1985); *U. Eckern* and *F. Pelzer*, Europhys. Lett. **3**, 131 (1987); *A. D. Zaikin* and *S. V. Panyukov*, Phys. Lett. **120A**, 306 (1987); Zh. Eksp. Teor. Fiz. **93**, 918 (1987) [Sov. Phys. JETP **66**, 517 (1987)]

[19] *B. G. Orr, H. M. Jaeger, A. M. Goldman*, and *C. G. Kuper*, Phys. Rev. Lett. **56**, 378 (1986); *H. M. Jaeger, D. B. Haviland, A. M. Goldman*, and *B. G. Orr*, Phys. Rev. **B34**, 4920 (1986); *H. M. Jaeger*, Ph. D. thesis, University of Minnesota 1987

[20] *L. Jacobs, J. V. José, M. A. Novotny*, and *A. M. Goldman*, Phys. Rev. **B38**, 4562 (1988); *L. Jacobs* and *J. V. José*, Physica **B152**, 148 (1988)

[21] *U. Eckern* and *A. Schmid*, Phys. Rev. **B39** (April 1989); *S. R. Shenoy*, to be published

[22] *S. Chakravarty, S. Kivelson, G. T. Zimanyi*, and *B. I. Halperin*, Phys. Rev. **B35**, 7256 (1987)

[23] *J. Kissner* and *U. Eckern*, unpublished

Quantum Effects and the Onset of Superconductivity in Granular Films

Wilhelm Zwerger

Technische Universität München, Institut für Theoretische Physik, James-Franck-Str., D-8046 Garching, Federal Republic of Germany

Summary: Recent experiments on granular films have shown an apparently universal maximum normal state sheet resistance of around $6.5\,k\Omega$ above which superconductivity cannot be established. A possible explanation for this phenomenon is presented which is based on the competition between quantum fluctuations in the phase of the superconducting order parameter and dissipative effects. The films are modelled by locally superconducting grains coupled through Josephson junctions. Quantum fluctuations in the phases of individual grains may lead to a state in which the Cooper pair number rather than its conjugate phase is a good quantum number and superconductivity is destroyed. Dissipative effects tend to suppress these fluctuations and lead to the appearance of a new type of superconducting state possessing only short range phase coherence. This occurs for normal state resistances smaller than the quantum unit $h/4e^2 = 6.5\,k\Omega$, in agreement with the experimentally observed threshold.

1 Introduction

The investigation of superconductivity in thin films started with the work of Shalnikov [1] and Buckel and Hilsch [2] on amorphous films which often result, when metals are evaporated on cold substrates. Later on thin films played a major role in the study of superconducting fluctuation effects [3] and the Kosterlitz-Thouless transition in two dimensional systems with a continuous symmetry [4]. More recently the interest in these systems has focussed on the interplay between localization and superconductivity [5] and the nature of the onset of superconductivity in very thin films [6-10]. It is the latter problem which will be addressed here and we will argue that it provides a unique example for the influence of quantum fluctuations and dissipation on the transition to superconductivity.

In the experiments [7-10] which initiated the recent interest in this problem different metals such as Al, Ga, In, Pb or Sn are evaporated onto silica-glazed alumina substrates at a temperature of about $15\,^0K$. The electrical resistance $R(T)$ is then measured as a function of temperature down to about $0.5\,^0K$ for different nominal film thicknesses d ranging between 10 and 50 Å. The qualitative behaviour is found to be identical in all the samples and is characterized by three different regimes (see Fig.1):

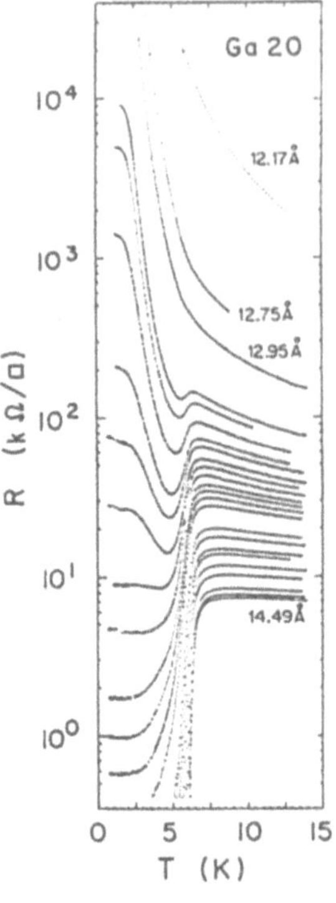
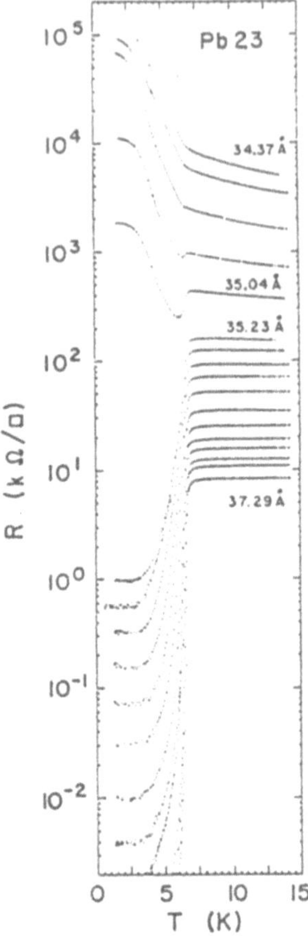

Fig.1

Resistances as a function of temperature for a series of Ga- and Pb-films of varying thicknesses.

a) For very small thicknesses the resistance increases exponentially with decreasing temperature over the whole range of temperatures considered. The films thus exhibit insulating behaviour but nevertheless there is a sign of superconducting order. Indeed in all but the very thinnest films, the activation energy $k_B T_0$ in $R(T) = exp(T_0/T)$ increases suddenly at a temperature T_c^l which is close to the bulk superconducting transition temperature T_c^b.

b) For somewhat larger thicknesses there is a drop in $R(T)$ at T_c^l, sometimes by several orders of magnitude. However, superconductivity is still incomplete since the resistance remains finite as $T \to 0$ ending either in temperature independent flat tails or rising again after going through a minimum.

c) For still larger thicknesses finally, the drop at T_c^l indeed leads to a superconducting state with vanishing resistance. In practice the resistance falls below the experimental resolution of around $10^{-3}\Omega$ which is more than six orders of magnitude smaller than the corresponding typical normal state values.

20

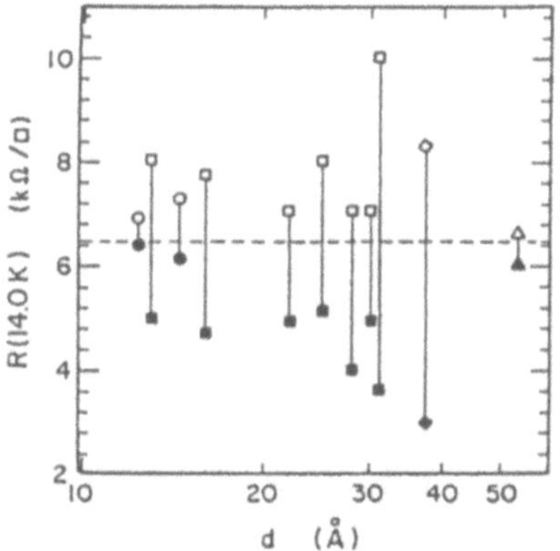

Fig.2

Pairs of connected points represent the normal state resistance of films at the threshold to superconductivity. Each pair brackets the value $\frac{h}{4e^2} = 6.5\,k\Omega$ denoted by the dashed line.

The surprising observation made in the context of these experiments now consists of the fact that independent of material or film thickness the boundary between regimes b) and c) always occurs at the same numerical value of the normal state resistance R_n [11]. This striking result is exemplified in Fig.2, where for different samples R_n is given as a function of thickness, with the open symbols denoting the last film in a series of successive depositions which had a finite resistance as $T \to 0$ and the full symbols denoting the corresponding first ones showing zero resistance. It is evident from the figure that this transition occurs at widely different values of thickness but within a narrow region of the normal state resistance around 6.5 $k\Omega$. Such a behaviour is consistent with results found earlier [12-14], however the apparently universal nature of the resistance threshold was only realized through the systematic work by Goldman and collaborators [7-10]. The accuracy of the critical value R_n^c of the normal state sheet resistance is limited by two factors: First of all, the normal state resistance of these films was operationally *defined* as $R_n = R(T = 14\,^0K)$. At this temperature the films are well above their local superconducting transition temperature, however due to the temperature dependence of $R(T)$ in this regime there is some uncertainty in the value of R_n. Secondly, the criterion for complete superconductivity used in the experiments involves a certain arbitrariness since it is defined by the resistance being smaller than a minimum observable value of around $10^{-3}\,\Omega$. Taking both factors together, one estimates that R_n^c is equal to 6.5 $k\Omega$ to within 15 % accuracy. Even in this restricted sense, however, the universality found is rather surprising and unexpected. In particular, assuming that the regimes b) and c) indeed correspond to different thermodynamic phases, one would naively expect that the phase diagram which is a *static* property of a thermodynamic system should not depend on a transport coefficient such as the normal state resistance. In fact in a classical system the commutativity between coordinates and momenta implies that the partition function - and thus the complete thermodynamics - does not depend on the dynamics of the considered degrees of freedom.

Quantum mechanically, however, this is no longer true which indicates that an explanation of the phenomenon discussed above must crucially involve quantum features in the superconductive ordering transition.

2 Superconductivity in Granular Films

2.1 Normal State Properties and Local Superconductivity

During the initial stages of film growth an inhomogeneous structure of isolated grains is formed. An example of an oxidized Sn film exhibiting similar behaviour as discussed above [15] is shown in Fig.3. The typical linear dimension L of grains in this case is of the order of a few hundred Å.

Electrical conduction in such a system may occur by quantum mechanical tunneling between close enough grains even below the limit where the film becomes geometrically connected. Due to the exponential dependence of the tunneling probability on the distance between grains, the spread in separations in any granular structure leads to very wide distributions of the local conductances. According to an argument given by Ambegaokar, Halperin and Langer [16], the actually measured conductance in such a situation is determined by the condition that the subset of all bonds with conductances larger then the measured one forms a percolating network.

An explanation for the observed thermally activated behaviour $R(T) = R_0 exp(T_0/T)$ of the resistance in thin granular films was given long ago by Neugebauer and Webb [17]. They argued that in order to transfer an electron between two originally neutral grains a charging energy $k_B T_0$ is necessary in order to account for the increase in electrostatic energy. The simple exponential form $ln(R) \sim T_0/T$ indicates that the analog [18] of *variable* range hopping between localized electronic states including Coulomb

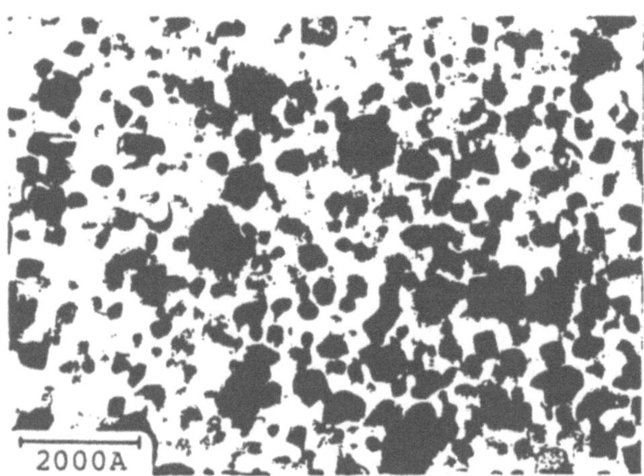

Fig.3

Electron micrograph of a granular Sn film with typical grain sizes of a few hundred Å.

effects, which would lead to $ln(R) \sim (T_0/T)^{1/2}$, does not occur in this system of mesoscopic grains [19]. With increasing film thickness and thus increasing conductance the activated behaviour becomes less pronounced and gives way to a metallic like regime with an almost temperature independent resistance in the normal state. Beyond the increase in typical grain dimensions which reduces the charging energy, this crossover may qualitatively be explained by the fact that the Coulomb-blockade is washed out with increasing conductance due to level-broadening or, equivalently, enhanced screening [20]. This effect was observed earlier [12] a quantitative theory, however, exists only at the level of a *single* tunnel junction [21].

Adding now an effective attractive interaction between electrons, it is well known that in granular systems superconductivity may be established locally. A theory for this phenomenon was given by Mühlschlegel, Scalapino and Denton [22]. They found that the essential parameter determining the existence of superconductivity in small grains is the ratio between the single electron level spacing at the Fermi-energy $\delta \simeq (n(0)V)^{-1}$ and the zero temperature gap Δ associated with a given attractive interaction (which may differ from its bulk value). Here $n(0)$ is the density of states at the Fermi-level per unit volume and $V \simeq L^3$ the volume of the grain. While a true thermodynamic transition is possible only in the limit of vanishing δ, it turns out that unless $\delta \geq 0.1\Delta$ even a small particle shows behaviour characteristic of a superconductor for instance in the specific heat or the spin susceptibility. For typical values of Δ/k_B of a few Kelvin this condition translates into a lower limit for the grain size L_{min} of order 100 Å. Sufficiently large grains therefore undergo a more or less smooth transition into a superconducting state at temperatures around a local condensation temperature T_c^l. The experimental fact that T_c^l is close to the bulk transition temperature T_c^b shows that within the grains the coupling strength is essentially unchanged and no microsopic disorder is present. Due to the opening of a superconducting gap below T_c^l the activation energy arising from the charging effects is enhanced further which explains the increase in T_0 at this temperature. As a result, the transition to local superconductivity in very thin films even *enhances* the insulating behaviour. In order to observe superconductivity in the familiar sense of a *decrease* and then vanishing of the resistance, it is necesssary to have a contribution to the current which arises from the coherent tunneling of pairs. It is then possible to establish global superconductivity in a network of disconnected, only locally superconducting grains [23].

2.2 Josephson Effect and Global Phase Coherence

If two isolated superconducting grains are close together, the tunneling of Cooper pairs between them gives rise to a nondissipative current flow (see the preceeding article by Eckern and Schön). With E_J as the transition amplitude (in units of energy) for the transfer of a single Cooper pair between adjacent grains, there is a correspondi ıg coupling energy

$$E(\varphi) = -E_J \cos(\varphi) \tag{1}$$

with φ the difference in the phases of the superconducting order parameter on both sides of the junction. The magnitude of E_J at zero temperature can be calculated microscopically within a tunneling Hamiltonian model and is given by

$$E_J = \frac{\alpha \Delta}{2} \qquad (2)$$

with the superconducting gap Δ assumed to be equal on both sides. The parameter

$$\alpha = \frac{R_Q}{R_n} \qquad (3)$$

is the dimensionless conductance of the contact in the normal state measured in units of $R_Q^{-1} = (2e)^2/h$. This parameter will play a crucial role in the following. In particular, measuring the *global* normal state conductance in the same units, the observed resistance threshold of around $6.5\,k\Omega$ just corresponds to a critical value $\alpha_c = 1$. The linear relation between E_J and the normal state conductance R_n^{-1} reflects the fact that R_n^{-1} is a measure of the tunneling *probablity* of a *single* electron which also gives the *amplitude* for *pair* tunneling. As was pointed out above, in any granular system there is a wide distribution of local conductances α and thus a corresponding one in the strength of the Josephson coupling. In the theoretical discussions of granular superconductivity this type of randomness is usually neglected and E_J is replaced by a fixed average value. This approximation seems to be justified in our case by recent experiments on regular arrays of Josephson junctions [24] which find a similar behaviour than in real granular systems, but certainly this problem needs further consideration.

In order to describe the transition to global superconductivity in a network of Josephson coupled grains, we assume that fluctuations in the magnitude of the order parameter may be neglected which is justified for temperatures well below T_c^l. The only remaining degrees of freedom are then the phases φ_l of individual grains l and the corresponding coupling energy is

$$H[\varphi_l] = -E_J \sum_{<ll'>} \cos(\varphi_l - \varphi_{l'}) . \qquad (4)$$

Here $<ll'>$ denotes a sum over nearest neighbor pairs with no double counting. The statistical mechanics of this model has been investigated in great detail. It exhibits a Kosterlitz-Thouless transition [25] at a critical temperature $k_B T_c \approx 0.9 E_J$ from a superconducting state with (algebraic) long range phase coherence below, to a non superconducting state with no phase order above T_c. For an estimate of the transition temperature we combine the BCS-relation $2\Delta = 3.5 k_B T_c^l$ for the local transition with eqn. (2) which leads to $T_J \approx \alpha T_c^l$ for the equivalent temperature describing the strength of the Josephson coupling and thus T_c. In this simple picture therefore, the assumption of a separate transition to global phase coherence well below T_c^l is consistent only if $\alpha < 1$ which means normal state resistances larger than $6.5\,k\Omega$.

24

In order to describe the experiments discussed above, the simple Josephson coupling model (4) is insufficient, however. Indeed, the high resistance samples do not become superconducting at all even below temperatures around αT_c^l where phase ordering should be present according to the above arguments. Nevertheless it seems that superconductivity in these systems is destroyed by phase fluctuations rather than a reduction in the magnitude of the order parameter. This may be inferred from the fact that the temperature where local superconductivity sets in is very close to the bulk transition temperature and thus disorder effects like an enhanced Coulomb repulsion or a reduction in the density of states [26] do not seem to be relevant. In order to explain the absence of phase ordering even at very low temperatures it is therefore necessary to consider fluctuations beyond the thermal ones.

2.3 Charging Effects

It was first pointed out by Abeles [27] that superconductivity in a Josephson coupled granular system may be suppressed by charging effects. Physically these are due to the Coulomb interaction between the Cooper pairs and they are particularly important in systems with small grains. On a mesoscopic level they are described formally by a capacitance matrix $C_{ll'}$ such that

$$\frac{(2e)^2}{2} \sum_{ll'} \left(C^{-1}\right)_{ll'} n_l n_{l'} \tag{5}$$

is the electrostatic Coulomb energy of a network of grains l with Cooper pair number n_l (more precisely n_l is the deviation of the number of Cooper pairs from the average value). A crucial simplification is obtained by assuming that due to the presence of the substrate and the possible existence of unpaired electrons the long range Coulomb interaction is screened. It is then a resonable approximation to replace the inverse capacitance matrix C^{-1} by a diagonal one. The Coulomb effects are thus characterized by a single effective charging energy $E_0 = (2e)^2/C$ and the total Hamiltonian is

$$H = \frac{E_0}{2} \sum_l n_l^2 - E_J \sum_{ll'} \cos(\varphi_l - \varphi_{l'}) . \tag{6}$$

The crucial point of introducing the charging effects lies in the fact that with the canonically conjugate variables φ and n generalized to quantum mechanical operators obeying

$$[\varphi_l, n_{l'}] = i\,\delta_{ll'} \tag{7}$$

we have now included quantum fluctuations in the phases of locally superconducting grains. Thus even at zero temperature phase ordering may be destroyed by zero point fluctuations. As was pointed out by Efetov [28], this effect can be understood easily by an analogy to the well known Mott transition in an interacting electronic system. Indeed the Josephson coupling corresponds to a tight binding model for hopping of Cooper pairs with amplitude E_J while the charging energy E_0 is a measure of the on-site Coulomb repulsion. The model (6) can therefore be thought of as a bosonic analog of the Hubbard model and in analogy with this problem one expects a Mott transition for pairs. For a large value of the ratio $J = E_J/E_0$ the Josephson coupling is dominant and leads to a ground state with delocalized pairs which is superconducting. Increasing the strength of the charging term, it will be destroyed at a critical value J_c of order one and for $J < J_c$ the ground state will be insulating with the Cooper pairs localized and strong quantum fluctuations in the phase.

For a qualitative estimate of whether these effects are relevant in the experiments discussed here, we recall that the equivalent temperature for the Josephson coupling is $T_J \approx \alpha T_c^l$. The charging energy may be inferred either from the activation energy $k_B T_0$ of the resistance in the normal state or from the geometrical capacitance of grains with typical sizes of order 100 Å. For the systems discussed above these estimates lead to equivalent temperatures T_0 of 30 ^0K or more which, at interesting values of α around one, are considerably larger than T_J. This indicates that charging effects are indeed strongly relevant and may explain the absence of phase ordering as $T \rightarrow 0$. However, using equation (2), the qualitative criterion $E_J \simeq E_0$ for the destruction of superconductivity by Coulomb effects even at $T = 0$, which is supported by various mean-field type calculations [29-31], leads to a critical conductance treshold $\alpha_c \simeq T_0/T_c^l$. Now the local transition temperature T_c^l was varied by a factor of about 8 in the experiments choosing different superconducting elements. Also the charging energy at the threshold to superconductivity differs widely for the various samples since they have rather different critical thicknesses (see Fig.2). There is therefore no reason to believe that α_c should be equal to one within 15 % accuracy for all the different samples. What is missing in this analysis is the possibility of screening the charging effects with increasing conductance which was mentioned above in connection with the observed crossover from insulating to metallic behaviour in the normal state. It is this phenomenon, first emphasized by Imry and Strongin [20], which turns out to be crucial for understanding the essential role played by the normal state resistance and leads to the universal threshold discussed above.

In order to describe these effects theoretically, it is necessary to introduce dissipative effects into the quantum mechanical model (6). Following the pioneering work by Caldeira and Leggett [32], an effective way of doing this at a phenomenological level is to use the Feynman path integral formulation of quantum mechanics. In this formulation all what is needed is the effective classical action in which dissipative effects may easily be incorporated by, for instance, giving a particle a frequency dependent mass.

26

3 Phase Coherence in Josephson Junctions as a Roughening Problem

3.1 Single Josephson Junction

In the absence of any dissipative mechanism a single quantum Josephson junction is described by the Hamiltonian

$$H = \frac{E_0}{2} n^2 - E_J \cos(\varphi) \tag{8}$$

which is equivalent to that of a particle in a periodic potential with φ playing the role of a coordinate. The corresponding eigenfunctions are extended Bloch states and thus the phase does not have a well defined value. Within this simple model, therefore, the charging effects completely destroy the coherent pair tunneling. In order to be able to include dissipative effects later on, it is convenient to study this problem within a functional integral formalism. Dividing the inverse temperature $\beta = (k_B T)^{-1}$ into $N_\tau = \beta/\varepsilon$ infinitesimal steps ε, the partition function $Z = Tr\, exp(-\beta H)$ in a representation in terms of number eigenstates takes the form [33]

$$Z = \sum_{n_j = 0, \pm 1, \pm 2, \ldots} exp\left(-\frac{\varepsilon E_0}{2} \sum_{j=1}^{N_\tau} n_j^2 - \frac{1}{2\varepsilon E_J} \sum_{j=1}^{N_\tau} (n_{j+1} - n_j)^2\right) . \tag{9}$$

This describes a discrete Gaussian roughening model of a one dimensional interface with height variable $n_j = 0, \pm 1, \ldots$ whose connection to the original Josephson junction problem has a simple physical interpretation. Each tunneling of a Cooper pair between two adjacent superconductors is represented by a discrete ± 1 step on an interface with average height equal to zero (see Fig.4).

The coordinate along the interface is a time like distance τ running from 0 to $\beta\hbar$. The junction is in a phase coherent state if the pairs are delocalized and can be exchanged freely between both superconductors. In terms of the interface this is equivalent to a state with strong fluctuations in the interface height and thus to a rough boundary between the two superconductors.

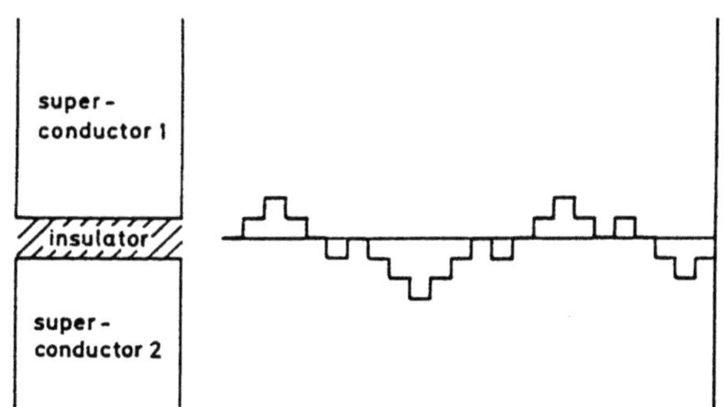

Fig.4

Cooper pair tunneling between two superconductors represented as a series of steps along a one dimensional interface.

Such a situation is, however, only realized in the absence of charging effects $E_0 = 0$. On the contrary, for any finite E_0 the first term in the exponent of (9) strongly pins the interface around $n = 0$ and leads to a smooth boundary which simply reflects the fact that the corresponding quantum particle has a ground state with an extended phase. In a single Josephson junction quantum fluctuations therefore completely destroy phase coherence as long as no mechanism for screening the charging effects is present.

A very simple model describing the screening of the charging effects is obtained by adding a shunting resistance R_n to each Josephson junction. In order to study this effect, we take the limit $\varepsilon \to 0$ and go over to a continuous height function $n(\tau)$. To mimic the original integer restriction, however, a term

$$K \int_0^{\beta \hbar} d\tau \cos(2\pi n(\tau)) \tag{10}$$

is added to the exponent S in (9) which favors integer n. Taking a Fourier transform with respect to τ, S takes the form

$$S = \frac{\hbar}{2E_J} \int_{-\infty}^{\infty} \frac{d\omega}{2\pi} (\omega^2 + \omega_0^2) \mid n(\omega) \mid^2 \tag{11}$$

with $\omega_0^2 = E_0 E_J / \hbar^2$ the square of the Josephson plasma frequency. The effects of a shunting resistor R_n can now easily be incorporated by replacing the capacity C in $E_0 = (2e)^2 / C$ by the frequency dependent generalization

$$C \longrightarrow C\left(1 + \frac{\gamma}{|\omega|}\right) \tag{12}$$

with $\gamma = (R_n C)^{-1}$ the inverse RC time. Thus ω_0^2 in (11) is replaced by $\omega_0^2 \frac{|\omega|}{\gamma + |\omega|}$ so that the effective charging energy vanishes at zero frequency. To determine under which conditions the interface will now be rough and thus when coherent pair tunneling is possible, one performs a simple renormalization group calculation. Eliminating fast varying fluctuations in $n(\tau)$ with frequencies in a range $(\omega_c - d\omega_c, \omega_c)$ one finds that the coefficient K in (10) changes as

$$\frac{dK}{dl} = -(\alpha - 1)K + O(K^3) \tag{13}$$

with $dl = -\frac{d\omega_c}{\omega_c}$ the infinitesimal reduction in the high frequency cutoff ω_c. Since K scales to zero if $\alpha > 1$ the interface is rough in this regime. Thus a phase coherent ground state for a single Josephson junction is realized for normal state conductances larger than the appropriate quantum unit $(2e)^2 / h$. For $\alpha < 1$, in contrast, the scaling (13) leads to increasing values of K and thus outside the range of perturbation theory. In this regime it is more appropriate to work with a functional integral using phase rather than number eigenstates [34-36]. One then finds that indeed the phase is a delocalized variable and thus the corresponding interface describing fluctuations in n will be smooth.

28

We have thus found that a single resistively shunted Josephson junction shows phase coherence only if the normal state resistance is smaller than $6.5 \, k\Omega$ independent of the values of E_J and E_0. This result is strongly reminiscent of the experimental findings discussed above and in the following we will discuss its relation to the behaviour of granular films which, as we have argued above, may be modelled as whole *networks* of Josephson coupled grains. Incidentally the model of a quantum particle in a periodic potential treated here can be analyzed in some detail even with respect to its *dynamical* properties [35,36]. It provides a unique example where the transition from coherent tunneling to incoherent hopping can be described analytically.

3.2 Josephson Junction Networks

To discuss the phase coherence properties of networks, it is convenient to start with a one dimensional chain of junctions. Being easier to treat formally, its phase diagram at zero temperature still displays all the essential features which are present in the actually interesting case of a two dimensional network. In the functional integral representation sketched above the partition function for vanishing dissipation is now of the form [33]

$$Z = \sum_{h_{lj}=0,\pm 1,\pm 2\ldots} exp\left(-\frac{\varepsilon E_0}{2}\sum_{lj}(h_{l+1j} - h_{lj})^2 - \frac{1}{2\varepsilon E_J}\sum_{lj}(h_{lj+1} - h_{lj})^2\right) \quad (14)$$

describing a two dimensional interface with height variable h_{lj} (here l enumerates the grains along the chain of Josephson junctions). Its relation to the pair number is given by $n_{lj} = h_{l+1j} - h_{lj}$, i.e. n is the spatial gradient of the interface height h. Chui and Weeks [37] have shown that a two dimensional interface as described by (14) has a Kosterlitz-Thouless like transition at $J_c \approx 1.2$ between a rough phase for $J > J_c$ and a smooth one for $J < J_c$. These phases may be distinguished for instance by the fluctuations in the interface width

$$c(l) = \langle (h_l - h_0)^2 \rangle = \langle (\sum_{l'=0}^{l-1} n_{l'})^2 \rangle \quad (15)$$

which diverges logarithmically as $l \to \infty$ if $J > J_c$ but remains finite otherwise. As is seen from (15) the rough phase in the interface problem corresponds to diverging charge fluctuations in the chain of Josephson junctions. Similar to the single junction case one expects that this state is superconducting. That this is indeed the case can be shown by calculating the frequency dependent dynamical conductivity $\sigma(\omega)$. In fact it is a unique feature of quantum phase transitions that static and dynamic properties are completely tied together [38]. In our present example, from the analytical continuation of the current-current correlation function in 'imaginary' time τ it is possible to show that $Re\,\sigma(\omega)$ has a singular $\delta(\omega)$ contribution at low frequencies if $J > J_c$ which is characteristic of a superconductor. The smooth phase in the interface problem, on the other hand, is characterized by a finite free energy of a step which is equivalent to a finite activation energy for the transport of charge in the corresponding chain of Josephson junctions. The phase at $J < J_c$ is thus a Mott insulating phase with a resistance which diverges exponentially as $T \to 0$. Thus we have found that even in one dimension there is a superconductor to insulator transition with a behaviour in agreement with the qualitative arguments by Abeles [27,39].

In order to include dissipation, which tends to wash out the charging effects, one proceeds in the same manner as above by phenomenologically introducing a frequency dependent capacity as in (12). In terms of the associated interface problem this leads to an anisotropy between the physical dimension along the chain and the 'imaginary' time direction related to the quantum nature of the problem. Indeed along the latter direction there is an effective long range interaction which may induce a state in which the interface is rough along it but remains smooth in the actual physical direction. This local transition can be analyzed using essentially the same renormalization procedure as above [33]. As a result it is found that for conductances $\alpha > 1$ the interface is rough along the τ-direction even if J is arbitrarily small. In terms of the chain of Josephson junctions this partially rough interface is equivalent to a well defined local phase on each individual grain but no global phase coherence. Indeed the broken symmetry in this state is the local discrete symmetry $\varphi_l \to \varphi_l + 2\pi$ for each separate l which may be broken in our case since the dissipation effectively introduces infinitely many degrees of freedom for each grain. In addition there is also the transition to a globally phase coherent state breaking the global continuous symmetry $\varphi_l \to \varphi_l + \delta\varphi$ for all l simultaneously. This transition corresponds to an essentially isotropic roughening of the interface. While the nature of this transition is different at finite α compared to $\alpha = 0$ the corresponding critical value $J_c(\alpha)$ is only a weakly decreasing function of the conductance α.

The complete phase diagram at zero temperature therefore consists of three different phases (see Fig.5): at small J and α there is a phase with finite number fluctuations (smooth interface) which is non superconducting. For J large enough there is one showing global phase coherence and thus superconductivity (rough interface). Finally a third phase occurs for $\alpha > 1$

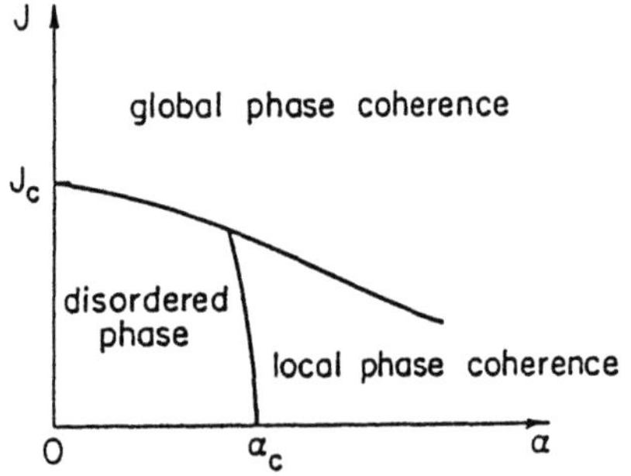

Fig.5

Qualitative zero temperature phase diagram for a chain of Josephson junctions including quantum fluctuations and dissipation.

30

and small J which has only a well defined local phase but no long range coherence (partially rough). Nevertheless this state has vanishing resistance at zero temperature. This may be indirectly inferred from a comparison with a model with long range Coulomb interactions showing zero resistance if $\alpha > 1$ even with stronger quantum phase fluctuations [40]. The two superconducting states may be distinguished by the Meissner effect which is complete only if long range phase coherence exists. The locally phase coherent state on the other hand has vanishing superfluid density and will expel a magnetic field only from the grains, i.e. partially. Its resistance is zero due to the fact that phase slips are frozen out, in analogy to a type II superconductor with pinned flux lines.

Using the same formalism as above, it may be shown that the actually interesting two dimensional network of Josephson junctions has a zero temperature phase diagram which is qualitatively the same as the one dimensional one shown in Fig.5. The transition to global phase coherence, however, is now of a different nature and has a smaller $J_c(\alpha = 0) \approx 0.1$. By contrast the transition to local phase coherence for small J again occurs at $\alpha_c \approx 1$.

Comparing our results with the experiments discussed in section 1, we see that our simple model indeed explains the observed universal conductance treshold $\alpha_c \approx 1$ assuming that the relevant values of J are smaller than the ones for global phase coherence. Now estimates of the actual values of J are of the order of 0.1 or smaller and are therefore consistent with this assumption. This is true in particular since the theoretical estimates of J_c are valid for *regular* arrays with no disorder. In real granular systems, however, with a distribution of E_J values the necessary J_c for global ordering will be larger. Due to the local nature of the transition at $\alpha_c \approx 1$ a sharp boundary between the non superconducting and the locally phase coherent superconducting state is present only at zero temperature. At any finite T, instead, there will only be a continuous crossover where the resistance begins to drop strongly. Thus, strictly speaking, it does never vanish although it may be unobservably small (see below). A prediction of the model presented here is the absence of a Meissner effect in the locally phase coherent regime and a transition to global phase coherence with increasing J. With the technology of fabricating well defined regular Josephson junction arrays in the parameter regime discussed here [24] these predictions should be amenable to experimental test.

4 Conclusion

We have discussed a very simple model which is able to explain the observed universal resistance threshold for the existence of superconductivity in granular films. It is based on a picture in which locally superconducting grains are coupled by the Josephson effect. The critical resistance arises from the competition between the quantum fluctuations due to charging effects and their screening with increasing conductance. At a conductance of the order of the quantum unit $(2e)^2/h$ for pairs this screening has suppressed the quantum fluctuations in the phase of the superconducting order parameter sufficiently that even for very small Josephson coupling a state with a well defined local phase is established. In analogy to a type II superconductor this state has vanishing resistance but does not possess long range phase coherence and thus will not show a complete Meissner effect. This distinguishes our model from other approaches to this problem [40-45] which we have not discussed and it allows a direct experimental test of the ideas presented here.

Beyond the calculation of the zero temperature phase diagram a more detailed understanding of superconductivity in granular films requires a theory which explains the resistance as a function of temperature for arbitrary normal state values, i.e. all the way from the insulating to the superconducting samples. At present there are no quantitative results in this direction and thus we can add only some qualitative remarks. In films with very small normal state conductances which, however, still show local superconductivity the low temperature resistance increases exponentially as $T \to 0$. Physically this may be explained by activated hopping of pairs between nearest neighbor grains which is consistent with a model in which due to charging effects the pairs form a Mott insulating ground state. Unfortunately our simplified model for the screening of the charging effects with increasing normal state conductance cannot explain this regime. Indeed it is clear that with simple parallel resistors R_n on each Josephson junction the actually measured resistance in a regular network can never be larger than R_n itself.

For films with intermediate conductances the experiments show the existence of a nonzero residual resistance as $T \to 0$ (see Fig.1). Qualitatively this is consistent with our model in which both normal and supercurrent flow are present and the latter becomes dominant with increasing normal state conductance where the quantum fluctuations in the Josephson current are quenched. Finally for conductances larger than the critical $\alpha_c \approx 1$ there is no measurable resistance at very low temperatures in agreement with theory. Now for classical superconductors it is known that in a dynamical sense there is no true superconducting transition. Indeed even below T_c there are resistive fluctuations due to thermally activated phase slips [46]. In the past few years it became clear that this behavior is not restricted to finite temperature but applies also to the flow of supercurrent at zero temperature [32]. There quantum fluctuations in the phase lead to the appearance of quantum phase slips and thus to finite resistances even at $T = 0$. In the superconducting state with *local* phase coherence discussed here, the analysis of the single junction indicates that no quantum phase slips are present at $T = 0$ and $\alpha > 1$. However the simplified model for dissipation used here tends to overestimate the suppression of quantum phase fluctuations. It is therefore conceivable that strictly speaking no zero resistance state occurs even at $T = 0$ and that the observed transition to superconductivity is a rapid crossover to a situation in which the rate for quantum phase slips becomes unmeasurably small.

32

Acknowledgement

I would like to to take this opportunity to thank Prof. W. Brenig for his continued support over many years. In addition I am very grateful to Prof. A.J. Leggett for many discussions and his kind hospitality at the University of Illinois.

References

[1]   *A.I. Shalnikov*, Zh. Eksper. i Teor. Fiz. **10**, 630 (1940)

[2]   *W. Buckel*, and *R. Hilsch*, Z. Phys. **138**, 109 (1954)

[3]   See e.g. the review by *W.J. Skocpol*, and *M. Tinkham*, in: Rep. Prog. Phys. **38**, 1049 (1975)

[4]   See e.g. the review by *J.E. Mooij*, in: Percolation, Localization and Superconductivity ed. by *A.M. Goldman*, and *S.A. Wolf* (Plenum Press, New York 1984), p. 325

[5]   *H. Ebisawa, H. Fukuyama*, and *S. Maekawa*, J. Phys. Soc. Jpn. **54**, 2257 (1985)

[6]   *M. Strongin, R.S. Thompson, O.F. Kammerer*, and *J.E. Crow*, Phys. Rev. **B1**, 1078 (1970)

[7]   *B.G. Orr, H.M. Jaeger*, and *A.M. Goldman*, Phys. Rev. **B32**, 7586 (1985)

[8]   *B.G. Orr, H.M. Jaeger, A.M. Goldman*, and *C.G. Kuper*, Phys. Rev. Letters **56**, 378 (1986)

[9]   *H.M. Jaeger, D.B. Haviland, A.M. Goldman*, and *B.G. Orr*, Phys. Rev. **B34**, 4920 (1986)

[10]   *H.M. Jaeger, D.B. Haviland, B.G. Orr*, and *A.M. Goldman*, Preprint, University of Minnesota (1988)

[11]   Note that in a two dimensional film the resistance is independent of the sample dimensions (assuming isotropy and a square geometry for simplicity). This peculiar scale invariance allows us to relate the local tunnel resistances between the grains to the overall film resistance.

[12]   *R.C. Dynes, J.P. Garno*, and *J.M. Rowell*, Phys. Rev. Letters **40**, 479 (1978)

[13]   *S. Kobayashi, Y. Tada*, and *W. Sasaki*, J. Phys. Soc. Jpn. **49**, 2075 (1980)

[14]   *A.F. Hebard*, and *M.A. Paalanen*, Phys. Rev. **B30**, 4063 (1984)

[15]   *S. Kobayashi*, and *F. Komori*, J. Phys. Soc. Jpn. **57**, 1884 (1988)

[16]   *V. Ambegaokar, B.I. Halperin*, and *J.S. Langer*, Phys. Rev. **B4**, 2612 (1971)

[17]   *C.G. Neugebauer*, and *M.B. Webb*, J. Appl. Phys. **33**, 74 (1962)

[18]   *B. Abeles, P. Sheng, M.D. Coutts*, and *Y. Arie*, Adv. Phys. **24**, 407 (1975)

[19]   For a recent discussion see: *R. Németh*, and *B. Mühlschlegel*, Z. Phys. **B70**, 159 (1988)

[20]   *Y. Imry*, and *M. Strongin*, Phys. Rev. **B24**, 6353 (1981)

[21]   *R. Brown*, and *E. Šimánek*, Phys. Rev. **B34**, 2957 (1986)

[22]   *B. Mühlschlegel, D.J. Scalapino*, and *R. Denton*, Phys. Rev. **B6**, 1767 (1972)

[23]   *G. Deutscher, Y. Imry*, and *L. Gunther*, Phys. Rev. **B10**, 4598 (1974)

[24]   *L.J. Geerligs*, and *J.E. Mooij*, Physica **B152**, 212 (1988)

[25]   For a recent review see: *P. Minnhagen*, Rev. Mod. Phys. **59**, 1001 (1987)

[26]   *J.M. Graybeal*, and *M.R. Beasley*, Phys. Rev. **B29**, 4167 (1984)

[27]   *B. Abeles*, Phys. Rev. **B15**, 2828 (1977)

[28]   K.B. Efetov, Sov. Phys. JETP **51**, 1015 (1980)

[29]   E. Šimánek, Solid State Comm. **31**, 419 (1979)

[30]   S. Doniach, Phys. Rev. **B24**, 5063 (1981)

[31]   P. Fazekas, B. Mühlschlegel, and M. Schröter, Z. Phys. **B57**, 193 (1984)

[32]   A.O. Caldeira, and A.J. Leggett, Ann. Phys. **149**, 374 (1983)

[33]   W. Zwerger, Europhys. Letters, in print (1989)

[34]   A. Schmid, Phys. Rev. Letters **51**, 1506 (1983)

[35]   M.P.A. Fisher, and W. Zwerger, Phys. Rev. **B32**, 6190 (1985)

[36]   W. Zwerger, Phys. Rev. **B35**, 4737 (1987)

[37]   S.T. Chui, and J.D. Weeks, Phys. Rev. **B14**, 4978 (1976)

[38]   J.A. Hertz, Phys. Rev. **B14**, 1165 (1976)

[39]   R.M. Bradley, and S. Doniach, Phys. Rev. **B30**, 1138 (1984)

[40]   M.P.A. Fisher, Phys. Rev. **B36**, 1917 (1987)

[41]   S. Chakravarty, S. Kivelson, G. Zimanyi, and B.I. Halperin, Phys. Rev. **B35**, 7256 (1987)

[42]   A. Kampf, and G. Schön, Phys. Rev. **B36**, 3651 (1987)

[43]   A.D. Zaikin, Physica **B152**, 251 (1988)

[44]   R.A. Ferrell, and B. Mirhashem, Phys. Rev. **B37**, 648 (1988)

[45]   S. Chakravarty, G.L. Ingold, S. Kivelson, and G. Zimanyi, Phys. Rev. **B37**, 3283 (1988)

[46]   J.S. Langer, and V. Ambegaokar, Phys. Rev. **164**, 498 (1967)

Pattern Formation in a Liquid Crystal

Ingo Rehberg, Bernhard L. Winkler, Manuel de la Torre Juarez, Steffen Rasenat, and Wolfgang Schöpf

Physikalisches Institut, PF 101251, Universität Bayreuth, D-8580 Bayreuth, FRG

Summary: Applying a voltage to a nematic liquid crystal sandwiched between two electrodes leads to convection when a critical threshold value of the driving voltage is exceeded. The patterns arising above this threshold are studied experimentally. Special emphasis is put on an instability leading to travelling waves and the role of defects on the route to turbulence.

1 Introduction

The problem of pattern formation in nonequilibrium systems obviously is the most important one in science: The formation of life on earth can simply be understood by assuming an ensemble of chemical compounds driven out of equilibrium by the radiation of the sun [1]. In order to understand some features of pattern formation processes, which are observed in many different fields including biology, chemistry and physics [2], one needs to deal with simpler systems. Within physics the motivation to study order-disorder transitions certainly has led to a renaissance of classical fluid dynamics during the last decade. The hydrodynamic instabilities offer the experimental advantage to allow reproducible measurements with well defined boundary conditions. For the theoretical physicist they offer the advantage to be based on well-defined theoretical grounds, the Navier-Stokes equations. The most popular hydrodynamic instabilities are circular Couette flow going unstable with respect to Taylor vortices [3] and Rayleigh-Bénard

Fig. 1   Image of Williams rolls (7 V)

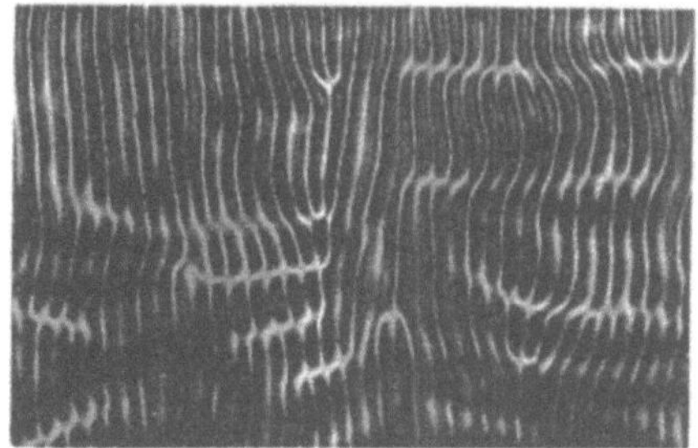

Fig. 2   Fluctuating Williams rolls (10 V)

convection [4]. A special case of the latter is thermal convection in a binary mixture, where the thermal expansion competes with thermodiffusion leading to an oscillatory instability [5]. In this talk we would like to propose a fluid dynamical system as useful for studies in pattern formation which can not exactly be called classical: Convection patterns in an anisotropic fluid, namely a nematic liquid crystal (Figs. 1 and 2). In the experiments presented here the convection is driven by an electric field, thus it is called electro-hydrodynamic convection (EHC) [6].

Many pattern forming instabilities including the ones mentioned above share a common scenario summarized in Fig. 3. The system is driven from equilibrium by means of a control parameter, for example a chemical concentration, a Reynolds number, a temperature difference or a voltage. The transition from the ground state to the structured state takes place at a well defined value of the control parameter. Above this threshold the order parameter of the system, which might be a velocity, a temperature or a director angle in the case of EHC, is different from zero. We concentrate here on systems where this upper state shows up as a periodic pattern. Above the transition point a wavenumber band exists as indicated by the upper part of Fig. 3. Theoretical descriptions of pattern forming instabilities generally start with a linear stability analysis of the spatially homogenous ground state. The amplitude of the most unstable mode might then be described by means of a weakly nonlinear analysis leading to an amplitude equation. The results

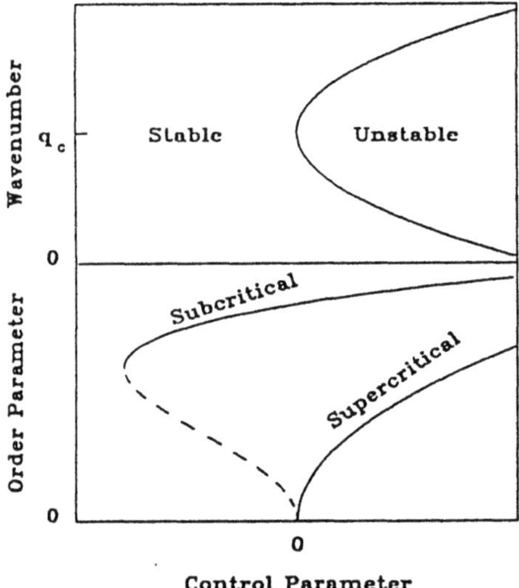

Fig. 3   Bifurcation scenario of pattern forming instabilities. At a well-defined value of the control parameter the system looses stability to a periodic structure with the wavenumber q_c. The bifurcation is called subcritical if the bifurcating branch extends to negative values of the control parameter, otherwise supercritical.

36

of the application of this concept to electro-hydrodynamic convection are given in [6].

In this paper we deal with experimental aspects of EHC. In section 2 the experimental methods used to explore EHC are described. The possible patterns arising above the convection threshold are then displayed in section 3. Increasing the voltage leads to a higher degree of disorder, which might manifest itself in more complicated, but regular patterns, or in an increasing number of defects as shown in Fig. 2. In section 4 we concentrate on more recent and special results: The observation of travelling waves, which are unexpected on the basis of the accepted theory for EHC, and the defect statistics in the weakly turbulent regime.

2 Experimental Set-up and Procedure

The experimental set-up is shown in Fig. 4: the nematic liquid crystal is sandwiched between two transparent, indiumoxid coated glass eletrodes. As the working fluid, we use either the nematic N-(p-methoxybenzylidene)-p-butylaniline (MBBA) or the Merck Phase V, a mixture of azoxy compounds. The former offers the advantage that the material parameters have been measured extensively. Phase V on the other hand is more convenient to observe oblique rolls, a different convection pattern which will be described in more detail in chapter 3. The spacing between the two electrodes ranges from 10 to 100 micrometer. The nematic is in the so called planar orientation: the director of the fluid is parallel to the plates. A preferred orientation (along the y axis) is achieved by rubbing the electrodes. Because all the material parameters of the liquid crystals are temperature dependent, it is neccesary to stabilize the temperature of the cell. We reach a stability of ±0.005 K by means of a water circuit. The cell is mounted on a polarizing microscope for flow visualisation.

Applying an ac-voltage across of the cell leads to convection when a certain value of that voltage is exceeded. The convection shows up in the form of parallel rolls as indicated in Fig. 4. Adjacent rolls have different sense of rotation. The deformation of the director field is also indicated in Fig. 4. If

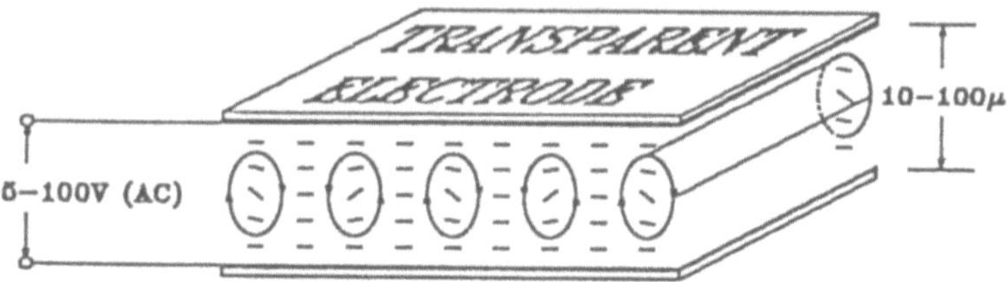

Fig. 4 The convection cell. The transparent electrodes have a typical spacing between 10–100 μm. The director orientation is indicated by (–); the streamlines by the circles.

light polarized parallel to the director is sent through the cell, an image of the convection pattern is formed as shown by the photograph 1. This visualisation method is a special case of the well known shadowgraph method: The light beams are bent inside the cell leading to intensity modulations. In the usual shadowgraph method, density gradients cause the deflection. In EHC, however, one deals with refraction of the extraordinary light inside the liquid crystal. A discussion of the differential equation describing the path of light inside the cell is given in [7]. If the optical anisotropy and the orientation of the optical axis (i.e. the orientation of the director) inside the cell are known, the path of the light and the intensity behind the cell can be calculated. Figure 5 shows the result of such a numerical calculation of the light path inside the cell. The most important result of the calculations presented in [7] is the fact that the angle of the outgoing light grows with the square of the director angle θ, but the light deviation inside the cell and therefore the intensity modulation measured behind the cell contain a term linear in θ. This linear term is responsible for creating different images of the upflow (broad intensity maxima in Fig. 1) and downflow (sharp maxima in Fig. 1). Figure 6 shows a measurement of this modulation as a function of the driving ac-voltage. From the measured intensities and the shape of the intensity curves a number proportional to the angle θ of the director field can be extracted, and the result is shown in Fig. 7. The solid line is a fit to the square root law expected for this supercritical bifurcation due to symmetry arguments.

The optical methods can be used for time dependent studies as well. In Fig. 8 a voltage step from a subcritical value without convection to a supercritical value and steps from a supercritical value to 3 different subcritical values have been applied. The intensity modulation along a line perpendicular to the axis of the rolls is plotted as a function of time. The amplitude grows or decays exponentially for small values. This growth or decay time increases when approaching the critical point. This behaviour is the analogue to the critical slowing down in equilibrium phase transitions.

$$\frac{\partial f}{\partial y} - \frac{\partial^2 f}{\partial y' \partial x} - \frac{\partial^2 f}{\partial y' \partial y} \cdot y' - \frac{\partial f^2}{\partial y'^2} \cdot y'' = 0$$

$$f(x,y,y') = n(x,y,y') \cdot (1+y'^2)^{1/2}$$

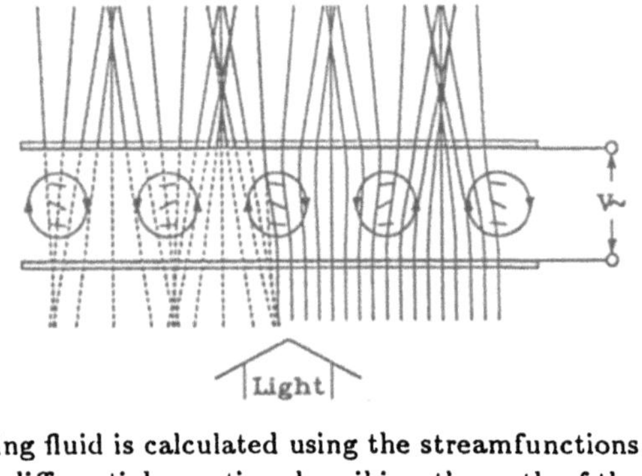

Fig. 5 The light path inside the convecting fluid is calculated using the streamfunctions obtained by linear stability theory and the differential equation describing the path of the extraordinary light.

38

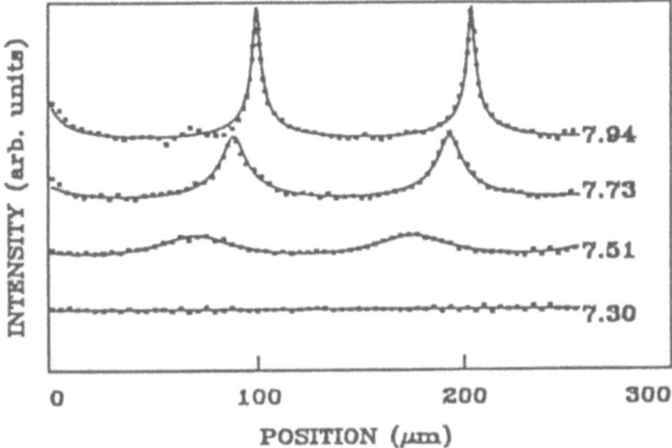

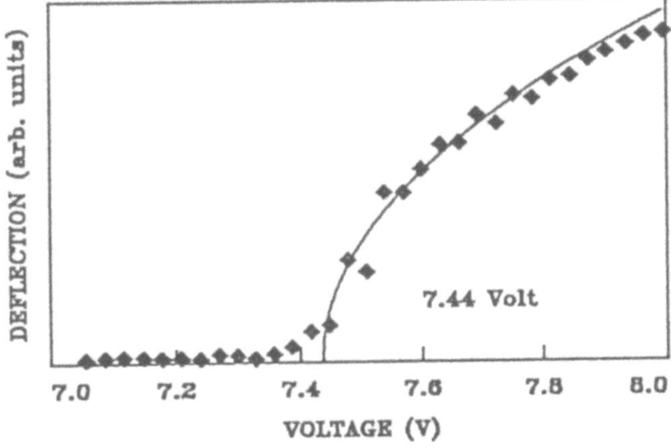

Fig. 6   The intensity measured for different driving voltages in a cell of 100 μm thickness.

Fig. 7   Order parameter extracted from the intensity measurements shown in Fig. 6. The absolute distance of the focal plane of the microscope from the cell is not known exactly, thus arbitrary units are given for the deflection of the light inside the cell. This value, however, is proportional to the tilt angle θ of the director and thus a suitable order parameter.

As an alternative to the optical methods described above, electrical methods can be used. From this point of view, the cell can be considered as a capacitance parallel to an electric resistance. If not only the voltage, but also the electrical current flowing into the cell and the phase between both is measured for different driving frequencies below the onset of convection,

39

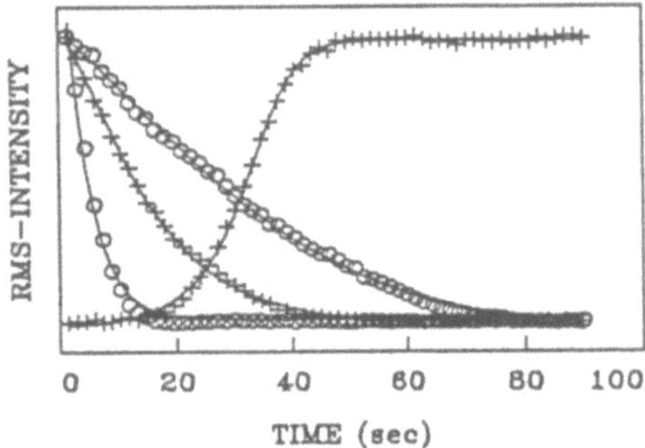

Fig. 8 Time dependence of the order parameter for different voltage steps, namely to 11.78 V (supercritical), 11.06 V, 11.03 V and 10.78 V (subcritical values), 30 IIz.

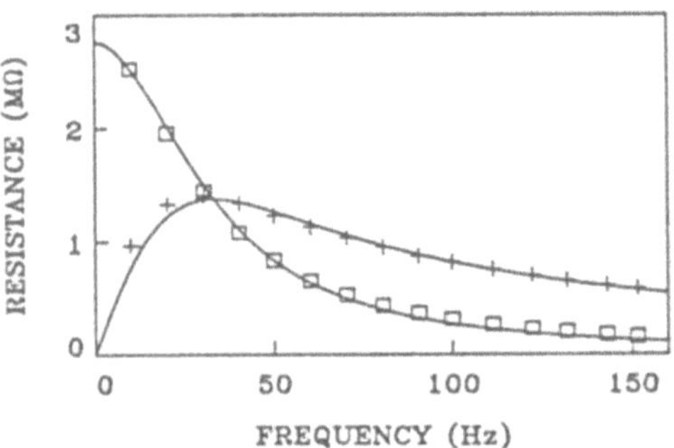

Fig. 9 Real (squares) and imaginary (plus) part of the resistance measured at a subcritical ac-voltage of 4.5 V RMS.

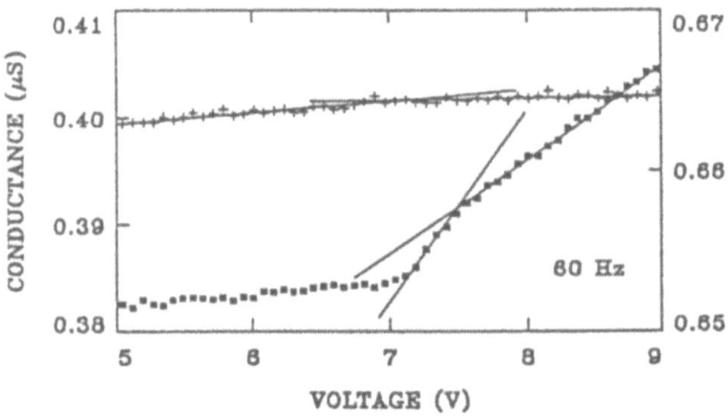

Fig. 10 The real (squares, left hand side labels) and imaginary (crosses) conductance measured as a function of the driving voltage at 60 Hz. The change of the slope indicates the onset of convection or a secondary instability.

40

both the resistive part and the capacitance can be extracted as demonstrated in Fig. 9. The onset of convection can be detected as an increase in the electrical conductance of the cell. This measurement is the analogue of the heat transport measurements in thermal convection. Figure 10 shows an example. Both the resisistive part (current and voltage in phase) and the imaginary part (phase difference 90 degrees) have been plotted as a function of the driving voltage. The increase in the resistive part reflects the fact that power is needed to drive the convection of the fluid. The change of the imaginary part does not have such a simple physical interpretation. It can be positive or negative depending on the driving frequency.

3 Pattern Formation

The convection pattern shown in Fig. 1 is not the only one possible in EHC. In fact, the kind of pattern appearing at threshold depends on the nematic used, its temperature, its electrical conductivity, the thickness of the cell and the frequency of the driving voltage. Figure 11 shows the

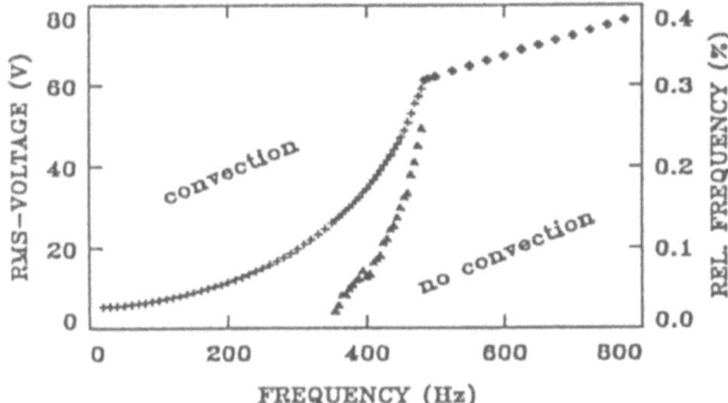

Fig. 11   Phase diagram for a 15 μm Phase V cell. Below 350 Hz convection sets in as a steady pattern, between 350 Hz and 420 Hz it starts as a moving pattern with the frequency given by the triangles; above 420 Hz dielectric rolls appear.

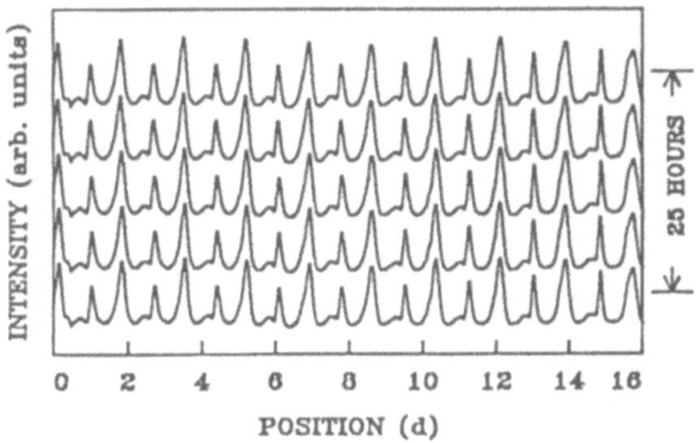

Fig. 12

Time-dependence of steady convection rolls.

threshold measurement of a Phase V cell of 15 μm thickness. Here both the threshold voltage and the kind of pattern appearing above threshold are a function of the driving frequency. The simplest pattern occurs at medium frequencies: The stationary Williams rolls shown in Fig. 1. Figure 12 shows a measurement of their time dependence which is a boring one: There is no hint for a movement within 25 hours.

Decreasing the frequency leads to another type of stationary roll pattern: the zig-zags shown in Fig. 13. Here the rolls are not normal to the director like in Fig. 1. They form a finite angle with respect to the normal axis, which becomes smaller when approching a critical frequency, the so called Lifschitz-point [8,9]. For frequencies higher than this critical frequency the rolls are normal to the orientation of the director.

Figure 14 shows a measurement of the angle of the rolls with respect to the axis normal to the director. On the basis of symmetry arguments, one would expect the angle to approach zero at the Lifschitz-point via a square root law. The measurement does not exactly support this statement. On the other hand it is not really contradictory, because all statements stemming from amplitude equations are local ones. The range of applicability of the square root law might be very small in this case.

Increasing the frequency above the normal roll regime leads to another type of instability: Travelling waves occur. They come in like a supercritical bifurcation as demonstrated in the Figs. 15,16. This travelling pattern is not explained by the stability analysis presented in [6]. We will discuss possible explanations in more detail in chapter 4.

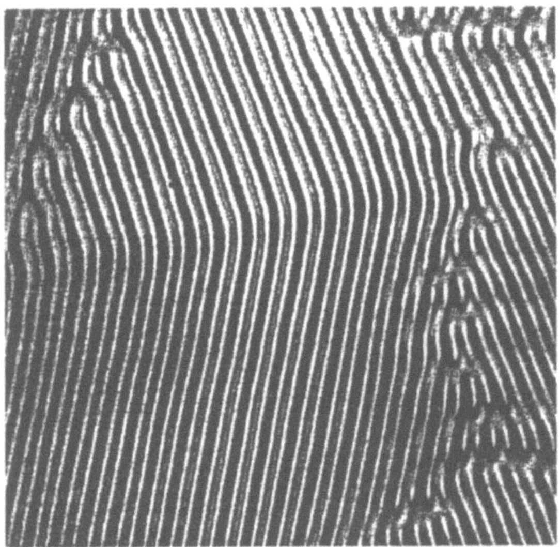

Fig. 13   Image of convection in the oblique roll regime.

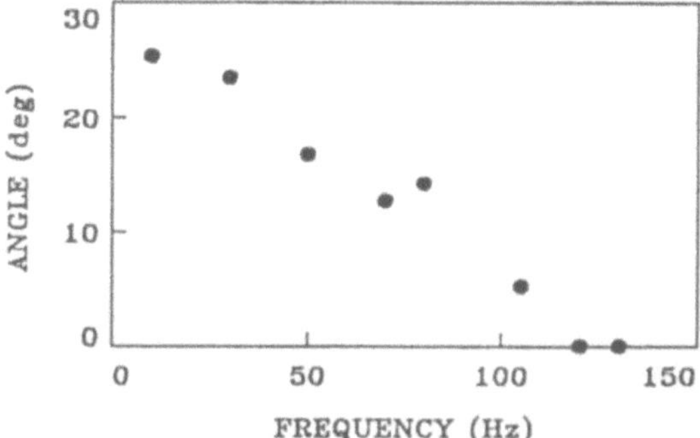

Fig. 14 The angle between "zig and zag" at the onset of oblique rolls.

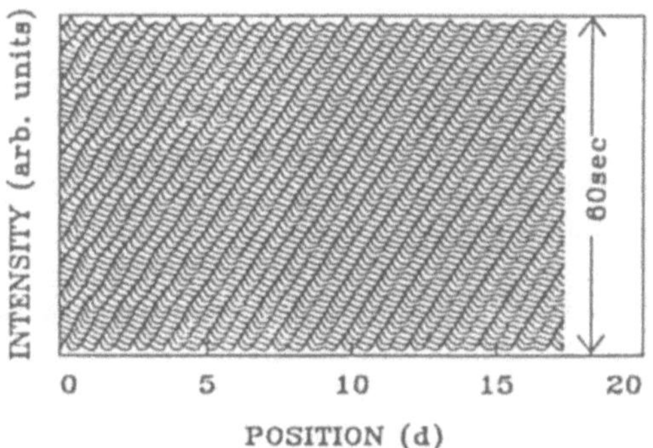

Fig. 15 Travelling waves. Intensity lines measured perpendicular to the roll axis are measured every second and plotted on top of each other.

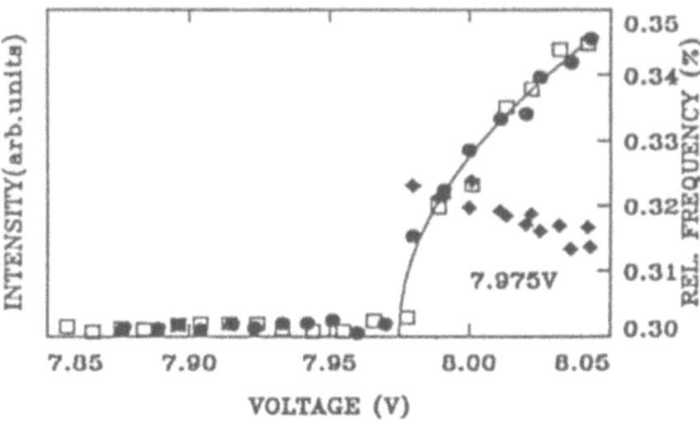

Fig. 16 Amplitude (squares: increasing voltage, circles: decreasing voltage) and frequency (diamonds) of TW. The amplitude grows continuously following a square root law; and the frequency comes in at a finite value.

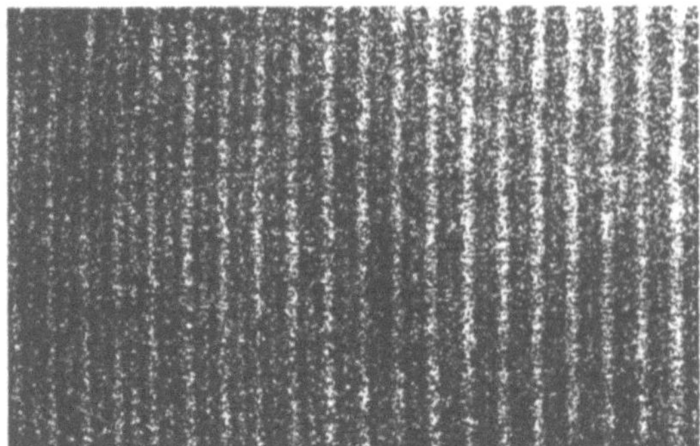

Fig. 17 Image of dielectric rolls obtained with stroboscopic illumination.

At higher frequencies the first instability to occur are the so called dielectric rolls. The difference between the low-frequency and the high frequency regime is illustrated nicely in [10]: In the first state, the director and flow fields are approximately stationary within one period of the external driving, while in the dielectric regime these fields oscillate following the external frequency. This fact makes dielectric rolls hard to observe, because the linear light modulation is averaged to zero in time. This difficulty can be overcome with stroboscopic illumination, which had been used to take the photograph 17.

The patterns shown up to now bifurcate directly from the homogenous ground state, i.e. it should be possible to describe them by means of a linear stability theory. If the driving voltage is increased beyond the critical one, one would expect the amplitude of the director field and the velocity to grow according to a square root law as already indicated in Fig. 7. Predicting the magnitude of this growth requires a nonlinear amplitude expansion as given in [6]. Experimentally, the optical method described above might be used to extract the tilt angle of the director field. Knowing the field of the optical axis within the liquid crystal, one can calculate the distance between the real and imaginary foci shown in Fig. 5. This distance can be measured pretty well with a microscope, and from this measurement the maximum angle of the director can be extracted (Fig. 18). The comparison with the nonlinear amplitude expansion (solid line) presented in [6] at the proper frequency seems pretty good, if one takes into account that no adjustable parameters are involved in the plot.

As demonstrated above, the order parameter increases with an increase of the control parameter. This growth of the amplitude does not go beyond some limit. At a second threshold the primary patterns go unstable with respect to the formation of new ones. For these patterns the sentence stated for the primary patterns has to be repeated: The kind of instability depends on many different parameters. We like to present two examples of patterns

44

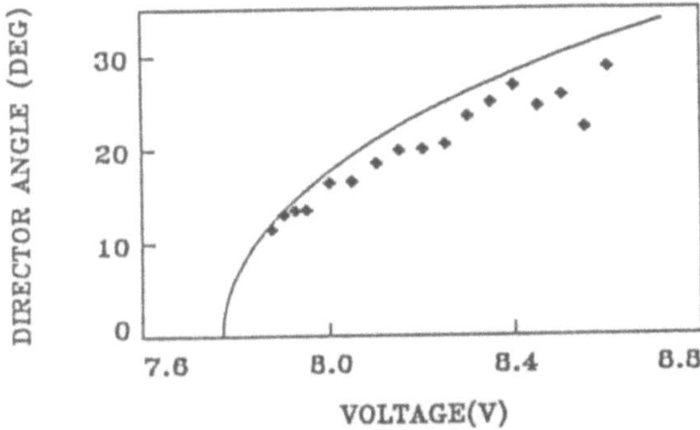

Fig. 18 Director angle obtained by measuring the distance between the real and imaginary foci. The solid line is the result of the nonlinear calculations presented in [6].

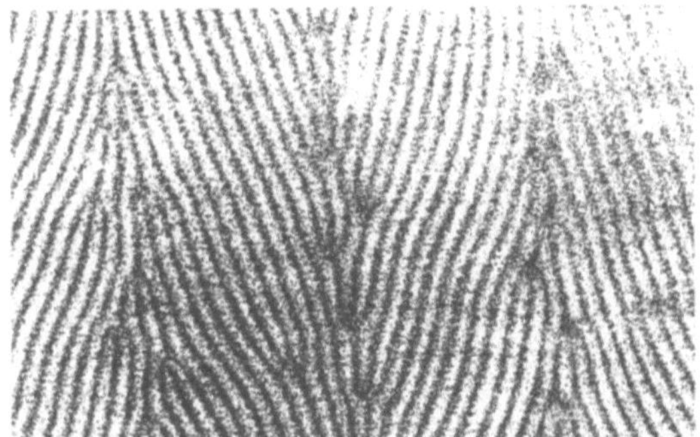

Fig. 19 The chevron pattern, the result of a secondary instability in the dielectric roll regime.

observed in the nonlinear regime: chevrons and defects. Increasing the voltage in the dielectric regime slightly above threshold leads to a characteristic flow pattern: the herringbone-like chevrons shown in the photograph 19. This pattern is unexplained theoretically, but it had been speculated [10] that its origin can be understood even by linear theory: Slightly above the first threshold linear stability analysis yields a second mode, and the superposition of these modes might be a possible explanation for the observed chevron pattern.

A characteristic feature of convection in the nonlinear regime is the existence of defects. In the following, we concentrate on the dislocations shown in Fig. 20. They are created by a secondary instability [11], which seems to have its reason in mean-flow effects. Sudden jumps in the amplitude or frequency of the driving voltage also creates this kind of imperfection of the roll

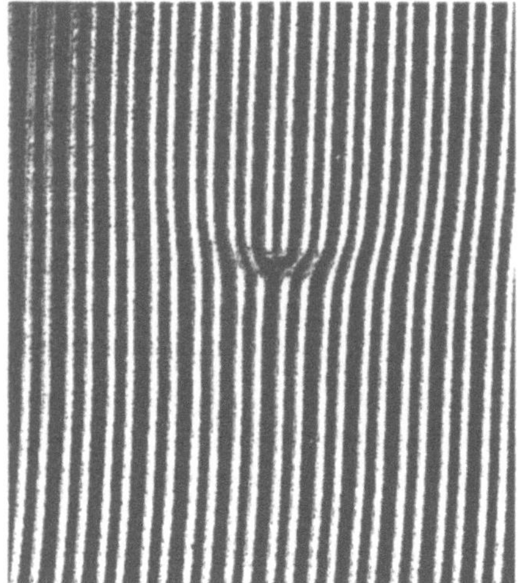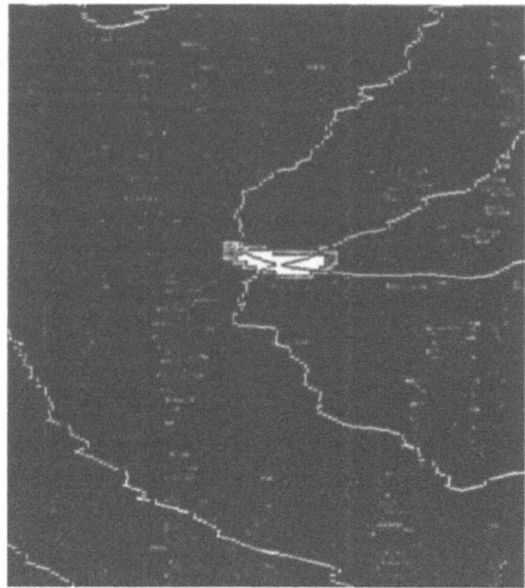

Fig. 20 The image of a defect is shown on the left hand side (virtual image). The slow modulation field extracted from this image is shown on the right hand side (black: large amplitude, white: small amplitude). The defect is located at the crossing of the zero lines of the real and the imaginary part of the slow modulation field.

pattern. In general, they are very robust and in fact hard to avoid-analogue to the process of crystal growing it is much harder to create perfect patterns than a structure accompanied by defects. Once created, the dislocations are stable at any driving voltage for topological reasons: they can annihilate in pairs only. Because they exist even slightly above threshold, they should be adequately described within the framework of a weakly nonlinear analysis, i.e. an amplitude equation. As an ansatz, one takes the linear unstable mode $\exp(ikx)$ and determines its slowly varying complex amplitude $A(x,y)$. The concept indeed leads to a realistic description of a defect. The experimentalist observes the real part of the field $A(x,y) \cdot \exp(ikx)$. The slowly varying amplitude can be extracted from this field by means of a demodulation procedure. Figure 20 shows a the magnitude of the $A(x,y)$-field, together with the lines where the real and imaginary part of this field are zero. The core of the defect is located where these lines cross.

4 Travelling Waves

Now we concentrate on one of the patterns shown in section 3, namely travelling waves (TW). As already mentioned above, they are unexplained by the theoretical calculations, but have nevertheless been observed by various experimental groups [12]. In order to discuss possible explanations for this kind of movement, let us recall a system where travelling waves are observed as well and theoretically well understood. The most popular example in the

46

field of fluid dynamics seems to be thermal convection in a binary mixture. Here a mixture of two fluids, say water and ethanol, is heated from below. Convection then sets in as an oscillatory instability. This is best demonstrated by pulse methods: A small (mechanical or thermal) distortion of the system leads to an oscillatory response as demonstrated in Fig. 21. A pulse of waves travelling out of the middle of the convection channel [13], where the heater is located, to the left and right of the cell is clearly seen. The reason for the oscillatory response in binary mixture convection is well understood: linear stability analysis predicts that the system goes unstable via a Hopf bifurcation, i.e. the linear growth rate contains an imaginary part.

When looking for possible explanations of the TW, one should compare to other systems as well: Taylor vortices between conical cylinders drift in a preferred direction [14], a broken left-right symmetry might induce drifting convection patterns [13], or convection patterns in a cylindrical convection cell might drift when the boundary conditions are spatially modulated [15]. We call this kind of movement drifting rolls instead of TW. They differ from the TW in the sense that they are induced by the broken symmetry of the system — they travel in one preferred direction. It is speculative but plausible to assume that non-parallel plates of the EHC-cell might lead to a drifting pattern similar to the drifting Taylor vortices between conical cylinders. In fact, it seems to be very difficult to prepare cells showing perfectly stationary rolls on the time scale presented in Fig. 12. Most often, the rolls drift with a speed of less than a wavelength/minute. One has to consider the question if the TW presented in Fig. 15 are of this drifting nature, i.e. caused by imperfections of the sample rather than by a Hopf bifurcation. One strong argument against this assumption is already the fact that the

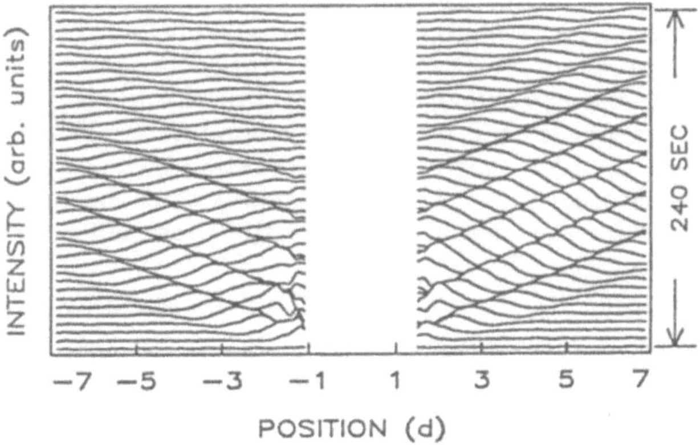

Fig. 21 Pulse of travelling waves caused by a heat pulse in a binary mixture of water and ethanol. The heater is located on the outside of the glass in the middle of the cell, where no intensity lines are visible.

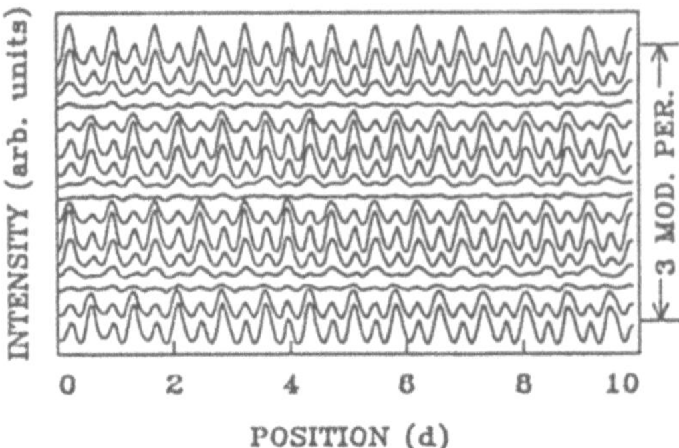

Fig. 22 Standing waves of oscillatory Williams rolls stabilized by an external modulation.

observed TW can travel in both directions and can change direction even spontaneously. Moreover, a theory concerned with the influence of a temporal modulation of the external driving on TW caused by a Hopf bifurcation supports this scenario. It was predicted, for instance, that TW should loose stability with respect to standing waves if the external frequency is close to double the linear frequency of the system [16]. This is indeed observed as demonstrated in Fig. 22. Roughly speaking, the stabilizing influence of the modulation on standing waves can be understood as follows: Both linear modes, the waves travelling to the left and the ones travelling to the right are phase locked and thus stabilized. Their superposition forms the standing wave.

Whether the external modulation leads to standing waves or not is a question of its frequency and amplitude. The phase diagram obtained with modulation is presented in Fig. 23. One has no convection on the left side of the line formed by crosses and plus signs. Convection may set in as a standing wave (diamonds) or as a travelling wave (plus). The solid line is a fit to the theoretical curve given by Riecke et al. [16]. This semiquantitative agreement between the experimental results and the ideas based on general symmetry arguments for the problem of a modulated Hopf bifurcation strongly supports the idea that a Hopf bifurcation is responsible for TW observed in EHC. Moreover, we like to point out that the modulation provides a useful tool for the measurements of the coefficents of the amplitude equation describing the TW [17].

The open circles in Fig. 23 indicate the upper bound of the standing wave state. Beyond this limit the system shows a disordered pattern consisting of nonperiodic changes in the direction of the travelling accompanied by defects [18]. This instability mechanism can be understood by a stability analysis of the coupled amplitude equations. The system is expected to go unstable with respect to the so-called Benjamin-Feir instability, which indeed has been shown to be able to produce chaotic spatio-temporal behaviour [19].

48

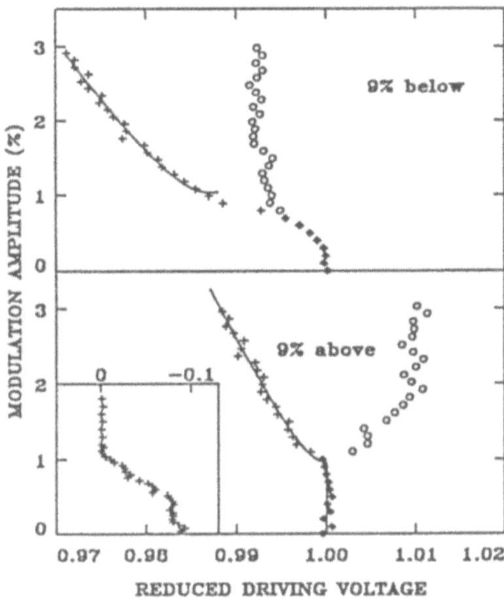

Fig. 23 Phase diagram with modulation.

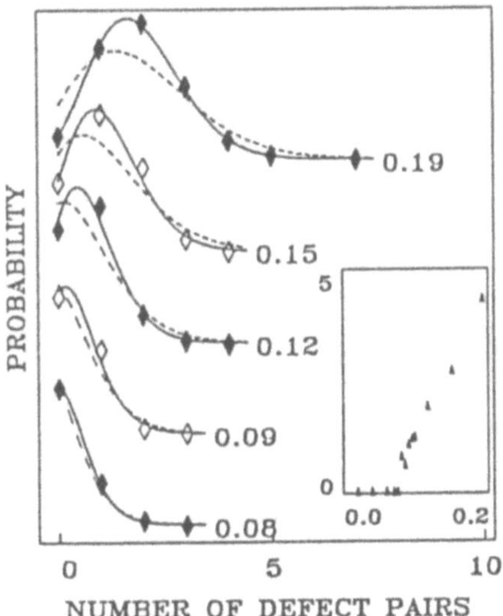

Fig. 24 Histogram of defect pairs for different driving voltages. The distribution seems well described by the solid line [20], while the dashed line (Poisson distribution) does not seem to be adequate. The inset demonstrates that the defects sets in 7.8% above threshold, thus they are very likely caused by a secondary instability.

Finally, we like to present another route to disorder. When increasing the
driving voltage beyond the threshold voltage in an ordered TW state, one
finds a well defined voltage where the correlation function starts to decay
[12]. This is accompanied by the observation of defects. Counting these
defects in a small spatial window as a function of time gives no hint for
any regularity within the time series. This leads to the idea of looking
for a theoretical description. Such an ansatz has been made by Gil et
al. [20]. With the idea that the rate of creation of defects is a voltage-
dependent constant, but the rate of annihilation processes is proportional to
the number of defects squared one ends up with a distribution looking like a
squared Poisson distribution for the number of defect pairs. This function is
shown as a solid line in Fig. 24 together with the histograms of the numbers
of pairs counted. If one takes into account that the distribution offers no
adjustable parameter provided that the mean value is given the agreement
can be called very good. This distribution, however, is not expected to be
valid for travelling waves only. An agreement of similar quality had been
found for steady convection as well [18].

5 Discussion

Obviously, electro-hydrodynamic convection in nematic liquid crystals is a
very suitable candidate for experiments dealing with pattern formation and
order-disorder transitions in nonequilibrium systems. This system, when
compared to other fluid dynamical systems, has at least two disadvantages:
There are more material parameters, and they are less well known when
compared to say thermal convection in water; and the mathematical description of the instability is more complex when compared to instabilities
in simple fluids.

- — The most striking advantage of the system is in the shorter time scale:
  reaching the instability point in the double diffusive convection experiment (Fig. 21) takes about a day — the same can be achieved in EHC
  in less than a minute.
- — The degeneracy of the wavevector present in an isotropic fluid is raised
  because of the preferred axis of the system. This, in principle, simpli-
  fies the theoretical description of the nonlinear state and in practice
  makes it simpler to perform reproducible measurements.
- — The instability is driven by an ac-voltage whose amplitude and frequency are experimentally easily controllable parameters. By changing
  the amplitude a few volts the complete transition scenario from highly
  ordered patterns to a very disordered two-dimensional turbulent state
  can be studied. Changing the frequency of the applied voltage offers a degree of freedom. Thus bifurcations of higher codimension are
  accessible.
- — The driving ac-field does not break the up-down symmetry of the system. This fact offers the possibility for the eigenmodes to be even
  or odd with respect to the middle of cell. Thus, two classes of solu-

tions (Williams rolls and dielectric rolls) become available adding to
the richness of the system.
— Cells with large aspect ratios can be built allowing the observation of
spatio-temporal disorder.
— Besides being well suited for studies in pattern formation the system
offers the challenge to be nontrivial in nature. The interaction of hy-
drodynamics and electrical forces in an anisotropic fluid is a beautiful
and fascinating field of physics with the additional advantage to be of
technical use.
— Most important, the fascination of EHC seems to attract especially
friendly people like E. Bodenschatz, L. Kramer, W. Pesch, and W.
Zimmermann. We like to thank them for many helpful suggestions,
patient explanations and enthusiastic support.

References

[1] *H. v. Ditfurth*, Am Anfang war der Wasserstoff. Hoffmann und Campe (Hamburg
     1973)

[2] Propagation in Systems far from Equilibrium, Proceedings of the Workshop in Les
     Houches, France, March 1987, ed. by *J.E. Wesfreid, H.R. Brand, P. Manneville,
     G. Albinet, N. Boccara*; Springer Series in Synergetics, Vol. **41** (Springer, Berlin
     1988) and other volumes of this series

[3] See e.g. the review of *R.C. DiPrima* and *H.L. Swinney*, in: Hydrodynamic Insta-
     bilities and the Transition to Turbulence, ed. by *H.L. Swinney* and *J.P. Gollub*
     (Springer, Berlin 1981)

[4] See e.g. the article of *F.H. Busse*, in: Hydrodynamic Instabilities and the Transition
     to Turbulence, ed. by *H.L. Swinney* and *J.P. Gollub* (Springer, Berlin 1981)

[5] See the contributions of *G. Ahlers, D.S. Cannell,* and *R.S. Heinrichs*, in: Chaos 87,
     International Conference on the Physics of Chaos and Systems far from Equilibrium,
     ed. by *M. Duong-Van* (North Holland, Amsterdam 1987) and Nuclear Physics B
     (Proc. Suppl.) **2** (1987); *P. Kolodner, A. Passner, H.L. Williams, C.M. Surko*, ibid.;
     V. Steinberg, E. Moses, J. Fineberg, ibid

[6] *E. Bodenschatz, W. Zimmermann,* and *L. Kramer*, J. Phys. France **49**, 1875 (1988)
     and references cited therein

[7] *S. Rasenat, G. Hartung, B.L. Winkler,* and *I. Rehberg*, Experiments in Fluids **7**, ?
     (1989)

[8] *W. Zimmermann* and *L. Kramer*, Phys. Rev. Lett. **55**, 402 (1985)

[9] *R. Ribotta, A. Joets,* and *Lin Lei*, Phys. Rev. Lett. **56**, 1595 (1986)

[10] *W. Thom*, Diploma Thesis, Bayreuth 1988; and *W. Zimmermann* and *W. Thom*,
     to be published

[11] *X.D. Yang, A. Joets,* and *R. Ribotta*, in: Propagation in Systems far from Equi-
     librium, Proceedings of the Workshop in Les Houches, France, March 1987, ed. by
     J.E. Wesfreid, H.R. Brand, P. Manneville, G. Albinet, N. Boccara (Springer, Berlin
     1988)

[12] *K. Hirakawa* and *S. Kai*, Mol. Cryst. Liq. Cryst. **40**, 261 (1977); *A. Joets* and *R. Ri-
     botta*, Phys. Rev. Lett. **60**, 2164 (1988); *I. Rehberg, S. Rasenat,* and *V. Steinberg*,
     Phys. Rev. Lett. **62**, 756 (1989)

[13] We used a 20% by weight concentration of ethanol in water and the experimental apparatus described in: *I. Rehberg, E. Bodenschatz, B.L. Winkler, and F.H. Busse*, Phys. Rev. Lett. **59**, 282 (1987)

[14] *M. Wimmer*, ZAMM **65**, T255 (1985); *M. Abboud*, ZAMM **68**, T275 (1988)

[15] *G. Hartung*, Diploma Thesis, Bayreuth 1988

[16] *H. Riecke, J.D. Crawford, and E. Knobloch*, Phys. Rev. Lett. **61**, 1942 (1988); *D. Walgraef*, Europhys. Lett. **7**, 485 (1988)

[17] *M. de la Torre Juarez and I. Rehberg*, in: New Trends in Nonlinear Dynamics and Pattern Forming Phenomena: The Geometry of Nonequilibrium, ed. by *P. Coullet* and *P. Huerre* NATO ASI Series (Plenum Press 1988)

[18] *I. Rehberg, S. Rasenat, J. Fineberg, M. de la Torre Juarez, and V. Steinberg*, Phys. Rev. Lett. **61**, 2449 (1988)

[19] *A.C. Newell*, in: Propagation in Systems far from Equilibrium, Proceedings of the Workshop in Les Houches, France, March 1987, ed. by *J.E. Wesfreid, H.R. Brand, P. Manneville, G. Albinet, N. Boccara* (Springer, Berlin 1988), and references cited therein. The fact that modulated TW go unstable via this mechanism has been demonstrated by *W. Zimmermann*, to be published

[20] *L. Gil, J. Lega, and J.L. Meunier* (to be published)

X-Ray Absorption and Reflection in Materials Sciences

Bruno Lengeler

Kernforschungsanlage Jülich, Institut für Festkörperforschung, Postfach 1913, D-5170 Jülich, Federal Republic of Germany

Summary: The X-ray absorption spectroscopy (XAS) is a local probe of the geometric and electronic structure of specific atoms in condensed matter. With the availability of synchrotron radiation from storage rings it has found widespread use in science and technology, giving interatomic distances, coordination numbers, electronic densities of state and the valence of the absorbing species. Like all X-ray techniques XAS can be made surface and interface sensitive by means of external total reflection with a probing depth as low as 20 to 70 Å . This new development is described in detail. Reflectivity measurements as a function of the angle of grazing incidence give the mass density in the surface layer, its thickness and the roughness of interfaces. The chemical composition and its depth profile can be determined in a non-destructive way by measuring the energy and angular dependence of the fluorescence intensity. A number of examples from materials science and technology will illustrate these spectroscopies.

1 X-ray Absorption Spectroscopy (XAS)

1.1 Introduction

When X-rays pass through matter, their intensity is attenuated. The relationship between the intensities $I_1(E)$ and $I_2(E)$ before and after passing a layer of thickness d reads

$$I_2 = I_1 \, exp(-\mu(E)d). \tag{1.1}$$

The linear absorption coefficient $\mu(E)$ shows edges at certain energies, which are characteristic for the absorber atom. At these absorption edges the photon energy is high enough for strongly bound electrons to be excited into empty states. At a K edge 1s electrons are excited, at L_1, L_2 and L_3 edges 2s, 2p1/2 and 2p3/2 electrons are excited into empty states. Figure 1.1 shows the linear absorption coefficient for amorphous and crystalline Ge in the vicinity of the Ge K-edge. μd shows oscillations above the edge. The oscillating contribution in μ is called EXAFS (extended X-ray absorption fine structure) and contains information on the geometric structure (interatomic distances, coordination numbers) around the absorbing species. The form and the position of the edge depend on the chemical bonding of the absorbing species. Especially, the higher the valence of the absorber the higher is the energy of the edge position. The details in the structure of the edge re-

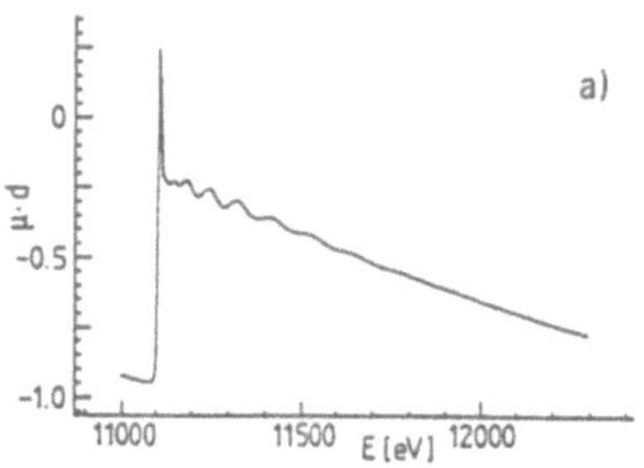
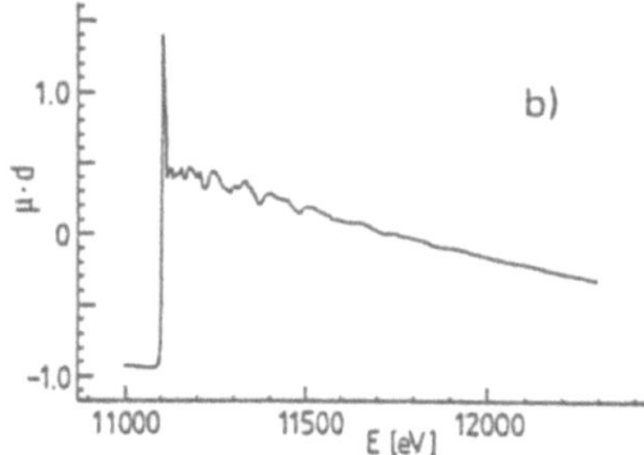

Fig. 1.1 X-ray absorption at the Ge K-edge in amorphous germanium (a) and in crystalline germanium (b) measured at 77 K.

flect the details in the empty electronic density of states. The references 1-8 are some recent review articles on the XAS. The XAS in transmission mode is a bulk probe. On the other hand, there is an increasing interest in making X-ray techniques surface and interface sensitive. This can be achieved by total external reflection. This interesting possibility will be discussed in the second part of this paper.

1.2 Experimental Set-Up in XAS

The most common procedure for measuring the X-ray absorption uses X-rays from an electron storage ring. The X-rays are monochromatized by a Si double crystal. At 10 keV an energy band about 1 eV wide is transmitted by the monochromator. At 3.7 GeV and 100 mA electron current about 10^{10} photons impinge per sec on a sample spot $1 \times 10mm^2$ in size. The photon flux can be increased by means of a focusing mirror or by using insertion devices instead of a bending magnet. The incoming intensity $I_1(E)$ and the transmitted intensity $I_2(E)$ are measured by ionization chambers (transmission mode). The hole created in a 1s, 2s, 2p ... shell will be filled up by an electron from an upper shell. This process is accompanied by the emission of a fluorescence photon or by the emission of Auger electrons. At high ionization energies fluorescence is the dominant decay channel, whereas at lower energies the Auger channel is favored. The intensities of the emitted photons and electrons show EXAFS oscillations just as the transmitted intensity does. The fluorescent radiation being characteristic for the absorbing species, the fluorescence mode is mainly used for dilute systems (from about 1 at% down to about 10 atppm). A great advantage of the transmission and fluorescence modes is the possibility to measure samples in-situ. The relatively high penetration of X-rays through matter allows for a sample chamber with appropriate windows (Kapton, beryllium, Al ...). The samples can be surrounded by a gas atmosphere or a liquid. This is especially interesting for high pressure experiments and for the investigation of catalysts. On the other hand the electrons emitted after X-ray absorption originate from a thin surface layer. This is a consequence of the short range

54

of electrons in condensed matter when their energy is in the range of a few
eV to about 1 keV. XAS in the electron yield mode is therefore a surface
sensitive procedure (SEXAFS). This detection mode will not be discussed
in this article. Details about this technique can be found in reference 8.

1.3 EXAFS

The relative change in the absorption coefficient μ is denoted in the following
by χ

$$\chi(k) = (\mu(E) - \mu_0(E))/\mu_0(E). \tag{1.2}$$

$\mu_0(E)$ is the absorption coefficient that would be measured if the absorbing
atoms had no neighbors by which the photoelectron could be scattered.
The photoelectron emerges from the absorber as a spherical wave, with
wave vector k given by

$$k = \{2m(E - E_0)/\hbar^2\}^{1/2}. \tag{1.3}$$

Other atoms at a distance r_j from the absorber scatter the photoelectron.
The incoming and the scattered wave can interfere. When the interference
is constructive the probability to find the photoelectron outside of the absorber is larger than in the case without neighbors, so that $\mu > \mu_0$. For
destructive interference $\mu < \mu_0$. Therefore one expects a periodicity of χ
with k, the periods being the double distances $2r_j$ to each shell j surrounding the absorber. The amplitude of the EXAFS $\chi(k)$ is proportional to the
coordination number N_j in the j-th shell. Thus, the periods of the EXAFS
oscillations give the interatomic distances measured from an absorber atom
and their amplitudes give the corresponding coordination number. The following two examples should demonstrate the capabilities of the technique.

Example 1: Amorphous germanium [9,10].

Figure 1.2 shows the EXAFS $\chi.k^2$ of a-Ge from figure 1.1. The signal
contains only one period. This is more clearly revealed by its Fourier transformation (figure 1.3). In contrast, crystalline germanium shows 8 coordi-

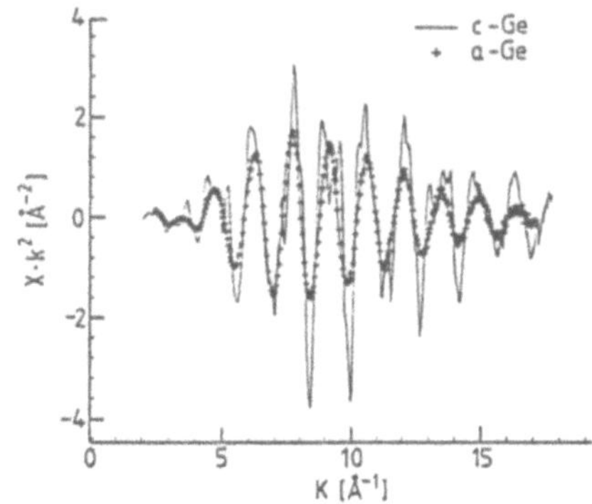

Fig. 1.2

EXAFS $\chi \cdot k^2$ of a-Ge (+) and
c-Ge (-) measured at 77 K.

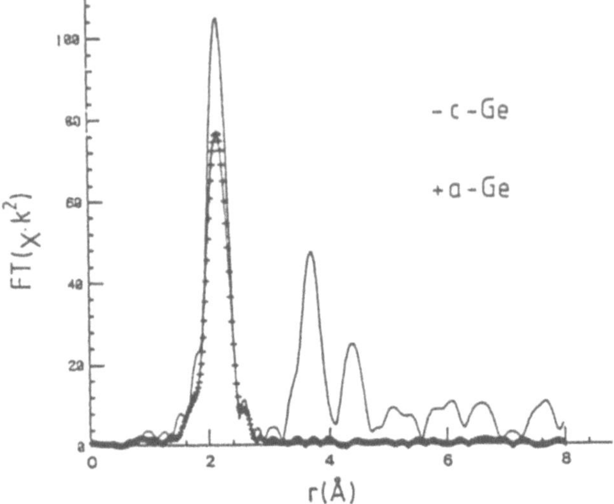

Fig. 1.3 Fourier transformation of the EXAFS $\chi \cdot k^2$ from figure 1.2

nation shells in the distance range up to 8 Å . The first interatomic distance in a-Ge deduced from the period in figure 1.2 is (2.448 ± 0.003) Å . This value is identical with the corresponding value 2.447 Å in c-Ge. The corresponding coordination numbers N_1 for the first shell are 4 ± 0.04 in a-Ge and 4 in c-Ge. According to the Polk model [11] the amorphicity of a-Ge can be understood in the following way. The elementary building bloc in a-Ge and in c-Ge is a tetrahedron in which each Ge atom is surrounded by 4 Ge atoms. Amorphicity is due to a spread of the tetrahedral angle of 10% about its crystalline value of 109.5°. This induces a small spread in the first interatomic distance r_1, which is visible as a reduction of the first shell contribution in the EXAFS signal of a-Ge compared to c-Ge (figure 1.3). The EXAFS analysis gives a Gaussian distribution with 0.1 Å spread (FWHM) around the average value. About 0.1% of the bonds in a-Ge are dangling bonds [12]. The EXAFS data for the first coordination number (4 within 1%) are consistent with this value. The spread in the tetrahedral angle of a-Ge induces strong variations in the second and all higher interatomic distances. They are so strong that only the first interatomic distance is visible in the EXAFS signal of a-Ge (figure 1.3). The EXAFS analysis gives a spread of at least 9° rms (root mean square) to the tetrahedral angle which compares well with the value of 10° deduced from X-ray scattering. A more detailed EXAFS analysis of a-Ge (including void structure and recrystallisation of a-Ge) is given in the references [9,10].

A note should be added about the accuracy of the data obtained by EXAFS. Accuracies of 0.01 Å for the interatomic distances are standard. The accuracy of 1% in the first coordination number of a-Ge is exceptionally

56

good. More usual are values of 5 to 10% . It is hard to analyse more than
the first three coordination shells. The EXAFS spectroscopy is indeed a
very local probe of the geometrical structure around the absorbing species.

Example 2: Lattice site location of hydrogen in metals by EXAFS [13,14].

Due to their small number of electrons the light elements H, Li, Be, B,
C, N, O, F are hard to detect by X-ray techniques. In EXAFS e.g. the
backscattering strength of hydrogen with only one electron is too weak to
be observable. Therefore hydrogen shows up in XAS mainly by the lattice
distortion it creates. This effect is not specific enough to determine the
lattice site occupied by the hydrogen. Nevertheless, EXAFS can be used in
many cases to determine the lattice site of hydrogen and other light atoms.
When the hydrogen is located on the line joining the absorber and the
backscatterer, it changes the phase and the amplitude of the photoelectron
on its way to the backscatterer and back, i.e. the hydrogen works as lens
for the photoelectron. This effect is demonstrated in figure 1.4 for Ni and
$NiH_{0.85}$. In both systems Ni forms an fcc lattice, in which the hydrogen
occupies octahedral sites. Those are the sites between an absorber and
its second neighbors. The hydrogen acting as lens increases the second
neighbor amplitude by 50% [13]. The lens effect is not limited to hydrogen.
But hydrogen shows only the lens effect. When the hydrogen is replaced by
a heavier element like oxygen in NiO, which also has the NaCl structure,
then the second neighbor contribution is enhanced even more (figure 1.4).
The oxygen shows up as backscatterers as well. The lens effect has turned
out to be very helpful in determining the lattice site of other light elements
(B in Pd [15]).

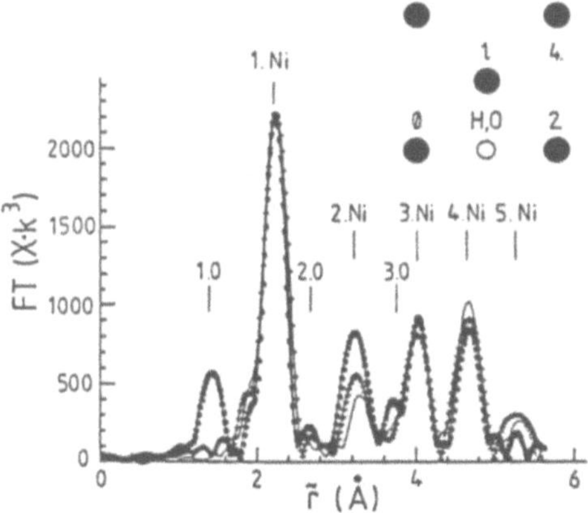

Fig. 1.4   Fourier transforms of the EXAFS $\chi \cdot k^3$ of Ni (–), $NiH_{0.85}$(+) and NiO(*).
The abscissa for the hydride and the oxide have been rescaled in order to eliminate the
influence of the lattice expansion. The inset shows the lattice position of H and O which
shadow a second Ni neighbor.

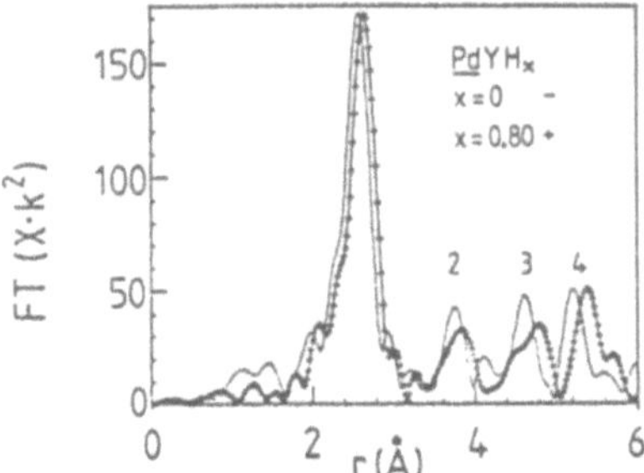

Fig. 1.5 Fourier transform of the EXAFS $\chi \cdot k^2$ at the Y K-edge in $\underline{Pd}$Y(2at%Y) (–) and in $\underline{Pd}YH_{0.80}$(+).

In the following the interaction of H with substitutional impurities in Pd will be analyzed by means of this effect [14]. Yttrium in Pd generates a strong lattice expansion. Y pushes the first shell of 12 Pd neighbors outwards by 0.06 Å , as deduced from EXAFS. When the hydrogen is dissolved in Pd the question arises whether the hydrogen prefers the vicinity of the Y atoms where the lattice has already been expanded, i.e. whether the Y traps the hydrogen. Figure 1.5 shows the Fourier transform of the EXAFS at the Y K edge in $\underline{Pd}$Y 2 at% and in $\underline{Pd}$Y 2% H 80% . Pd having an fcc lattice, it is again the second neighbor shell that is of interest here. It is obvious that in the hydrogen loaded sample the second shell is not enhanced by the hydrogen. This leads to the conclusion, that Y does not trap the hydrogen. In view of the large hydrogen concentration one must even conclude that the Y expels the hydrogen from its neighborhood. If the hydrogen were distributed statistically, 4.8 out of 6 nearest octahedral sites would be occupied with hydrogen for an alloy with 80% hydrogen. This is not the case as shown in figure 1.5. In other words, the hydrogen avoids the vicinity of the Y atoms. This conclusion is supported by an analysis of the lattice distortion around Y. Fig. 1.6 shows the interatomic distances r_1 of the first shell around Y and Pd in Pd. The hydrogen dissolved in pure Pd expands the lattice (full line in fig. 1.6). Substitutional Y in Pd also expands the lattice. The addition of a small amount of hydrogen (10-20%) to the dilute alloy expands the lattice according to the sum of the local expansions induced by H and Y alone (broken line in figure 1.6). If the interaction between Y and H were attractive, the value r_1 would be above the broken line. The addition of a large amount of hydrogen to the dilute alloy (73 and 80%) results in a local lattice expansion around Y which is less than the sum of the individual H and Y induced expansions. This implies that the H is expelled from the vicinity of Y. It also implies that at large H concentrations the H exerts a pressure that shifts the Pd atoms in the nearest Pd shell inwards towards the Y by 0.04 Å . So, there is indeed a repulsive interaction between Y and H in Pd. The same observation was

58

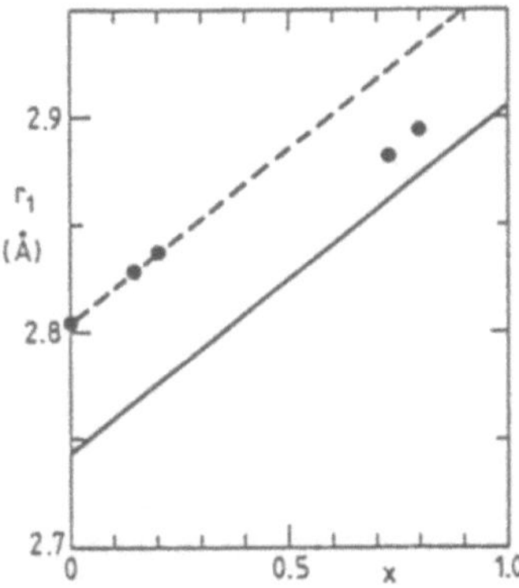

Fig. 1.6 First interatomic distance from Pd in PdH_x (–) and from Y in $\underline{Pd}$ Y2 at% H_x(•). Above the broken line the Y-H interaction is attractive, below it the interaction is repulsive.

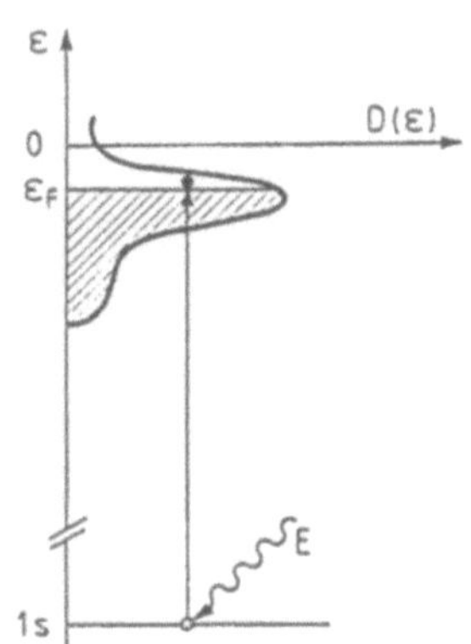

Fig. 1.7

Energy diagram for the absorption of a photon with energy E by a 1s electron.

made for a great number of other oversized impurities in Pd (rare earths, Th, Zr ...) [14].
It is noteworthy that the XAS is especially suited for this type of question. The analysis took full advantage of the atom specific character of XAS, so that even dilute alloys can be investigated.

1.4 Structure of the Absorption Edge

In figure 1.7 is shown an energy diagram of the absorption process. Only those photons can be absorbed by, say, a 1s electron whose energy is larger than its binding energy. Due to the conservation of angular momentum only those final states are allowed which have p-character about the absorber atom. In general, the difference Δl in electron angular momentum has to be $|\Delta l| = 1$. When the photons are absorbed by p-electrons (e.g. L_2 and L_3 edges) only electron transitions into empty s- and d-states are possible. However, the cross section for transition into empty d-states is about 50 times larger than that into empty s-states [2]. The near edge structure $\mu(E)$ can be written as a product [16]

$$\mu(E) = M(E) \cdot \rho_\ell(E) . \tag{1.4}$$

The matrix element M(E) gives the probability for a transition from an occupied, say, 1s state into an empty p state. It is only weakly energy dependent [16]. The projected (ℓ-dependent) density of states $\rho_\ell(E)$ gives the probability to find a final state of appropriate symmetry. The structure observed in the absorption edges reflects the structure in the projected densities of states $\rho_\ell(E)$. There is a number of calculations of $\mu(E)$ for elements and for stoichiometric compounds which confirm this statement [16-19]. An example should show that this is also true for dilute alloys. Since the X-ray absorption is specific to individual species it is possible to probe the projected densities of states locally around the impurity in a dilute alloy.

Example 1: K-edge of dilute Ge in Ni [20]

Figure 1.8 shows the measured K-absorption edge of Ge in a dilute $\underline{Ni}$ Ge alloy with 2 at% Ge. The experimental curve reflects the empty p-like density of states in the Ge cells of this alloy. The result of a band structure is shown for comparison. The p-like density of states is multiplied by the matrix element M(E) and then convoluted with a Lorentzian that takes into account the finite lifetime of the 1s hole, of the final states and the finite energy resolution of the spectrometer. Phase shifts up to ℓ=2,3 and 4 have

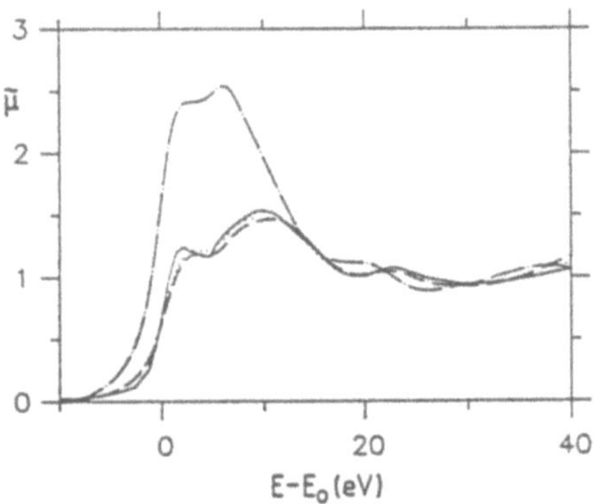

Fig. 1.8 Ge K-edge in $\underline{Ni}Ge$ 2 at% (experiment (—), band structure calculation with $l \leq 2$ (---), $l \leq 3$ ($\cdots$) and $l \leq 4$)(- -).

been included in the KKR calculation. It is obvious that at least phase shifts up to $\ell=3$ have to be included and that higher phase shifts do not improve the calculation. The main results are the following: (1) the structure in the absorption edges reflects the structure in the density of states. (2) The measurement of absorption edges is a simple and stringent test of the quality of band structure calculations. (3) The atom specific character of XAS is an advantage of XAS compared to other complementary spectroscopies like photoemission or BIS (bremsstrahlung isochromate spectroscopy).

In recent time XAS has been extended to the investigation of spin-split empty states. This possibility will be demonstrated in the next example.

Example 2: Spin Dependent X-Ray Absorption [21-24]

When the absorbed photon is circularly polarized the photoelectron becomes spinpolarized. Such photons are available in synchrotron radiation above and below the plane of the storage ring (about 0.1 mrad at 5 GeV electron energy). When, in addition, the final electron states are spin-split as in magnetized substances, the absorption process becomes spindependent. This effect has been observed recently by Schütz et al. in a number of metals and dilute alloys [21-23]. Figure 1.9 shows the absorption coefficient at the Pt L_2 and L_3 edges in a magnetized Fe foil with 3 at% Pt. The difference in absorption for the spin of the polarized photoelectron parallel and antiparallel to the magnetic moment of the sample can be as high as 20%. Relativistic band structure calculations by Ebert et al. [24] give good agreement with the experiment (figure 1.10). The spin splitting of the empty states is strong in the range of the Pt d-bands. The difference between the Pt L_2 and L_3 edges is due to spin-orbit coupling. XAS is a simple test for the quality of the band structure calculation and is applicable even to dilute systems as the present one, a feature that is due to the atom specific character of X-ray absorption.

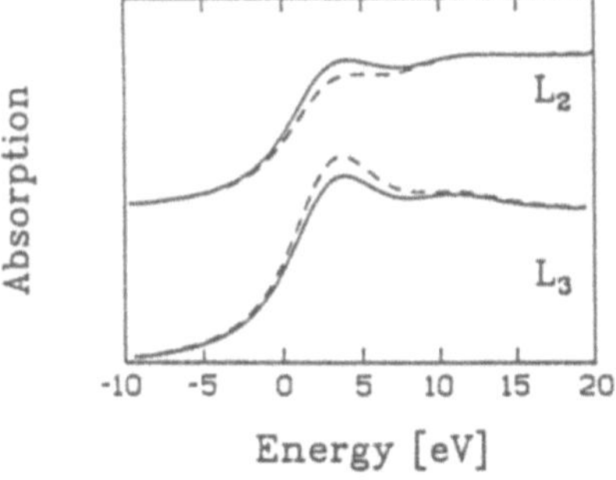

Fig. 1.9 Pt L_2 and L_3 edges in $\underline{Fe}$Pt 3 at% with the photon spin parallel (——) and antiparallel (- -) to the spin of the magnetic electrons in the sample [23].

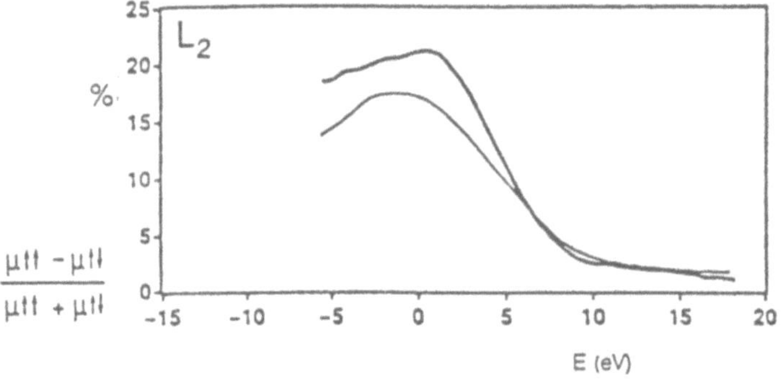
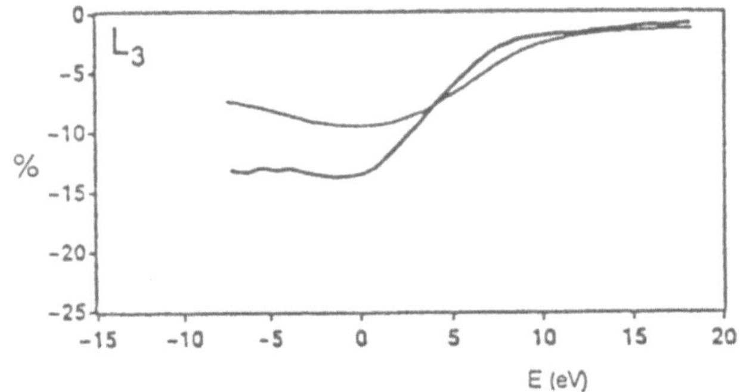

Fig. 1.10 Comparison of the calculated (thick line) [24] and experimental (thin line) [23] spin-dependent absorption coefficients at the Pt L_2 and L_3 edges in $\underline{Fe}Pt$ 3at%. E=0 is the Fermi level.

In many cases of practical relevance there are no band structure calculations available which could help to interpret the measured edge data. A fingerprint procedure might help in these cases. It is based on the experimental observation that the position and the form of the edges depend in a characteristic way on the valence of the absorber and its coordination. This procedure is most helpful when strongly electronegative ligands (like F and O) surround the absorber. In the following a few examples should illustrate the effect. Figure 1.11 shows the Fe $K-edge$ in $GeFe_2O_4$ with divalent Fe^{2+} and in Fe_2O_3 with trivalent Fe^{3+}. The edge is shifted by 4 eV when going from Fe^{2+} to Fe^{3+}. The spinell Fe_3O_4 which contains 1 Fe^{2+} and 2 Fe^{3+} per formula unit is located between the edges of Fe^{2+} and Fe^{3+}. This finger print technique can be used to determine the valence of metal atoms in high T_c superconductors. Figure 1.12 shows the Bi L_1 edge of a BiSrCaCu oxide which contains 2/3 of the 110 K phase and 1/3 of the 80 K phase. The edges of trivalent Bi_2O_3 and of pentavalent $NaBiO_3$ are shown for comparison. It is obvious that the superconductor contains mainly trivalent Bi.

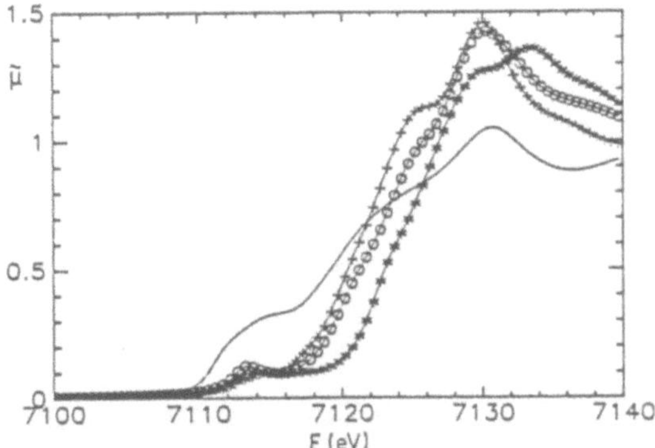

Fig. 1.11 $Fe\ K - edge$ in $GeFe_2O_4$ (+), in Fe_2O_3 (*) and in Fe_3O_4 (O). The first one contains Fe^{2+}, the second one Fe^{3+} and the third one Fe^{2+} and Fe^{3+}. The edge of metallic iron is shown for comparison (−).

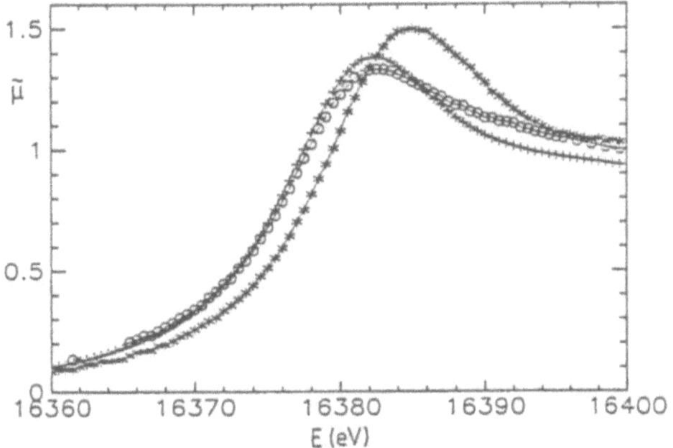

Fig. 1.12 $Bi\ L_1$ edge in Bi_2O_3 (+), in $NaBiO_3$ (*) and in the superconductor $BiSrCaCu$ oxide (O). The superconductor contains trivalent bismuth.

Strong shifts in the position of the edges are also observed in Rare Earth metals and compounds with variable or mixed valence [25]. The shift in the $L\ edges$ is 7 to 8 eV when the f count changes by one electron. The valency of mixed valent systems has been determined by decomposing the measured edges in an f^n and f^{n+1} contribution. An extensive review of this interesting field of research has been given recently by Röhler [25].

Finally, the influence of the type of coordination on the form of the edge should be demonstrated. In WO_3, hexavalent W^{6+} is coordinated octahe-

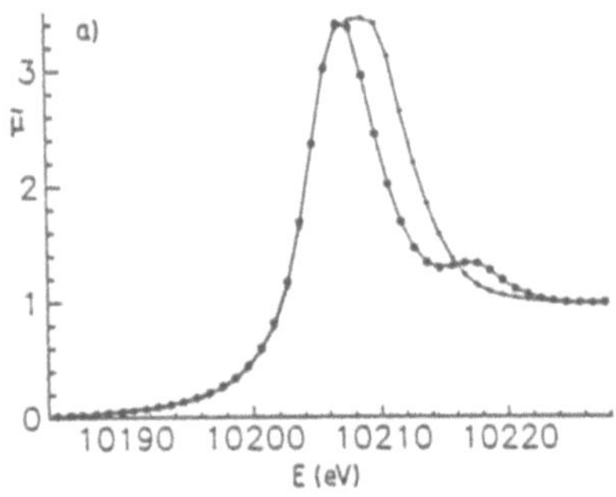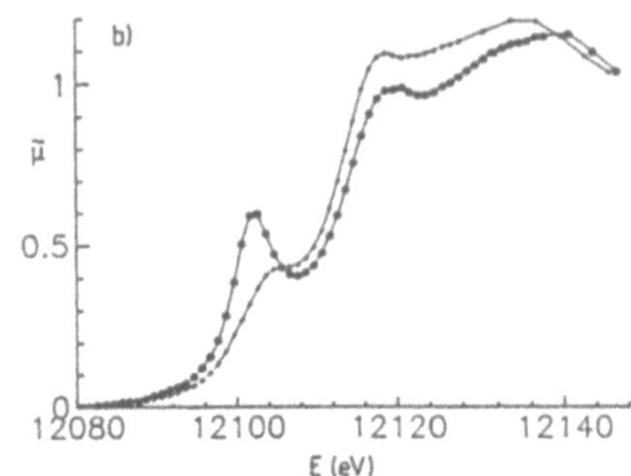

Fig. 1.13 W L_3 edge (a) and L_1 edge (b) of octahedrally coordinated W^{6+} in WO_3 (+) and of tetrahedrally coordinated W^{6+} in Na_2WO_4 (•).

drally by 6 oxygens whereas in Na_2WO_4 it is coordinated tetrahedrally by 4 oxygens. The figure 1.13 shows the L_1 and L_3 *edges* of these two hexavalent W compounds. The position of the edge is not shifted but their forms differ in a characteristic way. This feature has turned out to be very valuable in highly dispersed systems like in catalysts, where X-ray diffraction fails. Since the absorption edges can also be measured at high temperatures this spectroscopy is very helpful for in-situ studies of catalysts which often are operated at elevated temperatures [26].

2. Total Reflection of Hard X-Rays and XAS under Grazing Incidence

XAS in the transmission and fluorescence mode is a bulk probe. This is a consequence of the relatively large penetration depth of hard X-rays into matter. There is an increasing interest in science and technology for analytical tools to probe the geometric and electronic structure of surfaces and internal interfaces. The whole spectrum of X-ray techniques (X-ray absorption, diffraction, fluorescence analysis, topography ...) can be made surface sensitive by means of total reflection. This interesting new development will be described in the following and demonstrated in the case of XAS.

2.1 Total Reflection

Just like in the case of visible light, X-rays are refracted when they cross the interface between two optically different media (Figure 2.1). For X-rays in the hard X-ray range the index of refraction n can be written as [27,6]

$$n = 1 - \delta - i\beta \ .$$

(2.1)

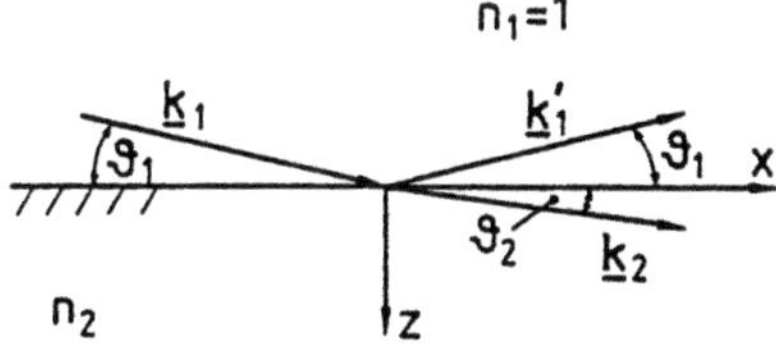

Fig. 2.1 Refraction and reflection of X-rays at the transition from vacuum ($n_1 = 1$) into matter with index of refraction n_2.

The dispersive and absorptive corrections δ and β are small positive quantities of the order 10^{-5} to 10^{-7}. So, for X-rays n is almost one and slightly smaller than 1. Thus, X-rays when entering into matter are refracted away from the normal. Consequently, there is a finite ange ϑ_{1c} for which $\vartheta_2 = 0$. From Snell's law

$$n_1 cos\vartheta_1 = n_2 cos\vartheta_2 \; . \tag{2.2}$$

the critical angle ϑ_{1c} can be deduced (no absorption).

$$\vartheta_{1c} = \sqrt{2\delta_2} \sim \sqrt{\rho(Z + f')\lambda} \; . \tag{2.3}$$

The critical angle is proportional to the photon wave length λ , to the square root of the mass density ρ and to the square root of the real part ($Z+f'$) of the atomic form factor in the forward direction. Typical values of ϑ_{1c} for hard X-rays are below 0.5°. Total external reflection of hard X-rays occurs only at grazing incidence.

2.2 Reflected Intensity at Grazing Incidence

The Fresnel reflectivity R of X-rays from a flat surface is 1 below the critical angle ϑ_{1c} when absorption can be neglected. Above ϑ_{1c} R vanishes like ϑ_1^{-4}. Absorption drastically reduces the reflectivity below ϑ_{1c} as demonstrated in figure 2.2 for the reflectivity of 13001 eV photons from a flat Au surface. The value $\sqrt{2\delta_2}$ is 5.84 mrads in this case.

When the mass density of a reflecting element is known, the reflectivity can be used to determine the energy dependence of the atomic form factor f'(E). In the case of figure 2.2 a value of (-7.9 $\pm$ 0.5) electrons is found for Au at 13001 eV. The same value was found by interferometry [28]. On the other hand, if f'(E) is known, the reflectivity can be used to determine the density of thin films which may differ substantially according to the way of production from that of bulk samples. An example is discussed in section 2.4.

65

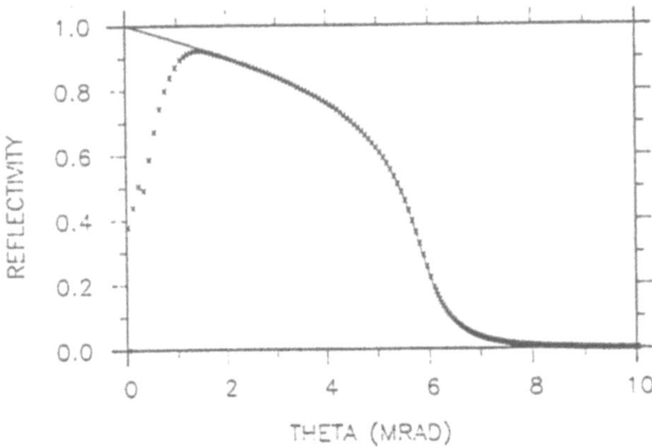

Fig. 2.2 Reflection of 13001 eV photons from a flat Au surface ($\times$) and fit to the measured curve giving a value (-7.9 $\pm$ 0.5) electrons for f' (—).

2.3 Transmitted Intensity at Grazing Incidence

X-rays can only be reflected at an interface when they penetrate by a certain amount into the lower medium. The amplitude of the transmitted electric field is damped exponentially with a damping constant z_0 that is given in figure 2.3 for Cu as a function of the angle of incidence. Above the critical angle ϑ_{1c} the absorption determines the penetration depth z_0. Below it, z_0 becomes independent of the absorption and saturates at a small value z_{00} which is independent of the photon wave length and mainly dependent on the mass density. For silicon with a density of 2.3 g/cm^3 z_{00} is 64 Å, whereas platinum with a density of 21.4g/cm^3 has a value z_{00} of only 23 Å. This small penetration depth z_{00} implies that all X-ray techniques like diffraction, absorption, fluorescence, topography ... become interface sensitive at total reflection, in the sense that the signal originates in a depth of 20 to 60 Å below the interface. This interesting feature opens new possibilities for the study of surfaces and interfaces.

For instance, multielement analysis by X-ray fluorescence is improved substantially when a solution is dried-up on a flat glas support and when the X-rays impinge on the glass at an angle smaller than the critical angle for the glass. In this way the background from the support is substantially reduced. A detection limit of 10^{-13} g for about 30 elements has been achieved in multielement analysis at Hasylab by this technique [29]. The technique was also successfully applied for detection of small amounts of impurities (10^{12} atoms/cm^2) at the surface of Si wafers for microfabrication [30]. When synchrotron radiation is used the technique can reach a lateral resolution better than 10 μm.

66

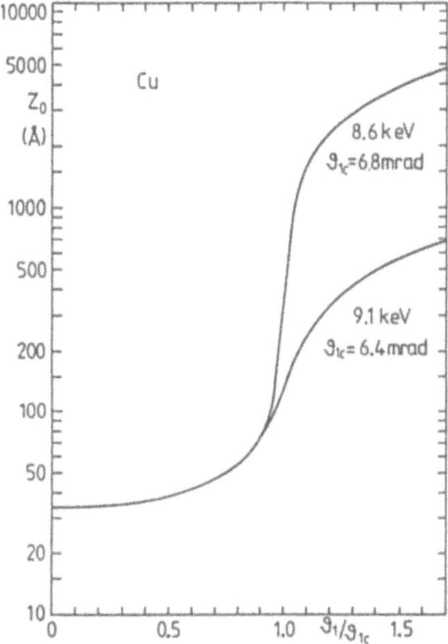

Fig. 2.3 Damping constant z_0 of the transmitted amplitude for 8.6 and 9.1 keV photons in metallic copper. The copper $K - edge$ is at 8980 eV. Note the small penetration depth of the X-rays below ϑ_{1c} which is the basis of making all X-ray techniques surface sensitive at total reflection.

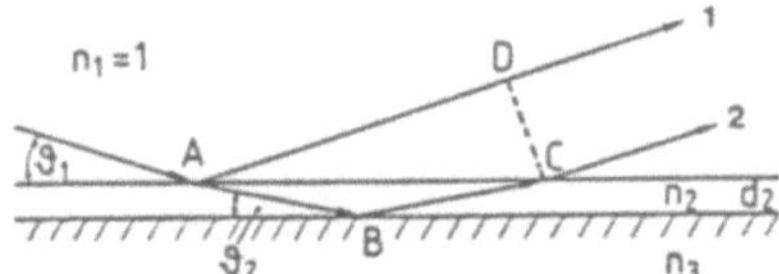

Fig. 2.4 Reflection of X-rays under grazing incidence from a layer (n_2, d_2) on a substrate (n_3).

2.4 Reflection from a Layer on a Substrate

Consider an X-ray beam falling on a flat layer on top of a substrate (figure 2.4). When the angle ϑ_{1c} of total reflection for the layer is exceeded, part of the beam will penetrate into the layer. The amplitudes 1 and 2 reflected from the interfaces can interfere. The difference in optical path is proportional to the thickness d_2 of the layer. The reflectivity shows oscillations, the period of which depends on the layer thickness [31]. Roughness at the interfaces reduces the amplitudes of the oscillations. When reflectivity

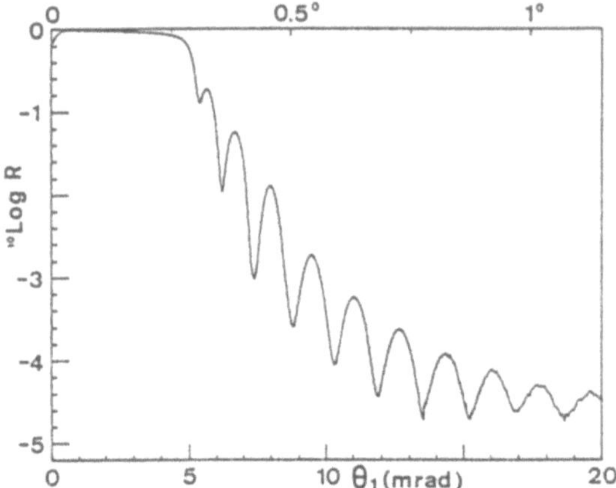

Fig. 2.5 Reflectivity of 9800 eV photons from a Ta oxide layer on Si deposited at $410^\circ C$ by CVD.

measurements on layers are combined with X-ray absorption measurement very useful information on the structure and the chemical composition of the layer can be obtained, as demonstrated in the next example.

Example: CVD (chemical vapour deposition) Ta Oxide Layer on a Si Substrate [32]

Figure 2.5 shows the reflectivity from a tantalum oxide layer deposited by CVD at $410^\circ C$ on a 4 inch silicon wafer. At 5.2 mrad the photon beam (9800 eV) penetrates into the oxide. Above this angle 10 oscillations due to the interference of the amplitudes reflected at the oxide and silicon interfaces are observed. When the square of the angular position of the maxima, $\vartheta^2_{1\nu}$, is plotted versus $(\nu + 1/2)^2$, ν being the consecutive numbers of the maxima, one expects a linear relationship, the slope giving the layer thickness and the intercept the critical angle ϑ_{1c} for the layer [6]. In the present case (figure 2.6) the slope gives a Ta oxide thickness of (352 ± 3) Å, whereas the intercept gives a critical angle ϑ_{1c} of 4.90 mrad. Note that the thickness determination is accurate and independent of optical constants (due to the fact that n is almost 1). From the critical angle the mass density of the Ta oxide layer can be determined, provided the chemical composition of the layer is known. This was obtained from X-ray absorption at the Ta L_3 edge. The result can be summarized as follows. The Ta L_3 edge of the CVD oxide is identical to that of crystalline $c - Ta_2O_5$. The EXAFS of the CVD Ta oxide gives the same local structure around Ta as in $c - Ta_2O_5$, and, finally, the compositon of the CVD oxide is homogeneous with depth. So, CVD Ta oxide deposited at $410^\circ C$, which is X-ray amorphous, is poorly crystallized Ta_2O_5. With this knowledge the mass density of the oxide can be deduced from the measured critical angle ϑ_{1c}. The value found is 94%

68

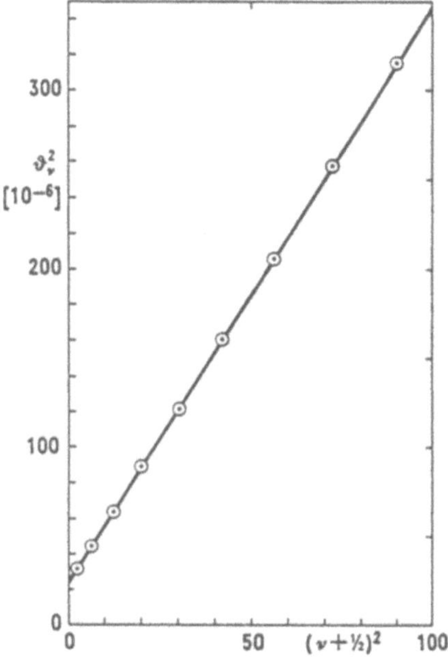

Fig. 2.6   Plot of the position of the maxima in fig. 2.5. The slope of the line gives the thickness and the intercept the density of the Ta oxide layer.

of the density of $c - Ta_2O_5$. Obviously, the Ta oxide deposited at $410^\circ C$ by CVD contains many structural defects. They are responsible for the low electrical breakdown field strength of these layers [32].

2.5 Reflection from Multiple Layers

Periodic multiple layers can show Bragg reflections in addition to the total reflection observed at small angles. Figure 2.7 shows the reflectivity from a systems of 75 double layers of 20 Å Si and 10 Å Ta, each. The layers have been deposited by sputtering on a 4" Si wafer. Total reflection for the 9800 eV photons is observed below 5 mrad. Above this angle the reflectivity drops by 3 orders of magnitude and increases again at 22.5 mrad to about 50 %. This reflectivity, limited to a small angular range, is a Bragg reflection from the 75 double layers. Correcting Braggs law for the refraction of the X-rays, the position of the Bragg reflection gives a periodicity D of 29.1 Å , compared to 30 Å as given by the deposition parameters. The oscillations in the angular range from 7 to 22 mrad are a consequence of the finite number of periods. Their angular spacing gives a number of periods of 75.2, in good agreement with the 75 layers known to have been deposited. These multiple layers are excellent monochromator crystals for X-rays. In addition they can

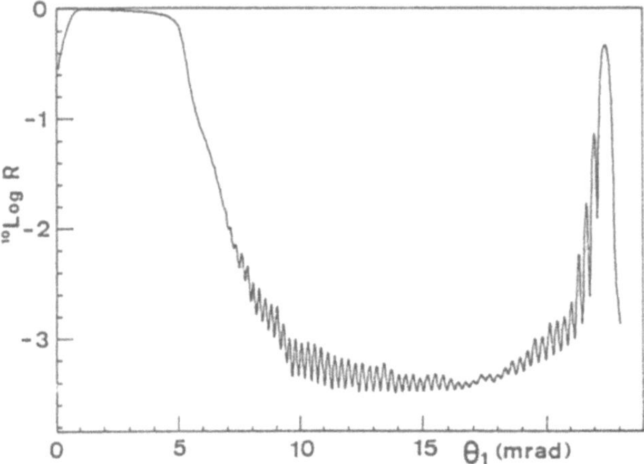

Fig. 2.7   Reflectivity of 9800 eV photons from 75 double layers of 20Å Si and 10Å Ta deposited by sputtering on a 4" silicon wafer. At 22.5 mrad a Bragg peak is observed with a reflectivity of 50 %.

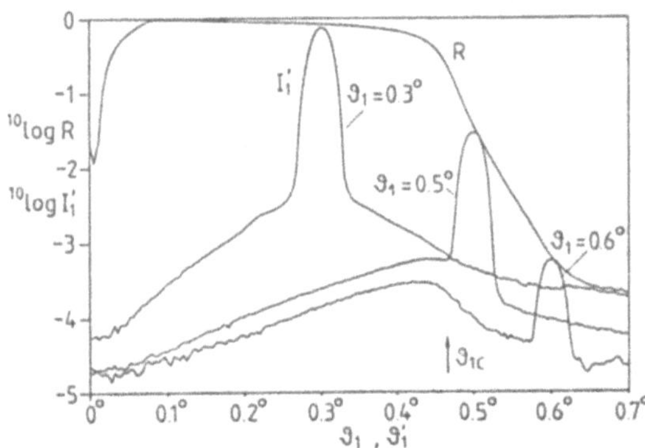

Fig. 2.8   Reflectivity R and intensity distribution $I'(\vartheta'_1)$ of reflected X-rays (9800 eV) at incident angles $\vartheta_1 = 0.3°$, $0.5°$ and $0.6°$ from a 3000Å Pt layer.

be used as mirrors for soft X-rays working even at normal incidence [33]. X-ray reflectivity measurements have turned out to be excellent for testing the quality of multiple periodic layers.

2.6 Surface and Interface Roughness

Surface roughness creates a halo of diffuse scattering around the specularly reflected beam. This is shown in figure 2.8 for the reflectivity of 9800 eV photons from a Pt layer, 3000 Å in thickness and deposited by evaporation

70

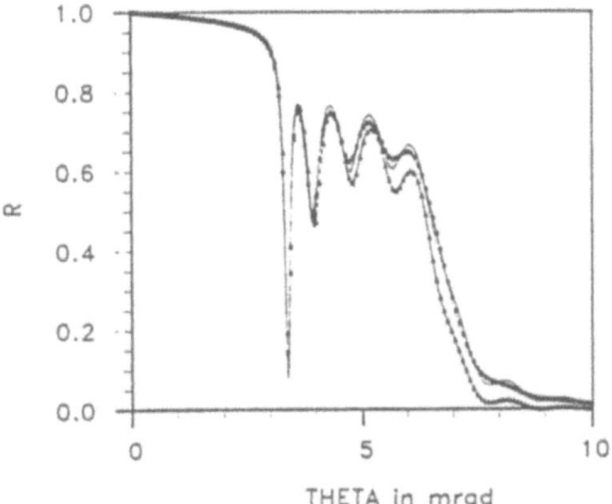

Fig. 2.9 Calculated reflectivity of X-rays (9800 eV) from a 500Å Si layer on top of a Ta substrate. $-$: both interfaces are assumed to be perfectly flat. $\bigcirc$: interface air-Si is smeared out with $\sigma = 20$Å. $\triangle$: interface Si-Ta is smeared out with $\sigma = 20$Å.

on a 4 inch Si wafer [34]. For angles above the critical angle the diffuse intensity can be as high or even higher than the specularly reflected intensity [34, 35]. The roughness of the two interfaces of a layer on a substrate shows up in different ways in the reflected intensity. Figure 2.9 shows the calculated, reflected intensity from a 500 Å Si layer on top of a Ta substrate. Below 3.2 mrad the beam is totally reflected from the Si layer. Below 7 mrad the beam is totally reflected from the Ta substrate, which is more dense than Si. The oscillations are due to the interference of the beam 1 and 2 in figure 2.4. Roughness of the Si-Ta interface reduces the intensity of the totally reflected beam, so that the average value of R decreases between 3.2 and 7 mrad. On the other hand, roughness of the air-silicon interface reduces the amplitudes of the oscillations in this angular range, without changing the average reflectivity. Thus, measurements of the reflectivity and of the diffuse scattering give the possibility to separate the amount of roughness of external and internal interfaces. More detailed investigations have to show how good these measurements can complement the information obtained on interfaces by other techniques like tunneling microscopy, atomic force microscopy, electron microscopy and other surface sensitive tools.

Summary

The XAS is a local probe for the geometric and electronic structure of specific species in condensed matter. The information obtained includes interatomic distances, coordination numbers, static and dynamic Debye-Waller factors, projected and spin-split empty electronic DOS and the valence of the absorbing species. It is especially suited for highly dispersed and diluted systems. In general, the probe is a bulk probe but it can be made surface sensitive by means of total external reflection with a probing depth

as low as 20 to 70 Å . An especially appealing feature is the possibility to
investigate internal interfaces which can be up to 1 μm below a flat surface.
The reflectivity of X-rays can be used to determine the thickness of layers,
indices of refraction, the mass density and the roughness of interfaces. The
measurements can easily be combined with a chemical analysis by X-ray
fluorescence, and depth profiles in chemical composition can be made by
measuring the angular dependence of the fluorescence.

XAS has a wide range of applications. It can be applied to crystalline,
microcrystalline, amorphous and liquid phases which contain the atomic
species of interest in concentrated and in diluted form. The sample preparation in XAS as bulk probe is simple. On the other hand, in the glanzing
angle version large and flat substrates are needed. The sample size should
exceed 1 cm^2. Silicon wafers and float glass substrates have turned out to be
excellently suited. XAS is a non-destructive probe and many investigations
can be made in-situ. This is especially interesting for the investigation of
liquid-solid and gas-solid interfaces. Since XAS is a local probe it does not
give information on long range order in condensed matter. Other techniques
(like X-ray, neutron or electron diffraction) have to be used for obtaining
this information. Since XAS needs the tunability of the X-ray energy, the
experiments are done in general with synchrotron radiation.

Acknowledgement

The author would like to thank Drs. H. Föll, H. Göbel, M. Schuster and M. Wilhelm from
Siemens AG for many helpful discussions and Mr. R. Baur and W. Rudnik from Siemens
AG for preparing the Ta oxide and Ta-Si films used as illustrating examples in this paper.
The author would also like to thank Mr. U. Dedek, F. Stanglmeier and W. Weber from
the Kernforschungsanlage Jülich for their invaluable help during the measurements at the
Hamburger Synchrotronstrahlungslabor.

References

[1] E.A. Stern, Contemp. Phys. 19, 289 (1978)

[2] P.A. Lee, P.H. Citrin, P. Eisenberger, B.M. Kincaid, Rev. Mod. Phys. 53, 796
     (1981)

[3] B.K. Teo, D.C. Joy (eds.), EXAFS spectroscopy Plenum Press, New York &
     London (1981)

[4] T.M. Hayes, J.B. Boyce, Solid State Phys. 37, 173 (1982)

[5] E.A. Stern, S.M. Heald, in: Handbook of Synchrotron Radiation Vol. 1B, D.E.
     Eastman, Y. Farge (eds.), North Holland Co. Amsterdam, New York, Oxford
     (1983)

[6] B. Lengeler, X-ray absorption and reflection in the hard X-ray range, Summer
     school "Enrico Fermi" 12-22.7.88 Varenna, Italy, eds. R. Rosei et al., North Hol-
     land Publ. Co (1989)

[7] B.K. Teo, EXAFS: Basic Principles and Data Analysis, Springer Verlag, Berlin,
     Heidelberg, New York, Tokyo, (1986)

[8]   D.C. Konigsberger, R. Prins (eds.) X-ray absorption: principles and techniques of EXAFS, SEXAFS and XANES, J. Wiley New York (1988)

[9]   B. Lengeler, Proc. 4th Intern. Conf. EXAFS and Near Edge Structure, J. Physique C8 47, 75 (1986)

[10] G. Stegemann, Röntgenabsorptionsmessungen an Fe, Ni, Cu Legierungen und an amorphem Ge, JÜL Report JUL-1075 (1986)

[11] D.E. Polk, D.S. Boudreaux, Phys. Rev. Lett. 31, 92 (1973)

[12] W. Paul, G.A.N. Connell, K.J. Temkin, Adv. Phys. 22, 529 (1979)

[13] B. Lengeler, Phys. Rev. Lett. 53, 74 (1984)

[14] B. Lengeler, J. Physique C8 47, 1015 (1986)

[15] B. Lengeler, Sol. State Comm. 55, 679 (1985)

[16] J.E. Müller, O. Jepsen, J. W. Wilkins, Solid State Comm. 42, 365 (1982)

[17] B. Lengeler, R. Zeller, Solid State Comm. 51, 889 (1984)

[18] L.A. Grunes, Phys. Rev. B27, 2111 (1983)

[19] G. Materlik, J.E. Müller, J.W. Wilkins, Phys. Rev. Lett. 50, 267 (1983)

[20] R. Zeller, G. Stegemann, B. Lengeler, J. Physique C8 47, 1101 (1986)

[21] G. Schütz, W. Wagner, W. Wilhelm, P. Kienle, R. Zeller, R. Frahm, G. Materlik, Phys. Rev. Lett. 58, 737 (1987)

[22] G. Schütz, M. Knüller, R. Wienke, W. Wilhelm, W. Wagner, P. Kienle, R. Frahm, Z. Phys. B 73 737 (1988)

[23] G. Schütz, R. Wienke, W. Wilhelm, W. Wagner, P. Kienle, R. Zeller, R. Frahm, Z. Phys. B (1989)

[24] H. Ebert, B. Drittler, R. Zeller, G. Schütz, Solid State Comm. 69, 485 (1989)

[25] J. Röhler, X-ray absorption and emission spectra, in Handbook on the Physics and Chemistry of Rare Earths, eds. K.A. Gschneidner et al. 10, 453 (1987)

[26] F. Hilbrig, H. Göbel, B. Lengeler, H. Schmelz , to be published

[27] R.W. James, The optical principles of the diffraction of X-rays, Cornell University Press (1967)

[28] R. Begum, M. Hart, K.R. Lea, D.P. Siddons, Acta Cryst. A 42, 456 (1986)

[29] W. Petersen, P. Ketelsen, A. Knöchel, R. Pausch, Nucl. Instr. Methods A 246, 731 (1986)

[30] V. Penka, W. Hub, Fresenius Z. Analyt. Chemie (1989)

[31] H. Kiessig, Ann. Phys. 10, 715 (1931)

[32] R. Baur, H. Föll, K. Hieber, B. Lengeler, to be pubslished

[33] T.W. Barbee, J.H. Underwood, Opt. Comm. 48, 161 (1983)

[34] F. Stanglmeier, (unpublished)

[35] R. Röhlsberger, Herstellung und Charakterisierung von dünnen Schichten im Hinblick auf die Monochromatisierung von Synchrotronstrahlung mit Hilfe des Mössbauereffektes. Interner Bericht, DESY Hasylab 88-08, Hamburg Nov. 1988

Propagation of Large-Wavevector Acoustic Phonons
New Perspectives from Phonon Imaging

James P. Wolfe

Physics Department, University of Illinois at Urbana Champaign, 1110 W Green St., Urbana, Illinois 61801, USA; and Physik-Department E 10, Technische Universität München

Summary: Within the last decade a number of attempts have been made to observe the ballistic propagation of large wavevector acoustic phonons in crystals at low temperatures. Time-of-flight heat-pulse methods have difficulty in distinguishing between scattered phonons and ballistic phonons which travel dispersively at subsonic velocities. Fortunately, ballistic phonons can be identified by their highly anisotropic flux, which is observed by phonon imaging techniques. In this paper, several types of phonon imaging experiments are described which reveal the dispersive propagation of large-wavevector phonons and expose interesting details of the phonon scattering processes.

1 Heat Pulses — The Basic Ideas

The heat pulse method, introduced in 1964 by von Gutfeld and Nethercot [1], measures the transit time for a packet of thermal energy to cross a single crystal at low temperature. A heat pulse can be viewed as a collection of quantized lattice vibrations, or phonons. Typical frequencies are in the gigahertz to terahertz range. The propagation of such phonons in solids is, in some respects, similar to the propagation of photons through matter. Thus, to introduce the basic ideas of a heat pulse experiment, I will begin with some comparisons to light propagation.

A heat pulse may be generated by passing a current pulse through a metal film which is evaporated onto one surface of the crystal. Alternatively, one may focus a pulsed laser beam or electron beam onto a metalized surface. Whichever method is chosen, the resulting pulse of thermal energy propagates radially away from the point source and can be detected on the opposite surface with a phonon detector.

Figure 1 depicts the spatial distribution of heat energy at some instant of time after the excitation pulse. Several non-spherical *shells* of thermal energy, corresponding to longitudinal and transverse modes, expand into the crystal. The anisotropic velocity and intensity of heat pulses is a principal topic to be discussed here. The detector (D) senses the arrival of the heat pulses, as shown in the figure.

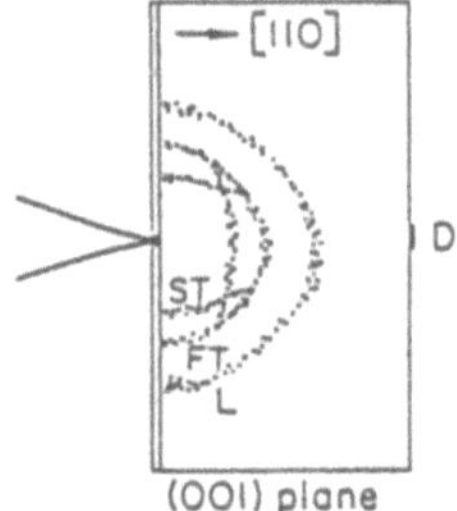

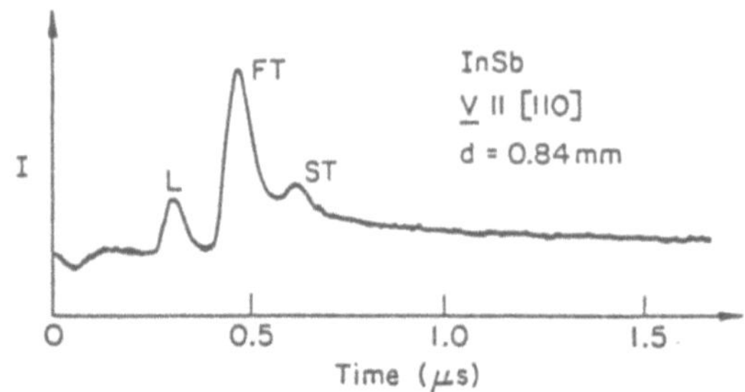

Fig. 1 (a) "Snapshot" of a heat pulse following a single laser pulse. Longitudinal (L), slow transverse (ST) and fast transverse (FT) modes propagate at different, direction-dependent velocities. (b) Heat-pulse signals detected in InSb by a PbTl tunnel junction (430 GHz) at a temperature of 1.6 K. V = group velocity. d = crystal thickness [2].

We assume for the moment that the frequency distribution of the generated phonons approximates a Planck distribution, just as the sun produces a Planck distribution of photons. On a clear day, photons from the sun can travel ballistically (i.e., without scattering) to our eyes. In a moderate fog we detect a diffuse source of scattered photons, but we may still observe some ballistic photons which sharply define a dim image of the sun. In a dense fog, only scattered photons are seen.

The phonons emitted from the heated region of the crystal can also scatter on their way to the detector. Common scattering centers in crystals are *mass defects*. These could be impurities, dislocated atoms, or even the random distribution of naturally occuring atomic isotopes. In short, anything that locally destroys the periodicity of the crystal lattice can scatter phonons. For such mass defects, the frequency dependence of the scattering rate is the same as for Rayleigh scattering of light, namely, ν^4. High frequency phonons scatter much more readily than low frequency phonons. For similar reasons, the sky is blue.

For a crystal of GaAs with a natural abundance of isotopes, the scattering length of 350 GHz transverse phonons is predicted to be about 3 cm [3]. For 1 Terahertz (10^{12} Hz) phonons, it diminishes to about 0.4 millimeter. Since the peak in the Planck distribution for a typical 10 K heat source is 2.8 kT = 600 GHz, this isotope scattering process plays a significant role in a heat-pulse experiment. That is, phonons can scatter out of the ballistically propagating *shell* to produce a *fog* of late-arriving phonons whose apparent source is quite diffuse.

An important question is whether the phonons actually retain their original frequency distribution as the heat pulse traverses the crystal. The isotope scattering process described above is an <u>elastic</u> scattering process, in which only the directions and polarizations of the phonons are changed. Thus, in the absence of <u>inelastic</u> processes, which change the phonon frequencies, the traveling heat pulse should retain its original frequency distribution.

76

Because the heat pulse is expanding into the crystal, its thermal energy
density must diminish with time. We may view the local phonon distribution as having a *color temperature*, just like the color temperature which
describes the frequency distribution of the sun in outer space or on earth.
The local temperature is much less than the color temperature because the
local photon (phonon) occupation numbers are much smaller than those of
the source. Indeed, we are dealing with highly *non-equilibrium* phonons
propagating through a relatively cold medium.

The deviation of the elastic crystal from perfect harmonicity leads to an
anharmonic decay rate which scales as ν^5 [4]. However, for phonons in the
sub-terahertz range, the strength of this inelastic scattering in most crystals
is much weaker than that of elastic scattering. Nevertheless, the influence
of this process becomes significant if higher frequencies are generated, say,
by directly photoexciting the crystal, or by highly heating a metal film
on the surface. The strong electron-phonon coupling and high density of
electrons in a metal cause a rapid thermalization which creates a local temperature; whereas, the relatively low density of optically excited electrons
in a semiconductor should produce a highly non-thermal distribution. This
photoexcitation case is not well characterized experimentally. More will be
said about this in Section 6.

This analogy between photons and phonons is useful but limited. Phonons
are, in several respects, much more complicated than photons. Phonons ex-

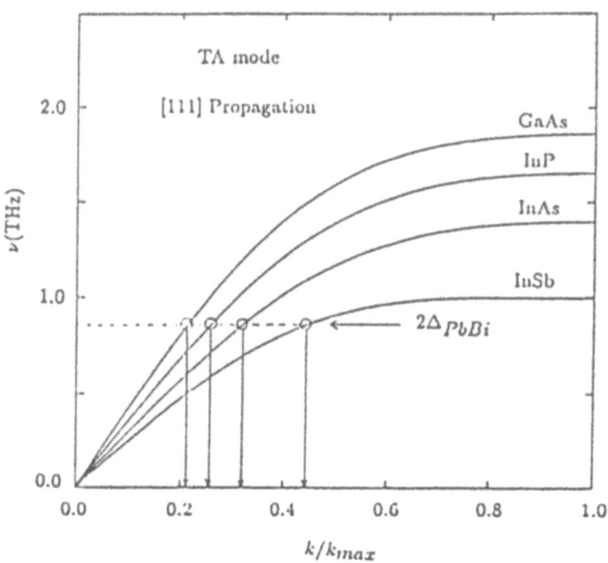

Fig. 2 Dispersion curves for several III–V semiconductors. The onset frequency of a
PbBi superconducting tunnel junction is shown for later comparison [2].

ist only in the *ether* of their crystal lattice, and they assume the complexity of this medium. In addition to the elastic anisotropy specific to each type of crystal, phonons have a minimum possible wavelength given by twice the atomic lattice spacing.

More precisely, phonon wavevectors are restricted to the Brillouin zone in wavevector space, whose exact shape depends on the particular crystal structure. As the wavevector approaches the zone boundary, the group velocity, $V = \delta\omega/\delta k$, of the wave departs from the velocity of sound. At the zone boundary this velocity of energy propagation is zero, signifying a standing wave. Examples of these non-linear dispersion relations, $\omega(k)$, for slow transverse waves in several semiconductors are shown in Figure 2.

Can one observe this reduced group velocity of *large-k* phonons using the heat-pulse method? Since higher frequency phonons scatter much more readily, it turns out to be very difficult to detect the ballistic propagation of these *dispersive* phonons, whose wavelengths are only a few lattice spacings. Part of the difficulty is in distinguishing a few dispersive phonons, traveling ballistically, from many late-arriving phonons which have scattered from defects. The problem is a little like detecting the dim image of the sun through a fog. Indeed, as we shall see, the key is to use imaging methods.

The velocity of a phonon also depends on its propagation direction. A useful way of displaying the dispersion relation of an anisotropic medium (i.e., any crystal) is to plot constant-frequency contours in wavevector space. Such a plot is shown for Ge in Figure 3. At low frequencies (the *continuum limit*) the shape of the contours are invariant to frequency and given by the static elastic constants of the crystal. At higher frequency, the constant-frequency contours distort due to the proximity of the Brillouin zone boundary, which the contours must intersect perpendicularly.

2 The Time-of-Flight Puzzle

In view of the tendency for large-k phonons to scatter, a report in 1980 claiming to observe ballistic propagation of nearly zone boundary phonons was received with more than passing interest. In this provocative paper, Ulbrich, Narayanamurti and Chin [7] claimed that photoexcitation of the semiconductor GaAs produced a rich distribution of large-wavevector phonons which propagated ballistically over several millimeters. The principal evidence was that the <u>peaks</u> of the heat pulses arrived with subsonic velocities, and their arrival times scaled linearly with the distance between source and detector. Diffusive propagation, in contrast, was expected to display arrival times which scale as the square of the distance. Stock, Ulbrich and coworkers subsequently showed that this linear scaling between time and distance could be observed in other III–V semiconductors such as GaP and InP [8–10].

The velocity of the peaks of the heat pulses observed in these crystals were as small as 1/3 of the sound velocity, which led these experimenters to

78

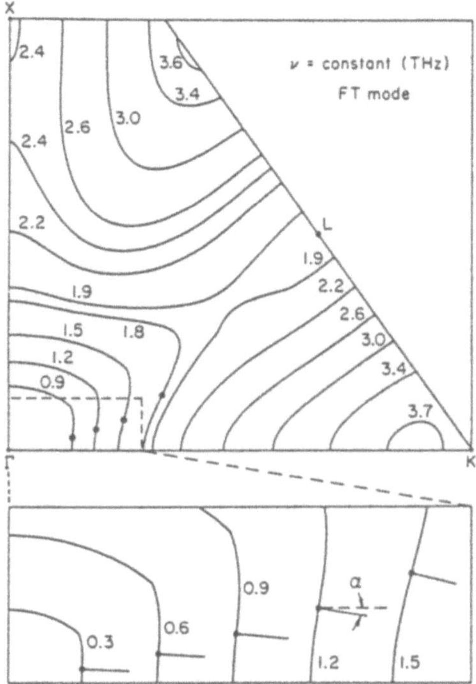

Fig. 3 Intersections of the constant-frequency surfaces for the FT mode in Ge with the (110) plane. As discussed later in the text, dots mark points of inflection which lead to caustics in flux along the surface normals, or group velocities, at these points. The enlargement shows how the group velocity along the caustic changes direction with frequency, as indicated by the increasing angle α [5,6].

conclude that the heat pulses contained a large fraction of phonons with frequencies of 1.5–2 THz and ballistic mean free paths of 1–2 millimeters. These results were incompatible with the commonly accepted theory of isotope scattering [3,4], which predicted scattering times at least an order of magnitude shorter than those *observed*.

A flurry of theoretical activity followed these reports. Lax et al [11–13] suggested that the isotope scattering rate in GaAs may be anomalously diminished for large wavevector phonons. In particular, if the ratio of Gallium to Arsenic displacements at the zone boundary is small, scattering from (isotopically impure) Ga atoms would be reduced. (Arsenic is isotopically pure.) Lattice dynamics calculations by Tamura [3], however, indicated that the scattering rate for dispersive phonons is one or two orders of magnitudes faster than suggested by the experiments of Ulbrich et al.

In an attempt to explain the experimental results within the framework of phonon scattering, Guseinov and Levinson [14] proposed that the linear scaling was the result of a combination of elastic scattering and anharmonic decay — termed *quasidiffusion*. Experiments by Northrop and Wolfe [15], using a broadband Al bolometer as a phonon detector, and numerical modeling of the quasidiffusion process supported this notion. Lax et al [16], however, argued that quasidiffusion could not account for the results of Ulbrich et al because they used a Pb tunnel-junction detector which selected only phonons with frequencies above 650 GHz. Based on the known dispersion curves for GaAs (see, for example, Figure 2), only one or two anharmonic decay events would lower the phonon frequencies to below the frequency threshold of the detector.

Lax et al [16,17] also carried out Monte Carlo calculations assuming only elastic scattering and taking into account the particular boundary conditions of the experiment. Their results showed that pure diffusion could lead to a nearly linear scaling because the proximity of the planar crystal boundaries significantly modified the shape of the detected heat pulse. This provides a compelling explanation to the time-of-flight data of Ulbrich et al.

How can one *experimentally* distinguish between these two alternatives — dispersion or scattering? The answer is an experiment that looks at the angular distribution of phonon flux: phonon imaging [18]. This technique is based on the fact that ballistic phonons emitted from a point source exhibit an extremely anisotropic energy flux [19].

Phonon imaging experiments [15,20,21] using similar excitation conditions and detectors as Ulbrich et al have shown that the heat pulses in GaAs contain both ballistic and diffusive components. The ballistic part, however, consists mainly of phonons with frequencies below 1 THz. Together with Monte Carlo simulations [21], these experiments are in accord with the existing theories of elastic scattering in III–V semiconductors. Some of the results will be discussed below in Section 5.

This lively controversy over large-k phonons has precipitated a deeper understanding of dispersion and scattering of phonons in these semiconductor crystals. One misconception was that direct photoexcitation was necessary for observing high frequency dispersive phonons. In fact, metal-film excitation produces a Planck distribution rich in high frequency phonons and has the advantage of producing a smaller phonon source, which is desirable for high resolution imaging experiments.

A second misconception was that time selection is an effective means of selecting phonon frequency, because slower group velocity means higher frequency. The existence of late-arriving scattered phonons, as well as the large anisotropy in velocity, make this idea impractical. Instead, it is found that the systematic tayloring of detector thresholds, combined with phonon imaging techniques, is an effective means of selecting ballistic phonons of a given frequency. Ironically, it is found that the rapid frequency dependence of elastic scattering can be a help, rather than a hindrance, in selecting a quasi-monochromatic band of ballistic phonons.

80

These newly gained insights have facilitated the observation of ballistic large-k phonons in InSb [22,23]. As seen in Figure 2 this crystal is somewhat softer than GaAs, allowing larger wavevectors to be sampled with a given detector frequency. Phonon images of InSb show striking changes with increasing phonon frequency. These results have been compared to predictions from several lattice dynamics models and some interesting differences have been found. Also, phonon imaging experiments have examined the phonon scattering processes, yielding new insights into elastic scattering and the phonon *hot spot* produced by inelastic scattering at the heat source. Highlights of these results are reviewed below.

3 Phonon Focusing of Large-k Phonons

A Planckian source of heat implies an isotropic distribution of phonon wavevectors. Remarkably, this does not mean that thermal energy is isotropically radiated from the heated point. The basis for this anisotropy in heat flux, known as *phonon focusing* [19], can be seen in Figure 3. The energy flux travels along the group velocity directions, $\mathbf{V} = \delta\omega/\delta\mathbf{k}$, which are normal to the constant-frequency contours. The non-spherical frequency surfaces are flatter in some regions than others. This means that the energy flux will be concentrated along those directions. More precisely, the energy flux along one of the normals, $\mathbf{V}$, is inversely proportional to the curvature of the constant-frequency surface at that point [24].

The shape of the constant frequency surfaces in the long wavelength limit (those contours nearer the origin in Figure 3) is given by continuum elasticity theory [25]. A traveling wave of the form $\mathbf{u} = \mathbf{e} \cos (\mathbf{k}\cdot\mathbf{r}-\omega t)$ must obey the Christoffel equations,

$$\left(\Gamma_{il} - \rho v^2 \delta_{il}\right) e_l = 0 \, , \tag{1}$$

where, for cubic crystals,

$$\Gamma_{il} = (C_{12} + C_{44}) n_i n_l \qquad (i \neq l) \tag{2}$$
$$= C_{11} n_i^2 + C_{44} \left(n^2 - n_i^2\right) \quad (i = l). \tag{3}$$

The C_{IJ} are the well known elastic constants, δ_{il} is the delta function, ρ is the density, and $n = k/|k|$. Summation over repeated indices is assumed. The solution to the Christoffel equations gives the phase velocity, $v(\theta, \phi) = \omega/|k|$, for a given $k = (k, \theta, \phi)$. The radial distance to a constant-frequency contour is $|k| = \omega/v(\theta, \phi)$, which explains why this contour has the shape of a *slowness* surface. Equations (1)–(3) do not depend on the magnitude of k; hence, in this continuum limit all constant-frequency surfaces have the same shape.

The black dots shown on the slowness surfaces in Figure 3 indicate inflection points where the curvature of the surface vanishes. Vanishing curvature implies that the corresponding group velocity normal is a direction of mathematically infinite flux, i.e., a *caustic*. In fact, the full three-dimensional slowness surface displays <u>lines</u> of vanishing curvature which separate convex, saddle and concave regions [26]. These parabolic lines give rise to a distinct pattern of caustics [27,28].

A simple experiment, shown in Figure 4, illustrates this highly directional flux for the crystal InSb [23]. When the laser beam is focused to point A there are three distinct heat pulses which are detected, corresponding to the longitudinal (L), fast transverse (FT) and slow transverse (ST) modes. Their velocities are the three solutions of the Christoffel equations. If the laser beam is translated only a fraction of a millimeter, a remarkable change

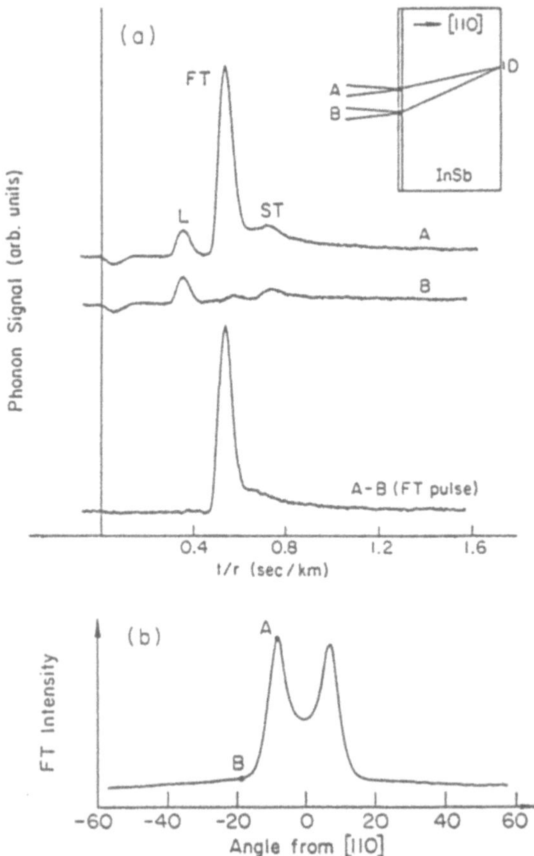

Fig. 4 (a) Heat pulses in InSb produced by exciting a 2500 Å copper film with a focused laser beam. As the beam is translated slightly a dramatic change in the FT pulse intensity is seen, due to phonon focusing. A subtraction of the two traces shows only the FT pulse. The detector is a PbTl tunnel junction (430 GHz onset). (b) Peak intensity of the FT heat-pulse as the laser beam is continuously translated in the (110) plane. The maxima correspond to caustics in phonon flux [23].

82

is observed in the heat pulses. The FT pulse almost completely disappears! A difference trace is computed which shows only the FT pulse, because the other two pulses hardly change for this small angular displacement.

The angular variation of the FT flux is directly obtained by *sitting* on the peak of the pulse (with a boxcar integrator) and continuously scanning the laser beam across the surface of the crystal. The result in Figure 4 b shows a double peaked structure occuring over only a few degrees of angular displacement. Similar effects in Ge were first recorded by Hensel and Dynes [29].

A more complete picture of this phenomenon is obtained by taking a series of line scans along x for different y positions. The result is a phonon image, where the intensity of the heat pulses is converted to brightness on a video screen. The phonon image of InSb shown in Figure 5 a is obtained for a time window which includes both of the transverse modes. The schematic diagram in Figure 5 b identifies the structures associated with each mode and labels the [110] and [100] propagation directions.

The bright lines observed in this phonon image are caustics in the heat flux, as predicted above. Many such patterns have now been observed for a variety of non-metallic crystals [18]. The shape of the phonon-focusing pattern depends on the particular elastic constants. The intensities of the various caustics can depend upon scattering centers in the crystal, but their angular positions are generally quite immutable. This is because most experiments observe the ballistic propagation of phonons in the long-wavelength limit. As indicated before, it takes special efforts to overcome the strong tendency of large-k phonons to scatter.

Such efforts have recently succeeded in InSb, as seen in the phonon images of Figure 5 c and d. These are frequency-selected phonon images taken by Hebboul [23]. The detectors were superconducting tunnel junctions consisting of Pb, Tl and Bi alloys, with varying concentrations to control the superconducting gap. To optimize the ballistic flux, undoped InSb crystals only 400 micrometers thick were used. To retain high angular resolution, junctions with just 10×10 μm^2 sensitive area were fabricated. How these detectors selected only a narrow band of phonon frequencies is described in the next section.

As seen in Figure 5 c and d, the pattern of caustics changes dramatically with increasing frequency. This is due to the changing shape of the slowness surface as the phonon wavevectors probe deeper into the Brillouin zone. The highest frequencies observed ballistically in these experiments correspond to wavevectors of about 40% of the zone boundary along [111], as indicated in Figure 2.

Accompanying this dispersive shift in the caustic pattern, there is a measurable decrease in the phonon velocities. Figure 6 shows FT heat pulses isolated by the method of Figure 4, for three different detectors. The onset of these pulses, with respect to the onset of the laser pulse, gives the ballistic velocity. At higher detection frequencies the pulses become broader mainly due to scattered phonons, so the <u>peaks</u> of the pulses are not representative of dispersive velocities (as assumed by Ulbrich et al for GaAs).

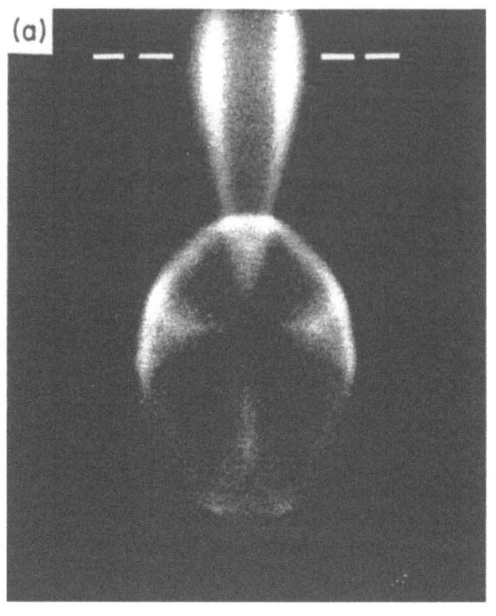

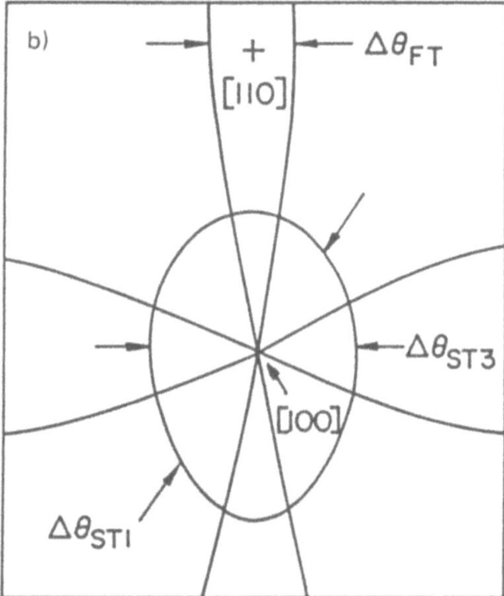

Fig. 5 (a) Phonon image of InSb using a PbTl tunnel junction detector (430 GHz onset frequency). Bright lines are phonon caustics. The dashed line corresponds to the line scan in Fig. 4. (b) Schematic defining angular sizes of caustic structures.

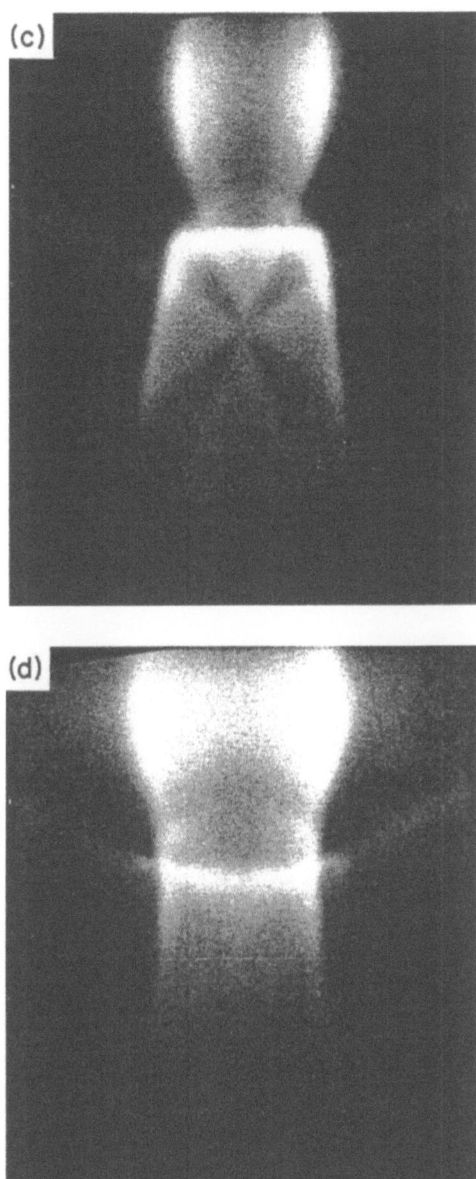

Fig. 5 (c) and (d): Images using PbTl (593 GHz onset) and PbBi (727 GHz onset) detectors. The caustic shift is due to phonon dispersion [23].

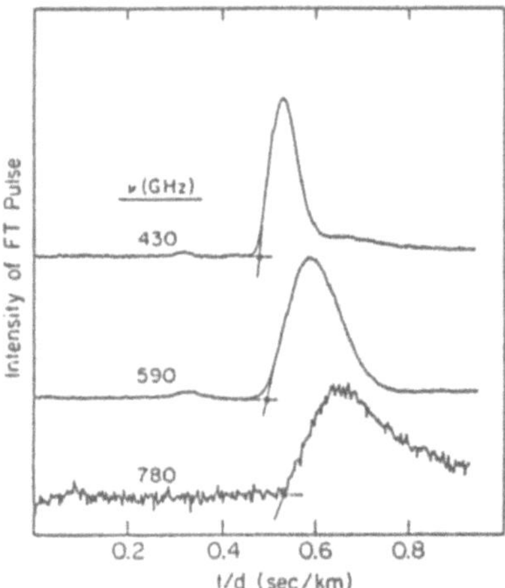

Fig. 6 FT heat pulses recorded by taking the difference between traces on and off the caustic, as in Fig. 4. Results for three different detector frequencies show the dispersion in the group velocity, which is determined from the time difference between the laser pulse onset and heat pulse onset. d is the source-to-detector distance [23].

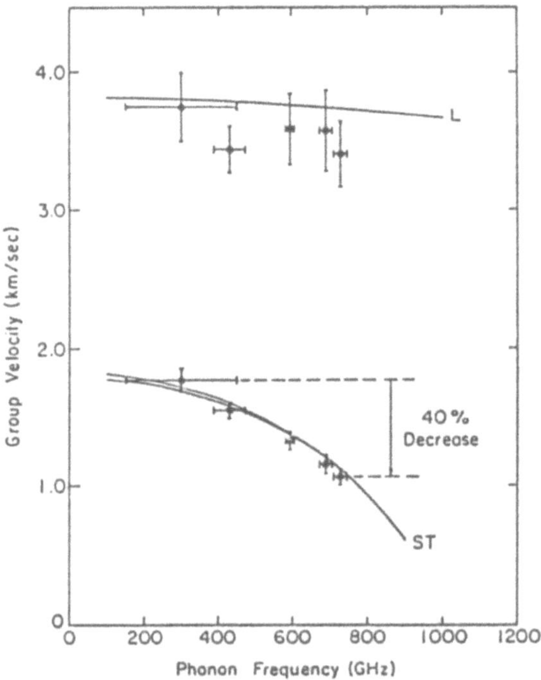

Fig. 7 Graph of group velocity dispersion in InSb along the [111] propagation direction. Predictions of the bond-charge model are shown as the solid curves [23].

86

The most dispersive velocities are observed for slow transverse phonons propagating along the [111] direction. The scattered phonon flux is dominant, but distinct onsets can be observed, and the resulting ballistic velocities are plotted in Figure 7. The longitudinal mode is not very dispersive but the ST velocities are found to decrease by 40% at 780 GHz. It is notable that the first experiments to demonstrate dispersion by time-of-flight were also performed on InSb, by Huet et al [30]. They observed a 6% reduction in velocity with a PbTl junction at 545 GHz.

4 Comparison with Lattice Dynamics Theories

The phonon images for InSb show well-defined caustic lines which shift as tunnel-junction detectors with larger superconducting gaps (2Δ) are chosen. Only phonons with $h\nu > 2\Delta$ can break a Cooper pair in the superconductor and cause a detectable current flow across the oxide barrier [31]. However, elastic scattering in the crystal greatly limits the ballistic propagation of high frequency phonons. The combination of these two effects — a sharp onset frequency of the detector and a rapid cutoff due to scattering — results in a rather narrow band of detected ballistic phonons, as shown in Figure 8. The peak in the ballistic flux occurs at the superconducting gap. This technique was first used by Dietsche et al [5] to observe the dispersive phonon focusing pattern of Ge at 650 GHz.

The scattered phonons are also detected in an image, but they mainly contribute to a background signal. As mentioned before, this ballistic selection is a little like seeing the sharp image of the sun through a fog. If one just measures arrival times of the phonons, it is difficult to separate the ballistic

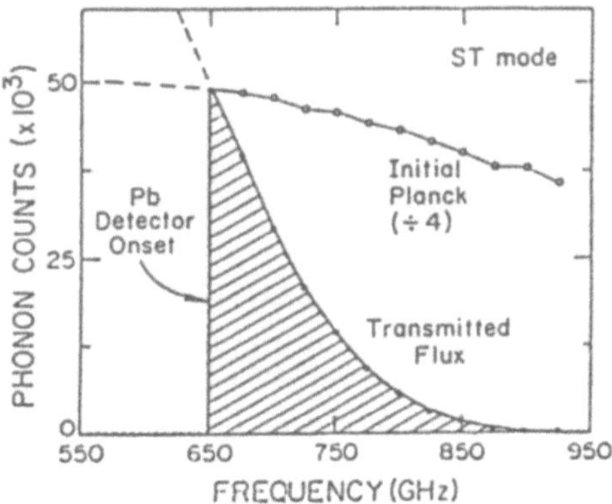

Fig. 8 Monte Carlo calculation of the ballistically transmitted flux multiplied by the detector onset response. InSb, $d = 0.5$ mm. Due to isotope scattering the crystal acts like a low-pass filter. Hence, a relatively narrow frequency distribution of ballistic flux can be selected [22].

and scattered flux. Both have a range of velocities. But phonon imaging is able to isolate the sharp caustics even if most of the phonons are scattered out of the ballistic beam.

What new physics is learned from such an experiment? The pattern of caustic lines is directly related to the shape of the slowness surface. These constant-frequency surfaces, as seen in Figure 3 for Ge, are determined by the basic forces between atoms in the crystal. The wave equation in the dispersive regime is formally quite similar to the Christoffel equation [32]. Again, we have a set of three equations for the displacement eigenvector,

$$\left(D(\mathbf{k}) - M\omega^2\right)\mathbf{e} = 0 \,, \tag{4}$$

but in this case the *dynamical matrix*, D(k), depends on the model chosen to describe the forces between atoms. (A monatomic lattice of atoms with mass M is assumed in this equation.) Now the phase velocity depends on the magnitude of the wavevector; i.e., $v = \omega/k = v(k, \theta, \phi)$. Also, different force models will predict constant-frequency contours of different shapes, suggesting that the validity of these models may be tested by phonon imaging.

A variety of lattice dynamics models have been devised over the years to simulate the motion of vibrational waves in crystals. For example, a basic shell model approximates the atoms as ionic cores located at the center of rigid electronic shells. Core-core, core-shell and shell-shell interactions are considered up to second-nearest neighbors, requiring 10 or more parameters. The bond charge model (BCM) was designed for crystals with covalent bonds. It assumes pointlike bond charges at the center of the covalent bonds which move adiabatically between neighboring atoms — requiring 6 parameters. Usually the free parameters are adjusted to fit inelastic neutron scattering data.

The phonon images provide information about lattice dispersion which is quite complementary to inelastic neutron scattering. While the latter method generally gives the phonon frequency vs wavevector along symmetry directions, the caustic patterns are extremely sensitive to the <u>curvatures</u> of the constant-$\omega(\mathbf{k})$ surfaces, generally away from symmetry axes. Of course, phonon imaging observes only the acoustic branches, whereas neutron scattering also measures the optical branches.

For InSb, a covalent semiconductor, a 10-parameter shell model, an 11-parameter rigid-ion model, a 15-parameter deformation dipole model, and a 6-parameter BCM model all give reasonably close representations of the inelastic neutron scattering data [33]. We may now ask how well these models do at predicting the phonon imaging data.

To make this comparison, Hebboul [23] performed Monte Carlo calculations of the heat flux for each of the above models, based on computational methods developed by Northrop [18]. First, a uniform grid of wavevectors k is chosen over the entire Brillouin zone. For each k, the frequency $\omega(\mathbf{k})$ is

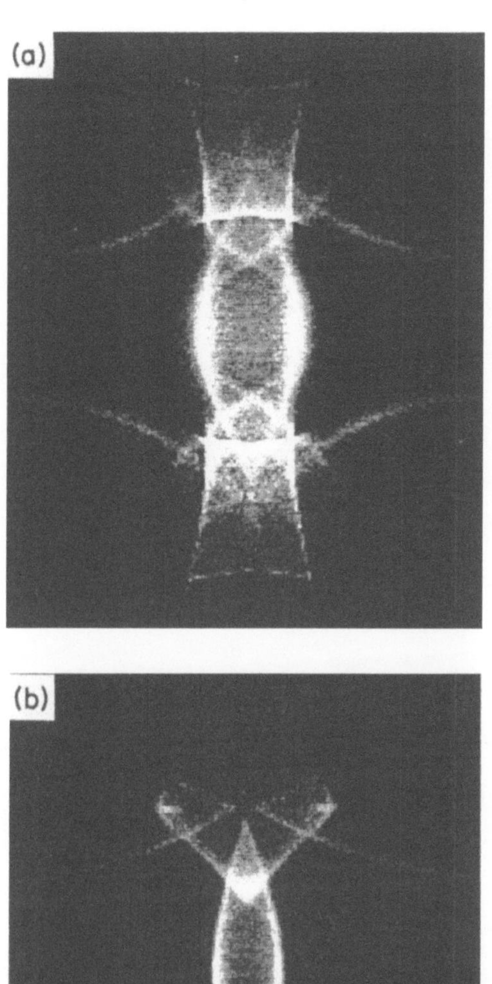

Fig. 9 (a) and (b): Calculated phonon-focusing patterns for bond-charge and rigid-ion models, respectively. These Monte Carlo simulations assume a detector onset of 730 GHz, a 10 K source, and include the effects of isotope scattering [23].

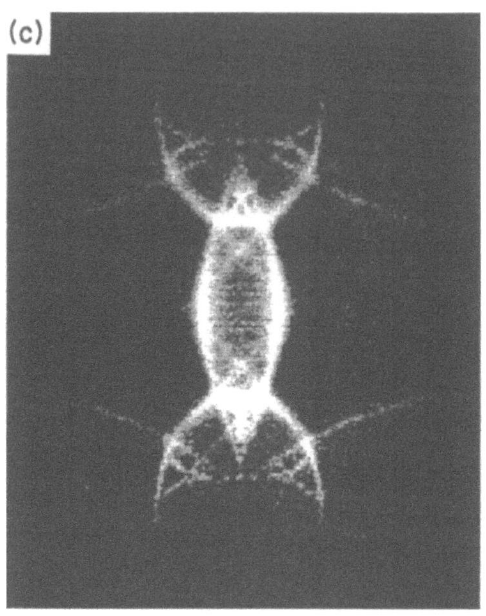

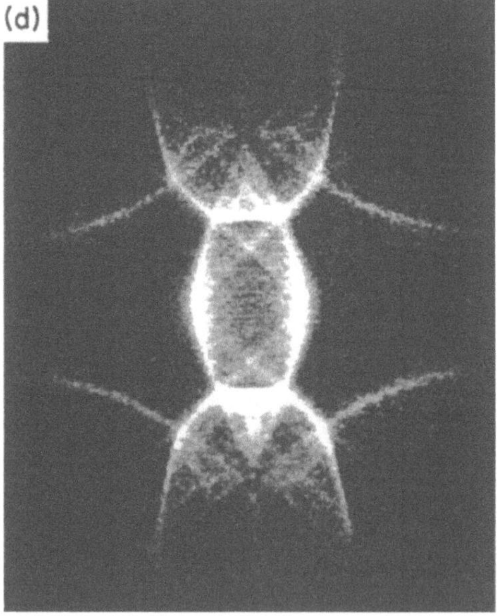

Fig. 9   (c) dipole model,   (d) shell model. All images include both ST and FT modes. The [110] direction is at the center of this scan across the (110) crystal surface.

then computed by solving Equation (4). (This is the time-consuming task, facilitated by an array processor.) In the neighborhood of each k on the grid, $\omega(\mathbf{k})$ is approximated by a local expansion using the derivatives of $\omega(\mathbf{k})$ calculated at this point. Once this table is constructed, phonon images of any frequency distribution can be simulated.

The resulting Monte Carlo images for the four force models are shown in Figure 9. These calculations use the best-fit parameters which were previously found from neutron scattering data on InSb [33]. The phonon-focusing calculations assume a detector onset $2\Delta = 730$ GHz and a ballistic transmission rate given by the theoretical isotope scattering in InSb. The caustic patterns for the three models are qualitatively quite different. The BCM calculation comes the closest to describing the data taken with a 730 GHz junction, which is shown in Figure 5 d. It is clear from these calculations that the phonon focusing pattern is a sensitive probe of the dispersive lattice dynamics.

A more quantitative comparison between experiment and theory can be made by defining some characteristic dimensions of the focusing structures. Figure 5 b shows three of the angles chosen in Hebboul's study. For each model the frequency dependence of these angles is determined numerically and is shown as the solid lines in Figure 10. Consistent with the qualitative conclusion drawn from the phonon images, the BCM model shows the best overall agreement with the data. The results for other angular dimensions and a refitting of the BCM parameters is reported in Ref. [23].

Particularly notable is the frequency dependence of the angle $\Delta\Theta(\mathrm{ST1})$, which clearly <u>decreases</u> with frequency for BCM and <u>increases</u> for the three other models. This parameter seems to be sampling a property of the lattice dynamics which is qualitatively different for BCM and the other models. Perhaps it is the ability of the BCM model to better represent the torsional motion of the covalent bonds in such semiconductors. Investigations of the relations between the phonon caustics and the atomic forces are required to resolve this interesting issue.

5 Channeling of Elastically Scattered Phonons

Let us now consider the *haze* of phonons which are scattered on their way from the heat source to the detector. It turns out that phonon imaging can provide us with some rather basic information about these bulk-scattered phonons. The reason is that phonons retain some of their highly directional focusing even after a few scattering events. These observations provide a bridge between purely ballistic and purely diffusive propagation.

We return, for the moment, to the claims of ballistic propagation of terahertz phonons in GaAs. Stock, Fieseler and Ulbrich [34,35] adopted phonon imaging methods to further examine this possibility. They found an interesting new structure in the phonon focusing pattern which could not be

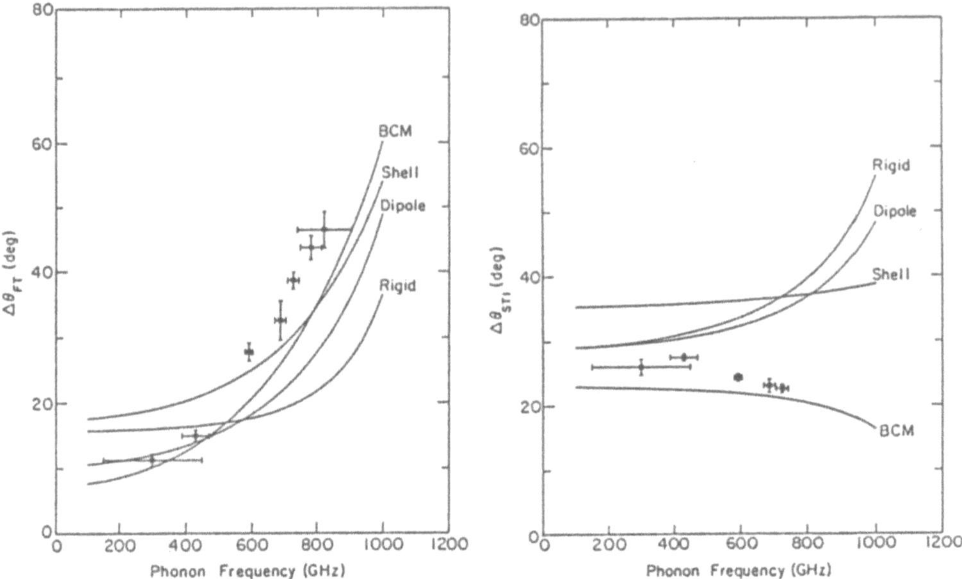

Fig. 10 Comparison of the force model predictions (curves) with the phonon-imaging data (dots) [23].

explained by ballistic propagation of sub-terahertz phonons. Their experimental results were reproduced by Ramsbey, Wolfe and Tamura [36], who obtained the data shown in Figure 11.

These traces are line scans, like that in Figure 4 c, but taken across the slow-transverse structure which is centered on the [100] direction. In addition to the usual caustic peaks, a new structure appears at later delay times: there is a pronounced peak at the center of the scan — along [100]. Stock et al [34] interpreted this new structure as the ballistic focusing of highly dispersive phonons.

Support for this idea came from a dispersive phonon-focusing calculation of Schreiber et al [37], who showed that the BCM model indeed predicted a pile-up of phonon intensity along [100] for high frequencies. A further study of this effect by Hebboul and Wolfe [21] indicated that the [100] pile-up occured for phonons at about 1.8 THz in the BCM model but was not present in the shell model. They also predicted radically shifted caustic patterns at these frequencies and strong pile-up effects along the [110] directions, none of which were observed experimentally.

It is interesting to consider the theoretical origin of this pile-up of ballistic flux. Cross sections of the slow-transverse slowness surface in the (010) plane are plotted in Figure 12. The 1.8 THz contours of the BCM model display a global flattening of the surface which leads to a concentration of

92

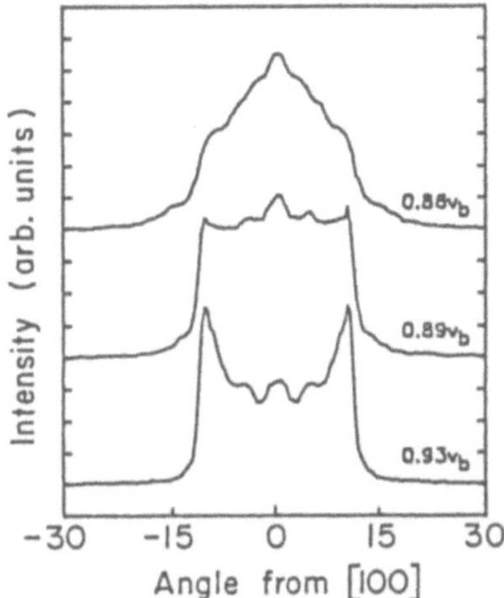

Fig. 11 Line scans of the ST focusing structure in GaAs for increasing delay times after the excitation pulse. Actually, the delay time is continuously changed during a scan to select a constant velocity. v_b is the sound velocity. The sample thickness is 1.82 mm. The laser excites a 2000 Å Cu film on the (100) surface [36].

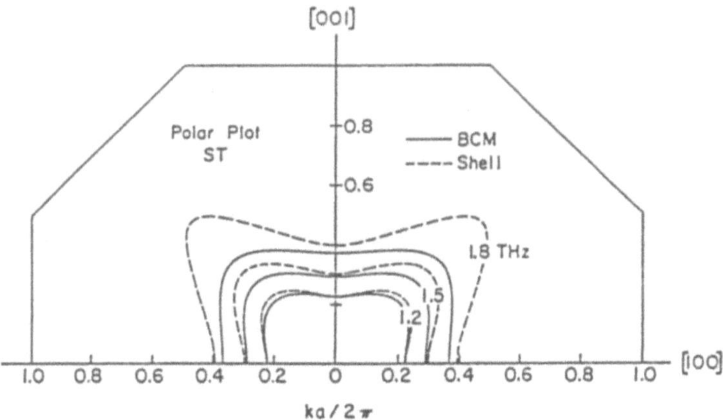

Fig. 12 Constant-frequency contours for GaAs in the (010) plane. Two lattice dynamics models are shown. BCM predicts a flattening near 1.8 THz which should lead to a concentration of phonon flux along [001] [21].

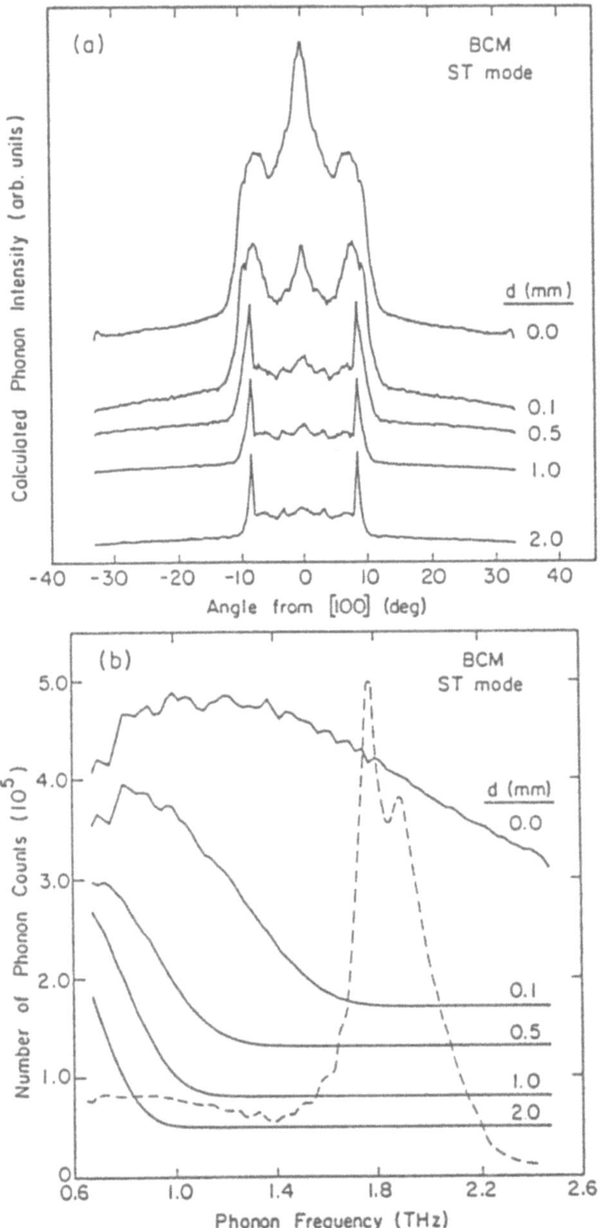

Fig. 13 (a) Calculated line scans of phonon intensity using the BCM model for GaAs. The calculation includes isotope scattering. Results for various crystal thicknesses, d, are shown. A 20 K heat source and 0.65 THz detector onset are assumed. (b) Corresponding frequency distributions [21]. Curves are displaced vertically for clarity.

group velocity vectors along the [100] directions. (A (110) slice of the BCM surface also shows this flattening.) As seen in Figure 12, the contours of the shell model do not display a similar flattening, nor is the [100] flux pile-up predicted.

Hebboul and Wolfe [21] next carried out a Monte Carlo simulation of the ballistically transmitted flux with isotope scattering included. They used the isotope scattering rate calculated by Tamura [2]. Plotted in Figure 13 is the spatial distribtuion of ballistically transmitted phonons for several sample thicknesses. A high frequency pile-up along [100] is predicted for samples with thickness less than 0.1 mm. That is, due to the strong ν^4 dependence of isotope scattering, the mean-free-path of 1.8 THz phonons is expected to be less than 0.1 mm.

Figure 13b shows how isotope scattering limits the frequency spectrum of the ballistically transmitted phonons. For comparison, the calculated phonon intensity along the [100] axis, assuming no scattering, is plotted as the dashed line. The phonons involved with the ballistic [100] pile-up are centered at 1.8 THz, and there should be little chance of observing this effect for samples thicker than 0.1 mm.

What, then, is the origin of the delayed pile-up of phonon flux along [100]? Ramsbey, Wolfe and Tamura [36] proposed a logical solution to this puzzle: the effect is due to sub-terahertz phonons, scattered just a few times, but retaining some semblance of the ballistic focusing. This idea was also adopted by Fieseler, Wenderoth, and Ulbrich [38], who estimated that the mean free path of 0.7 THz phonons is very close to that predicted by the isotope scattering rate calculated by Tamura [6].

Figure 14 illustrates this *channeling* effect for the fast-transverse (FT) phonons. This mode exhibits narrow ridges of high ballistic flux intensity along (100) planes, represented schematically in the figure. An FT phonon com-

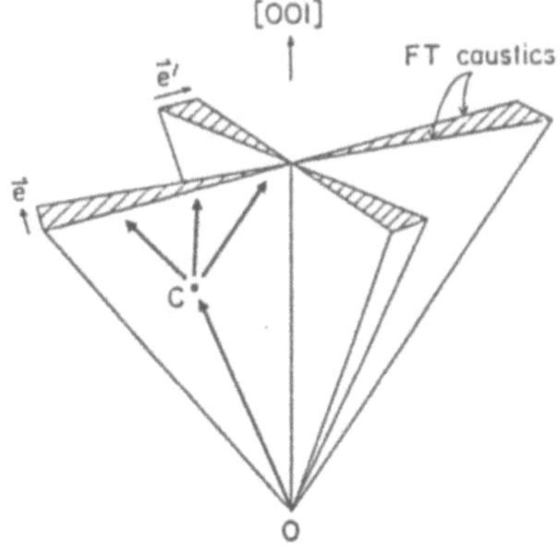

Fig. 14 Schematic diagram showing a portion of the FT phonon-focusing ridges. A phonon scattering at point C inside the ridge will tend to stay within the ridge [36].

ing from the heat source may elastically scatter at point C. For elastic scattering, the rate has the form,

$$\tau^{-1} \propto |e_1 \cdot e_2|^2 \,, \tag{5}$$

where $e_{1,2}$ represent the polarizations of the initial and final phonons. The polarizations of phonons in the FT ridges are shown in the figure. From the above equation one can see that the probability of an FT phonon scattering from one ridge to another orthogonal ridge is effectively zero. Due to the large phase space occupied by a ridge (hence large intensity) and the above selection rule, the FT phonons will tend to scatter into the same ridge. Thus, the distribution of phonons scattered a few times will tend to retain some of the ballistic anisotropy.

Figure 15 shows the Monte Carlo distribution of once-scattered FT and ST phonons [36]. They retain a large degree of the ballistic anisotropy. Of course, the arrival times of these scattered phonons will be later than those of the ballistic flux. When one considers a horizontal line scan across the center of Figure 15 a, one can see how this could be interpreted as a pile-up of delayed phonon flux along [100].

The scattered phonon need not be the same mode as the incident phonon. The above equation still holds for mode-converted phonons, but the scattering probability depends on the density of states of the final mode. Channeling of the mode converted phonons is reported in Ref. [36]. Figure 16 a shows a comparison of the predicted flux, including all modes, for up to three scattering events per phonon. This channeling of elastically scattered phonons is seen to provide a good explanation of the delayed-flux pile-up along the [100] axis.

A graphic illustration of this channeling effect was found in a recent experiment on Si, which displays a similar ballistic focusing pattern to GaAs. To separate the ballistic and scattered flux, Shields, Wolfe and Tamura [39] cut a narrow slot in the crystal as shown in Figure 17. The position of the slot with respect to the phonon-focusing caustics is shown in part (b) of this figure. To the left of the dashed line (slot boundary) the slot obstructs the *line of sight* between the laser source and the detector. Phonon signals detected when the laser beam is *behind the slot* could only be from scattered phonons.

The resulting phonon image in Figure 18 shows part of the ballistic phonon-focusing pattern at the right. The intense FT ridges end where the ballistic flux intercepts the edge of the slot. Behind the slot are weaker extensions of the ridges which cross at the (100) direction. (The intensity on the left side of this image is enhanced in the printing.) This is the channeling of elastically scattered FT phonons which was discussed above. The FT channeling is the principal source of the delayed flux along [100].

It is interesting that the phonon focusing effect alone is not sufficient to explain the intensity of the channeled FT ridges. The $e_1 \cdot e_2$ factor in the elastic scattering formula helps to concentrate the scattered flux within the ridges.

96

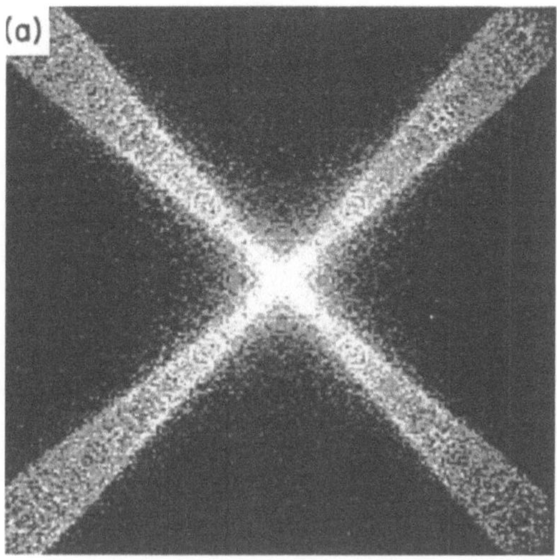

Fig. 15 (a) Monte Carlo calculation of the once-scattered FT phonons in GaAs [36].

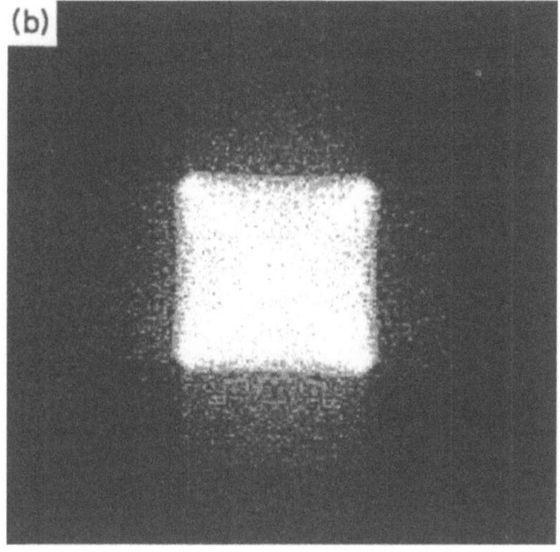

Fig. 15 (b) Monte Carlo calculation of the once-scattered ST phonons in GaAs [36].

Therefore, this experiment shows the anisotropy of the elastic scattering process. Further experiments of this sort have potential for quantitative measurements of scattering rates.

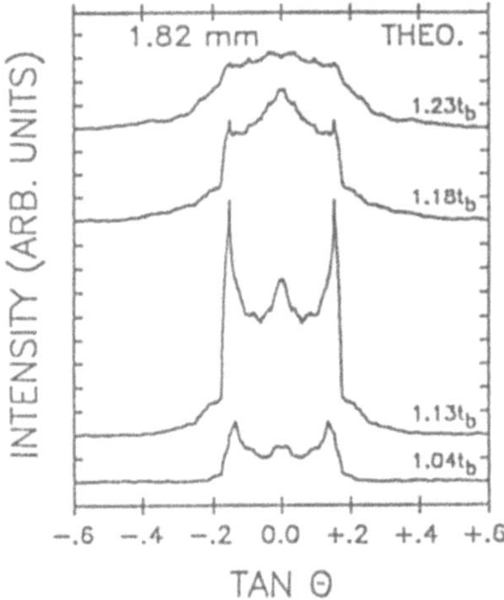

Fig. 16 Theoretical line scans of phonon intensity in GaAs for several delays after the excitation pulse. The central peak is due to scattered phonons. Compare to Fig. 11 [36].

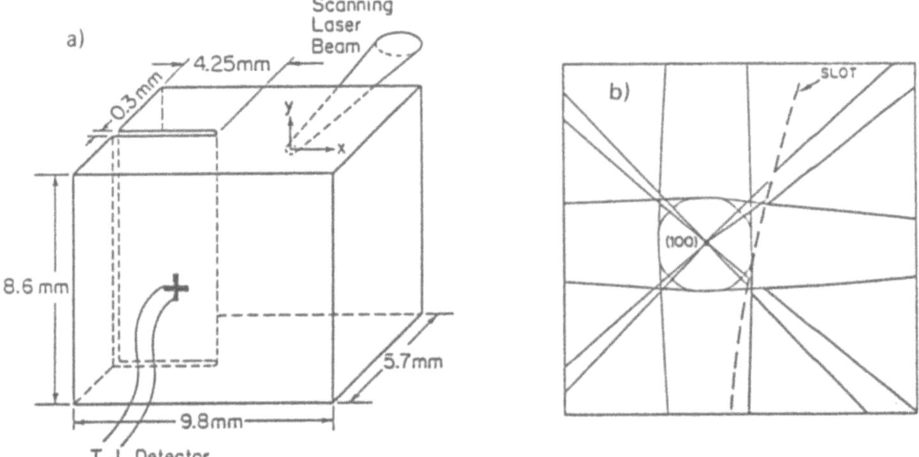

Fig. 17 (a) Slot imaging experiment designed to separate the ballistic and scattered flux in Si. (b) Positions of the slot boundary and focusing caustics [39].

98

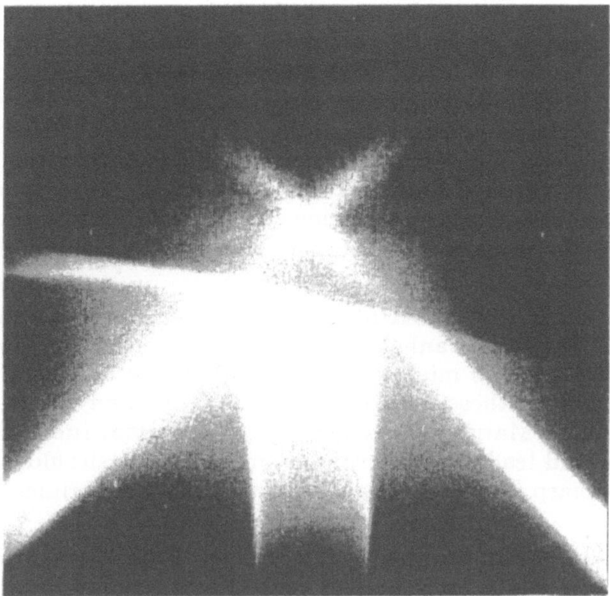

Fig. 18 Phonon image corresponding to the geometry of Fig. 17. Ballistic caustics are seen to the right of the slot. The weaker continuation of the FT ridges behind the slot is due to the channeling effect described in the text [39].

6 The Phonon Source

Earlier it was mentioned that for metal film excitation the frequency distribution of the phonon source could be regarded as Planckian [40], and that the heat pulse retained this same *color temperature* throughout its motion. This hypothesis is true only for frequencies low enough that the ν^5 anharmonic decay rate may be ignored. The down-conversion process is expected to be especially important when the semiconductor itself is directly illuminated with a light pulse, which creates energetic electron-hole pairs that thermalize by emission of very high frequency phonons.

It should be mentioned that the large-k phonon imaging experiments described earlier generally employed metal-film excitation, but they were also tried with direct photoexcitation. Generally, the heat pulses were somewhat broader in that case, but no significant differences in the ballistic flux (e.g., the caustic patterns) were observed. This result implies that there is an effective mechanism (or mechanisms) for down-converting the high energy phonons.

One possibility for an effective down-conversion of high-frequency phonons is the formation of a *hot-spot*, as originally postulated by Hensel and Dynes [41] to explain their observation of broad heat pulses in laser-excited Ge. This idea has been examined theoretically by Kazakovtsev and Levinson [42,43] and seems to be a viable process. Basically, if the energy density deposited by the laser pulse is high enough, phonon-phonon interactions become important, and an effective temperature is set up in the crystal near the excitation region. Low frequency phonons most easily escape this hot spot and they are able to propagate ballistically through the remainder of the crystal.

The hot spot can be a much more efficient down-conversion mechanism than spontaneous decay. For Si the anharmonic decay process for 1 THz phonons is predicted to be about 6 microseconds [44]. Because the elastic scattering time is much shorter, these phonons would undergo many elastic scattering events. Indeed, as Maris pointed out [45], the normal inelastic and elastic decay rates would lead to a diffuse phonon source. It should be very difficult to observe sharp phonon-focusing caustics. But experiments show otherwise.

The size of the phonon source can be directly determined from the sharpness of the phonon caustics, which for a perfect crystal and a point source should be extremely sharp, perhaps limited only by diffraction [46]. A high resolution experiment of this sort was recently performed by Shields and Wolfe [47]. A 10×10 μm^2 detector was evaporated on a $2 \times 2 \times 2$ cm^3 crystal of undoped Si, which was immersed in a superfluid helium bath and excited with a focused laser beam of diameter about 15 μm.

The resulting phonon image is shown in Figure 19. A line scan across the ST caustic indicates that the width of this structure is about 30 μm, corresponding to only 0.09° propagation angle. This is much smaller than one expects from the theoretical anharmonic and elastic scattering processes [45,47].

A further result is that the sharpness of the caustic depends on the energy of the excitation pulse, as shown in Figure 20 a. This result implies that there are some important scattering processes which depend on the density of the phonons at the source. This is just what is expected for a phonon hot spot. In this particular experiment, the crystal is immersed in superfluid helium, which may play a significant role in the kinetics of a hot spot.

The width of the caustic gives the size of the phonon hot spot. In these first experiments, the dimension of the hot spot is found to increase approximately as the cube root of the excitation energy, as shown in Fig. 20 b. That is, the volume of the phonon source increases linearly with excitation energy.

Further experiments are needed to characterize the properties of a hot spot produced by surface excitation of a semiconductor. For example, one would like to know the temporal development of the hot spot, and look for an energy threshold for its formation. A number of predictions have been made by Levinson and coworkers [42,43] which may be tested by time-resolved phonon imaging experiments.

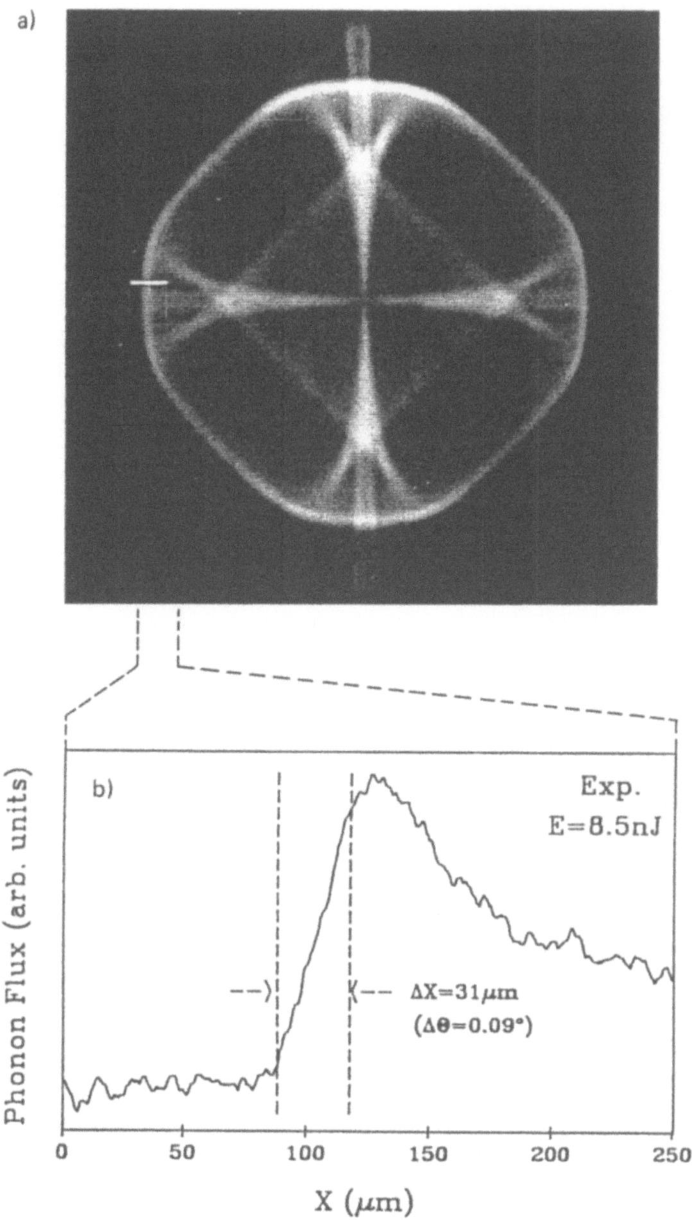

Fig. 19 (a) High resolution phonon image of photoexcited silicon. The center of the image corresponds to the [100] direction. The width of the image is 19.5 degrees from left to right. (b) Line scan of the ST caustic as indicated by the line on the image. The horizontal scale has been greatly expanded [47].

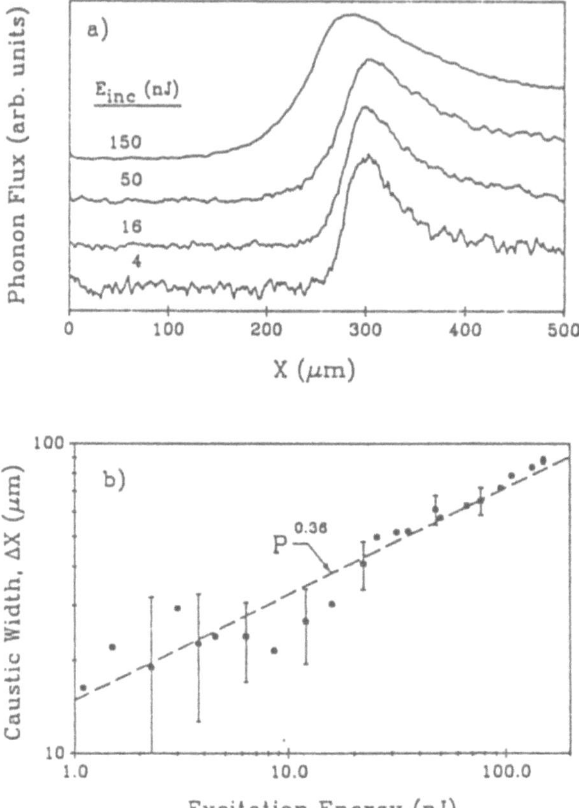

Fig. 20 (a) Line scans across the ST caustic indicated in Fig. 19 a at various incident laser energies. The width of the caustic varies significantly with incident energy, indicating an expansion of the ballistic-phonon source. (b) Power dependence of the caustic width. The experimental resolution has been deconvolved from these data.

7 Conclusions and Future Perspectives

Much can be learned from the angular distribution of phonon flux emitted from a point source. Phonon imaging is sensitive to the curvatures of the phonon slowness surfaces. By appropriate choice of experimental conditions, it is possible to observe the ballistic propagation of acoustic phonons in InSb with wavevectors up to 40% of the Brillouin zone boundary. The experimental caustic patterns contain new information about the interatomic forces, as underscored by the fact that the existing force models predict significantly different caustic patterns. Quantitative methods have been developed to compare phonon-imaging experiments with lattice-dynamics

102

theories. The microscopic interpretation of these results remains as an interesting problem.

InSb was chosen for this first study because high quality undoped crystals were available, and because dispersion of the transverse acoustic branches matched well with the superconducting gaps of Pb-based alloys. The success of this work hinged upon the ability to fabricate tiny junctions with continuously variable superconducting gaps. The present Pb-based detectors are applicable to other moderately dispersive crystals. Ionic crystals would provide an interesting contrast to the covalent semiconductors.

The theoretical studies of dispersive phonon focusing in GaAs have revealed striking variations in the large-k phonon flux, for example, the flux pile-ups along the [100] and [110] axes. The predictions are very model dependent; thus, observation of such effects would further probe the validity of various force models. The detection of such highly dispersive acoustic phonons requires a significant extention of the present technology. The smaller phonon mean free paths require micron-sized detectors and microscopic imaging techniques. Development of larger gap tunnel junctions or other types of frequency selective detectors may also be required.

Elastic and inelastic phonon scattering are basic processes for which few quantitative measurements have been made. Isotope scattering and anharmonic decay rates still need to be determined, not to mention the effects of impurities [48]. The new phonon-imaging experiments described in the latter part of this article seem to hold significant promise for revealing the microscopic details of the scattering processes. For example, it should be possible to observe the evolution of a phonon hot spot in time and space.

An important practical reason for understanding the phonon scattering processes in semiconductors has arisen recently. Single-particle detection of nuclear events based on ballistic phonon propagation and imaging has been proposed [See, for example, 45,49,50]. Absorption of α-particles, γ-rays, x-rays, neutrinos, etc. produce energetic electrons which relax, as in the photoexcitation case described above, by emission of high frequency phonons. The subsequent down-conversion and scattering processes of the phonons will dictate the possible configurations of such particle detectors.

No doubt, the study of large-wavevector acoustic phonons in solids will continue to provide some interesting surprises.

Acknowledgement

The Illinois work was supported by the National Science Foundation under the Materials Research Laboratory Grant DMR 86–12860. I wish to thank the Alexander von Humboldt Foundation, H. Kinder, W. Dietsche, P. Berberich, L. Koester, and the students and staff at the Physics Department of the Technische Universität München for their kind support this year.

References

[1] R.J. von Gutfeld, A.H. Nethercot Jr., Phys. Rev. Lett. **12**, 641 (1964)

[2] S.E. Hebboul, Ph.D. Thesis, University of Illinois, 1988; see also Ref. 23

[3] S. Tamura, Phys. Rev. **B 30**, 849 (1984)

[4] P.G. Klemens, Proc. Phys. Soc. London, Sect. **A 68**, 1113 (1955)

[5] W. Dietsche, G.A. Northrop, J.P. Wolfe, Phys. Rev. Lett. **47**, 660 (1981); G.A. Northrop, Phys. Rev. **B 26**, 903 (1982)

[6] S. Tamura, Phys. Rev. **B 28**, 897 (1983); S. Tamura, T. Harada, Phys. Rev. **B 32**, 5245 (1985)

[7] R.G. Ulbrich, V. Narayanamurti, M.A. Chin, Phys. Rev. Lett. **45**, 1432 (1980)

[8] R.G. Ulbrich, V. Narayanamurti, M.A. Chin, J. Phys. (Paris) **42** C6, 226 (1981)

[9] B. Stock, R.G. Ulbrich, Physica **117 B − 118 B**, 540 (1983)

[10] B. Stock, R.G. Ulbrich, M. Fieseler, in: Phonon Scattering in Condensed Matter, ed. by W. Eisenmenger, K. Lassmann, S. Dottinger (Springer Berlin, Heidelberg, New York, 1984) p. 97

[11] M. Lax, P. Hu, V. Narayanamurti, Phys. Rev. **B 23**, 3095 (1981)

[12] M. Lax, V. Narayanamurti, P. Hu, W. Weber, J. Phys. (Paris) **42**, 161 (1981)

[13] V. Narayanamurti, J. Phys. (Paris) **42**, 221 (1981). The observation of near zone-boundary phonons in (isotopically pure) solid helium was reported by T. Haavasoja, V. Narayanamurti, M.A. Chin, Phys. Rev. **B 27**, 2767 (1983)

[14] N.M. Guseinov, Y.B. Levinson, Sol. State Commun. **45**, 371 (1983)

[15] J.P. Wolfe, G.A. Northrop, in: Phonon Scattering in Condensed Matter, ed. by W. Eisenmenger, K. Lassmann, S. Dottinger (Springer Berlin, Heidelberg, New York 1984), p. 100

[16] M. Lax, V. Narayanamurti, R.C. Fulton, N. Holzwarth, in: Phonon Scattering in Condensed Matter V, ed. by A.C. Anderson, J.P. Wolfe, (Springer Berlin, Heidelberg, New York 1986), p. 335

[17] M. Lax, V. Narayanamurti, R.C. Fulton, in: Laser Optics of Condensed Matter, ed. by H.Z. Cummins, A.A. Kaplyanskii, (Plenum Press 1988) p. 229

[18] G.A. Northrop, J.P. Wolfe, in: Proc. on a NATO Advanced Study Institute of Nonequilibrium Phonon Dynamics, Les Arcs, France 1984, ed. by W.E. Bron, Plenum Press, New York (1985)

[19] B. Taylor, H.J. Maris, C. Elbaum, Phys. Rev. Lett. **23**, 416 (1969)

[20] G.A. Northrop, S.E. Hebboul, J.P. Wolfe , Phys. Rev. Lett. **55**, 95 (1985)

[21] S.E. Hebboul, J.P. Wolfe, Z. Phys. **B 74**, 35 (1989)

[22] S.E. Hebboul, J.P. Wolfe, Phys. Rev. **B 34**, 3948 (1986)

[23] S.E. Hebboul, J.P. Wolfe, Z. Phys. **B 73**, 437 (1989)

[24] H.J. Maris, in: Nonequilibrium Phonons in Nonmetallic Crystals, ed. by W. Eisenmenger, A.A. Kaplyanskii, (North-Holland, Amsterdam 1986), p. 51

[25] B.A. Auld, Acoustic Fields and Waves in Solids (Wiley, New York, 1973)

[26] G.A. Northrop, J.P. Wolfe, Phys. Rev. Lett. **43**, 1424 (1979)

[27] G.A. Northrop, J.P. Wolfe, Phys. Rev. **B 22**, 6196 (1980)

[28] A.G. Every, Phys. Rev. **B 24**, 3456 (1981); ibid, **B 34**, 2852 (1986)

[29] J.C. Hensel, R.C. Dynes, Phys. Rev. Lett. **39**, 969 (1977)

[30] D. Huet, J.P. Maneval, A. Zylbersztejn, Phys. Rev. Lett. **29**, 1092 (1972)

[31] W. Eisenmenger, in: Physical Acoustics, Principles and Methods, Vol. 12, ed. by W.P. Mason, R.N. Thurston, (Academic Press, New York 1976), p. 79

104

[32] *M. Born, K. Huang*, in: Dynamical Theory of Crystal Lattices (Oxford University Press, New York 1954)

[33] *D.L. Price, J.M. Rowe, R.M. Nicklow*, Phys. Rev. **B 3**, 1268 (1971)

[34] *B. Stock, M. Fieseler, R.G. Ulbrich*, in: Proc. of the 17th Int. Conf. on Physics of Semiconductors, ed. by *J.D. Chadi, W.A. Harrison* (Springer Berlin, Heidelberg, New York 1985), p. 1177

[35] *R.G. Ulbrich*, in: Nonequilibrium Phonon Dynamics, ed. by *W.E. Bron* (Plenum Press, New York 1985), p. 101

[36] *M.T. Ramsbey, J.P. Wolfe, S. Tamura*, Z. Phys. **B 73**, 167 (1988)

[37] *M. Schreiber, M. Fieseler, A. Mazur, J. Pollman, B. Stock, R.G. Ulbrich*, in: Proc. of the 18th Int. Conf. on Physics of Semiconductors, ed. by *O. Engstroem* (World Scientific, Singapore 1987), p. 1373

[38] *M. Fieseler, M. Wenderoth, R.G. Ulbrich*, to be published in: Proc. of the 19th Int. Conf. on Physics of Semiconductors (August 1989, Warsaw)

[39] *J.A. Shields, J.P. Wolfe, S. Tamura* (preprint)

[40] *O. Weis*, Z. Angew. Phys. **26**, 325 (1969)

[41] *J.C. Hensel, R.C. Dynes*, Phys. Rev. Lett. **39**, 969 (1977)

[42] *D.V. Kazakovtsev, Y.B. Levinson*, Sov. Phys. JETP **61**, 1318 (1985)

[43] *Y.B. Levinson*, in: Nonequilibrium Phonons in Nonmetallic Crystals, ed. by *W. Eisenmenger, A.A. Kaplyanskii* (North-Holland, Amsterdam 1986), p. 91

[44] *S. Tamura*, in: Phonon Scattering in Condensed Matter V, ed. by *A.C. Anderson, J.P. Wolfe* (Springer, Berlin, Heidelberg, New York 1986), p. 288

[45] *H.J. Maris*, in: Phonon Scattering in Condensed Matter V, ed. by *A.C. Anderson, J.P. Wolfe* (Springer, Berlin, Heidelberg, New York 1986), p. ?

[46] *H.J. Maris*, Phys. Rev. **B 28**, 7033 (1983)

[47] *J.A. Shields, J.P. Wolfe*, Z. Phys. **B 75**, 11 (1989); Phonon focusing and source size from electron-beam excitation of Ge has been examined by *W. Metzger, R.P. Huebener*, Z. Phys. **B 73**, 33 (1988)

[48] *E. Held, W. Klein, R.P. Huebener*, Z. Phys. **B 75**, 17 (1989)

[49] *Th. Peterreins, F. Probst, F. von Feilitzsch, R.L. Mossbauer and H. Kraus*, Phys. Lett. **B 202**, 161 (1988)

[50] *B. Cabrera, J. Martoff, B. Neuhauser*, Nuclear Instruments and Methods in Physics Research **A 275**, 97 (1988)

Theory of Dynamical Surface States and Reconstructions at Crystal Surfaces

Wolfgang Ludwig

Westfälische Wilhelms-Universität Münster, Institut für Theoretische Physik II, Festkörperphysik, Wilhelm-Klemm-Str. 10, D-4400 Münster, Federal Republic of Germany

Summary: During the spring meeting of the DPG in 1988 J.P. Toennies reported on experimental methods and results on the dynamics of crystal surfaces. In this contribution theoretical investigations on this topic will be reviewed. After some remarks on the dynamics in general and on the elastic limit (Rayleigh-waves) lattice dynamical calculations and results are discussed. Most of the investigations are based on phenomenological models which give quite good agreement with experiment for the localized as well as for the resonant surface states. It turns out further, that a number of results is independent of the special assumptions for the models but is rather determined by the mere existence of surfaces. These are e.g. statements on the occurrence of soft modes and surface reconstructions. Microscopic (ab initio) calculations will be referenced, too, but these are just in an early stage of discussions.

1 Introduction

Surface waves have been investigated theoretically since 1885, when Rayleigh [1] published his famous paper on this subject. In the following decades the discussion of surface waves in the elastic or long wave limit has been continued including waves in layered structures and the coupling to electromagnetic waves.

The first lattice-dynamical calculations have been done about 15–20 years ago by different groups [2–5]. Though the principles of such calculations and the main features can be explained rather simply, quantitative calculations are tedious and need a large amount of numerical work. Thus these early discussions suffer from the lack of large enough computer facilities as well as experimental data for comparison. As far as I see, the first surface mode found experimentally is that of Ibach [6]. But since then the situation has changed drastically. Quantitative calculations are possible due to the growth in computers, and experimental methods have been developed which allow for more detailed measurements. These have been reviewed at the meeting last year by J.P. Toennies [7]. Therefore I will restrict to the theory here.

Nearly all the calculations so far have been done for high symmetry surfaces of cubic or nearly cubic crystals. Though there has been also a lot of work on surfaces with surface layers or with defects or rough surfaces or interfaces or on the coupling to electromagnetic waves (surface polaritons) in this report I will concentrate on plane surfaces of semi-infinite crystals. Such investigations have been done by several groups using different methods and models.

In most calculations the *slab method* is used, solving the equation of motion for a crystal slab consisting of a certain number of layers parallel to the surface instead of treating a semi-infinite crystal. Others use a Green function technique which treats the surface as a twodimensional perturbation of an infinite crystal. Parallel to the surface a translational periodicity is maintained. Calculations of the surface dynamics can also provide insight into structural questions. The theory of phase transitions combines the occurrence of a soft mode, i.e. a zero for the surface phonon dispersion, with the formation of a surface reconstruction.

The models discussed reach from purely phenomenological ones with fitted force constants to the use of potentials and nowadays also to microscopic (ab initio) calculations. Nearly all kinds of substances have been investigated: semiconductors, ionic crystals and metals. In many cases there is good agreement with experimental results; especially statements on soft modes and surface reconstructions seem to be very realistic though the models are very simple in some cases.

2 General Dynamics in Harmonic Approximation

Some remarks on the dynamics of a harmonically vibrating system seem to be necessary. The motions of such systems are completely described by the equation

$$M^m \ddot{u}_i^m = -\sum_{n,j} \Phi_{ij}^{mn} u_j^n = -M^m \omega^2 u_i^m \, . \tag{2.1}$$

Because of the harmonicity the time dependence can always be replaced by the set of 3N frequencies, the eigenfrequencies of the system. u_i^m describes the displacement in i-th Cartesian direction of an atom m with mass M^m. Eq. (2.1) is valid for all systems of masses M^m provided there exist

  i)   equilibrium (*rest*) positions for the masses,
  ii) only harmonic couplings between the masses.

There is no need for an introduction of a potential energy, though in general the coupling (force) constants Φ_{ij}^{mn} can be derived from a potential (adiabatic approximation). Eq. (2.1) does not require the existence of a lattice or an ordered structure rather it holds for any set of coupled masses. Thus the solutions of Eq. (2.1) are not directly dependent on lattice distances, positions or the geometric structure. The positions enter only via the equilibrium (*rest*) positions, at which the coupling constants or the curvatures of the potential have to be taken.

As a consequence small changes of positions near defects or surfaces influence the frequency spectrum only to that extent with which the force constants are changed. In reconstructed surfaces, e.g. there are changes in

108

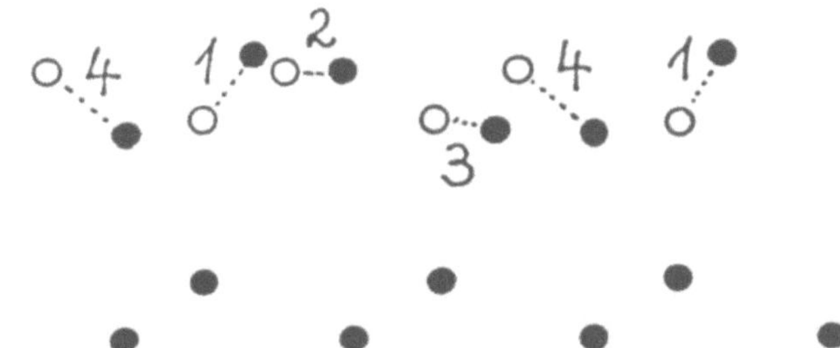

Fig. 1 Si-surface according to the π-bonded chain model (Pandey [8]) with undisplaced (open circles) and displaced atoms (full circles) in the uppermost layer.

the positions (Fig. 1), but one does not know the forces very well; consequently, there is an arbitrariness about the force constants entering into the calculations even if the positions are precisely known. In other words, it is not necessary to describe the geometric structure of a reconstructed surface but rather the force constants at the surface. We found that in many cases force constants like f_1 or f_2 between ions 4–1, 2–3, or 1–2, 3–4, respectively in Fig. 1 can be assumed to be equal to those of the ideal surface in a first discussion of surface modes.

The situation is somewhat different, if there are holes or lines of missing ions at the surface. Then, of course, corresponding force constants have to be put equal to zero.

For the calculation of surface modes at crystal surfaces, of course, it is suitable to start using the translational symmetry of the crystal. This implies the existence of a lattice constant, a Brillouin zone and so on. But one should have in mind that this is only a convenient starting point for the calculations. The essential ingredients are the force constants.

Another point is the number of force constants used and their determination. In many cases the force constants are not known, especially those at the surface. Often one works using them as free parameters determined by fitting the eigenfrequencies or some of them to experimental data. In general this gives reasonable results. But one should realize, that such a procedure never gives unique values for the force constants (see Szigeti et al. [9,10]). On the other hand many results are very insensitive against the special choice or fitting of force constants. This gives a certain confidence in such results and I will concentrate myself to statements which to a certain extent are independent of special assumptions.

The situation is different and, of course, the better if one starts with a reasonable given potential or even more, with microscopic (ab initio) calculations. Then the force constants can be calculated, but this requires a large extent of numerical work. Discussions of such methods are in progress now and first quantitative results have been obtained [11–15].

3 The Elastic Limit

As already mentioned in the introduction, surface waves at surfaces of elastic media have been discussed since 1885. The localized states at plane surfaces of semi-infinite elastic media are the Rayleigh- or generalized Rayleigh-waves. If they exist (there may be certain orientations of surfaces and of wave propagation in non-isotropic crystals without localized surface states) their frequencies lie below the elastic bulk wave frequencies and they are the lowest frequencies in ideal solids with a plane surface.

The lattice theory of surface waves for plane surfaces of semi-infinite crystals has to coincide with the results of the elastic theory in the limit of long wavelengths. Thus the dispersion curves have to merge into the Rayleigh waves. Consequently we use the term Rayleigh waves only for the surface states with the lowest frequency lying below all the bulk vibrational states.

Some features of the Rayleigh waves should be mentioned. As all localized states they have exponentially decreasing amplitudes with increasing distance from the surface. The decay constants of the amplitudes depend on the elastic properties of the media and on the wavevector $q_\parallel$ parallel to the surface. The exponential decrease can be superposed by an oscillatory behavior in more complex cases (generalized Rayleigh waves in anisotropic media). Rayleigh waves have a mixed polarization with a longitudinal and a transverse component perpendicular to the surface. The relative contributions change with distance from the surface.

These properties have to correspond to the long wave limit of the lowest lattice surface mode. For higher frequencies the phenomena will be even more complicated. Surface waves above the Rayleigh-wave might also be of shear (transverse) horizontal character.

4 Methods

The method most widely used in calculating surface phonons is the slab-method, in which the equation of motion is solved for a slab of about 7 to 50 layers of atoms. The slab is infinitely extended in two directions, in which it has the translational symmetry of the surface under discussion. Taking into account this translational symmetry one has to deal with a 100- to 1000-dimensional eigenvalue problem depending on the number of atoms in the surface unit cell.

This procedure has its advantages if tedious numerical calculations are necessary, e.g. if the unit cell contains a large number of ions (perovskite-type crystals, see [16–18]) or if calculations for reconstructed surfaces or if microscopic calculations are to be done. In the latter case one has to use the density functional formalism, firstly calculating the electronic structure of the slab. The frequencies of surface phonons can then (at q-vectors of high symmetry) be calculated by freezing in the displacements and comparing the energy of the electronic system with the ground state energy. In an alternative approach the atomic force constants are determined from the response of the electronic system to displacements of the atoms. This leads to the dynamical matrix of the slab [13,14].

However, the slab method has disadvantages, too. In any case one gets for a given $q_\parallel$ a discrete spectrum contrary to that of a semi-infinite crystal. Apart from slight quantitative differences, this difference is not very serious for proper localized states which lie outside the region of the continuous bulk band states. Resonance states are identified with the accumulation of eigenstates (always for fixed $q_\parallel$). That this might be misleading can be seen from Fig. 2. The band edges of bulk states also show an accumulation of frequencies and only a thorough comparison of bulk states and additional states due to the surface leads to an unique identification of resonance states.

If the number of layers in a slab is not sufficient, errors might also occur from an interference between the two surfaces of a slab. Even for states at the zone boundary (Fig. 3) the decrease of the amplitudes as function of the layers is rather slow. Both surfaces are not independent of each other. This can be checked relatively simple because all states in a slab must be twice degenerated (from both surfaces) and the splitting of such states can be taken as a measure of accuracy.

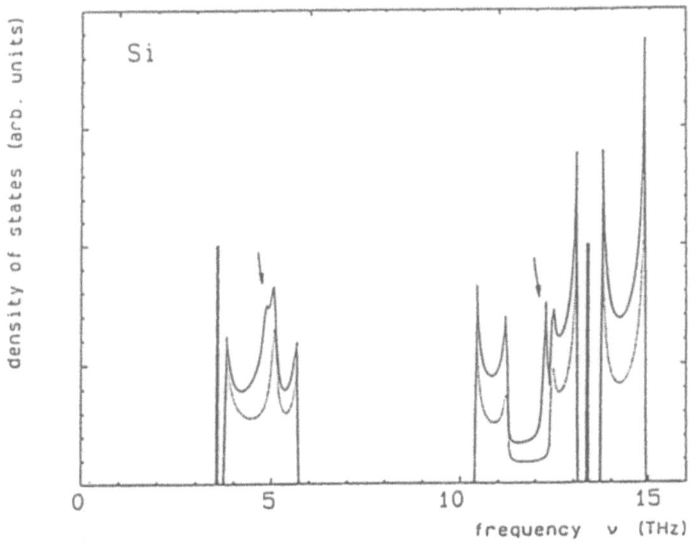

Fig. 2 Density of frequencies for bulk (dotted) and surface states (solid line) showing localized (fat lines) and resonance states (arrows) and band edges. The resonance states can be identified only by detailed comparison with bulk states. Si-(111) surface, $q_\parallel = \overline{M}$-point.

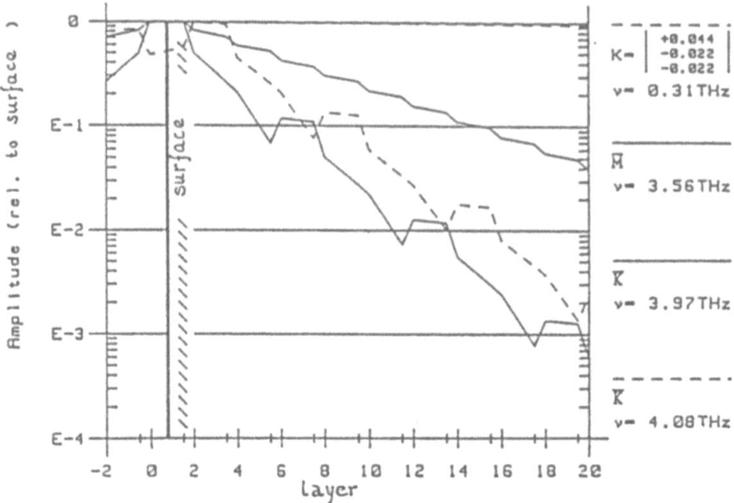

Fig. 3 Amplitudes of localized surface states as a function of the layers for special $q_{\parallel}$-values for an ideal Si-(111) surface. The amplitude scale is logarithmic!

The Green function technique treats the surface as a perturbation of an otherwise ideal crystal. The equation of motion of the ideal infinite crystal is (in matrix notation).

$$M\ddot{u} + \Phi u = 0 \tag{4.1}$$

and that of the semi-infinite crystal

$$M\ddot{u} + \Phi' u = 0 \quad \text{or} \quad M\ddot{u} + \Phi u = Ju \tag{4.2}$$

with the perturbation

$$J = \Phi - \Phi'. \tag{4.3}$$

J contains all the changes which are due to creating a surface by cutting all the force constants between two adjacent semi-crystals but also changes in the force constants near the created surface. The dimension of the non-zero part of the defect matrix J is determined only by the range of the defect. This is automatically limited to a few atomic layers within models employing short range force constants. But even in problems involving long range electrostatic interactions it was found to be sufficient to take into account 4–8 layers in the defect matrix. This is due to the fact that consideration of the translational invariance parallel to the surface leads to summation of the contributions parallel to the surface. Thus layer-to-layer couplings are important which decay much more rapidly than atom-to-atom couplings. This limited range of the defect is of crucial importance for the calculations because it determines the dimensions of the matrix equations that have to be solved.

112

The formal solution of (4.2) is

$$u = u_h + G_0 J u \tag{4.4}$$

with the Green function G_0 of the ideal infinite crystal and a solution u_h of the unperturbed (homogeneous) Eq. (4.1). Taking into account the translational symmetry parallel to the surface the Green function is expanded with respect to the polarization vectors $e_i(^{m_3}_{\beta})$ of the ideal lattice.

$$G_{0ij}(^{m_3 n_3}_{\beta \ \tau}) = \frac{1}{sN\sqrt{M_\beta M_\tau}} \sum_{q_3 \sigma} \frac{e_i^\beta e_j^\tau(q_{\|}, q_3, \sigma)}{\omega^2(q_{\|}, q_3, \sigma) - \omega^2 - i\varepsilon} e^{iq_3(R^{m_3} - R^{n_3})} \tag{4.5}$$

where i, j denote the Cartesian components, m_3 and n_3 the layers and β, τ the basis indices. $q_{\|}$ and q_3 are wave-vector components parallel and perpendicular to the surface, σ denotes the polarization. Localized surface modes do not have a homogeneous contribution u_h, thus their frequencies are obtained from

$$\det\left(1 - G_0(q_{\|}, \omega) \cdot J(q_{\|})\right) = 0 \tag{4.6}$$

where $J(q_{\|})$ is the Fourier-transformed perturbation. In this case $\varepsilon = 0$ and G_0 is Hermitean. The calculation of G_0 can be limited to about 50 q_3-values, which leads to errors of about 1%. The frequencies $\omega_s(q_{\|})$ lie outside the projected bulk band frequencies ω_b, i.e. they satisfy

$$\omega_s(q_{\|}) \neq \omega_b(q_{\|} + \gamma g_3), \qquad -1/2 < \gamma \leq 1/2; \tag{4.7}$$

g_3 is a reciprocal lattice vector perpendicular to the surface.

Resonance states lie inside the bulk bands. Their frequencies are obtained best from the density of states of the perturbed crystal:

$$3sN\pi g(\omega^2) = \frac{3sN\pi}{2\omega} z(\omega) = \mathrm{Im} \sum_{m_3 \beta i} M_\beta G_{ii}(^{m_3 m_3}_{\beta \ \beta})(\omega^2 + i\varepsilon). \tag{4.8}$$

Here G is the Green function of the perturbed crystal (with surface) obtained via the Dyson-equation

$$G = G_0 + G_0 J G \quad \text{or} \quad G = (1 - G_0 J)^{-1} G_0. \tag{4.9}$$

A single term in the sum of (4.8) gives the contribution of the m_3-th layer to the density of states. For resonance states the Green function is no longer Hermitean. A reasonable choice is $\varepsilon \approx 1\%$ of the squared Raman frequency, the summation in (4.5) then needs about 600 q_3-values.

In the Green function method the main numerical effort lies in the calculation of the Green functions whereas the diagonalization of (4.6) is that of a matrix with a dimension of about 20 to 50 depending on the perturbed region (about 4–8 layers) and the number of atoms in the surface cell. Both methods have been used in the calculation of the same potential model.

5 Models

The simplest potential model, but even now often used, is just describing the interaction by force constants which are fitted to some experimental data, but sometimes also calculated from very simple assumptions on the interaction potential. It is generally known as rigid ion- or Born-von Kármán-model. In many cases the potential can be described as

$$
\delta\Phi = \frac{1}{2}f\sum_{1.n.}(\delta r)^2 + \frac{1}{2}g\sum_{2.n.}(\delta r)^2 + \frac{1}{2}h\sum_{3.n.}(\delta r)^2 + \frac{1}{2}\ell\sum_{4.n.}(\delta r)^2 + \ldots
$$
$$
+ \frac{1}{2}\sum_{val.angl.} \sigma\, r_0^2(\delta\vartheta)^2 + \sum_{val.bonds} \varphi\, r_0(\delta r \cdot \delta\vartheta) + \ldots
$$

$$(5.1)$$

The terms of the first line describe central forces up to fourth neighbor interactions, those of the second line non-central (many body) forces up to second neighbors. Of course, both kinds of interactions can easily be extended to higher order neighbors. But this does not make very much sense because the number of unknown parameters increases rapidly (see the problems mentioned in Sect. 2). (5.1) can be looked upon as a model with short range forces. Higher order terms would also introduce effects of long range forces depending on the nature and structure of the crystals. Especially for metals long range effects are important. The determination of the corresponding force constants, however, may cause severe problems in practice. For semiconductors it may be fruitful to take into account particularly important interactions in addition to the short range force constants as for example fifth neighbor interactions in silicon [47].

Long range interactions are better described by appropriate models. We have used the dipole model first proposed by Wallis and Wanser [19,20] for calculating surface phonons in zincblende structure crystals. In this model it is taken into account that the motion of ions in a crystal leads to a relative displacement of positive ions and the negative electronic charge distributions, thus to dynamically induced dipoles

$$
d_i\binom{m}{0} = \sum_j P_{ij}\binom{m\ m}{0\ 0}u_j\binom{m}{0} + \sum_j\sum_{n\beta} P_{ij}\binom{m\ m+n}{0\ \ \beta}u_j\binom{m+n}{\beta} + \ldots \qquad (5.2)
$$

where n, β run over neighbors of $m, 0$. (5.2) takes into account that a displacement $u_j\binom{m}{0}$ of $m, 0$ does not only induce a dipole moment $d_i\binom{m}{0}$ on the ion itself, but also on neighboring ions (non-local effects). In our model we have taken the sum over n, β up to nearest neighbors. The parameters in the matrices P have to be determined in accordance with the requirements of infinitesimal translational invariance, which in this case means that no dipole moment is induced if the crystal is displaced as a whole. In contrast

114

to local models [48–50] this can be done unambiguously within the non-local approach. The dipole interaction energy is simply

$$W_{\beta\,\tau}^{mn} = \frac{1}{\varepsilon} \sum_{ij} \Omega_{ij}({}^{mn}_{\beta\,\tau}) d_i({}^{m}_{\beta}) d_j({}^{n}_{\tau}) \tag{5.3}$$

with

$$\Omega_{ij}({}^{mn}_{\beta\,\tau}) = \frac{\delta_{ij}}{|\mathbf{R}_\beta^m - \mathbf{R}_\tau^n|^3} - \frac{3\left(\mathbf{R}_i({}^{m}_{\beta}) - \mathbf{R}_i({}^{n}_{\tau})\right)\left(\mathbf{R}_j({}^{m}_{\beta}) - \mathbf{R}_j({}^{n}_{\tau})\right)}{|\mathbf{R}_\beta^m - \mathbf{R}_\tau^n|^5} . \tag{5.4}$$

The total dipole-dipole interaction energy then is

$$\Phi^{dd} = \frac{1}{2} \sum_{\substack{mn \\ \beta\tau}} W_{\beta\,\tau}^{mn} . \tag{5.5}$$

Force constants are obtained by taking the second derivatives of (5.5), the occurring lattice sums can be calculated by Ewald's method.

In case of diamond type lattices (5.2) contains two independent parameters described by (n_1: neighbors to 0)

$$P_{00}^{0n_1} = \begin{pmatrix} P_1 & P_2 & P_2 \\ P_2 & P_1 & P_2 \\ P_2 & P_2 & P_1 \end{pmatrix} . \tag{5.6}$$

In zincblende type crystals (GaAs) there are four parameters depending on whether "0" is a Ga- or an As-ion. The short range forces being present additionally are again described by (5.1). The parameters of these models are determined by a least square fit to the experimental phonon dispersion curves. Figs. 4 and 5 show the results for the ideal crystal in case of Si. The typical flattening of the TA-branch in Δ-direction is reproduced only by the long range dipole model.

The long range forces can be described also by the shell model which takes into account polarization effects like the dipole model and also by the bond charge model. Since these are very well-known models we will not describe their details here. The latter has been used by G. Benedek and his group [21,22] for calculations of surface phonons in diamond type crystals. The results are very similar to those of the dipole model (see Sect. 6).

In calculations of some cubic (alkali halides, MgO, PbS) and Perovskite-type crystals ($KZnF_3$, $KMnF_3$) by the Regensburg-Stuttgart-Austin-group [16–18,23,24], also the shell model together with short range forces between the shells of Born-Mayer-type

$$V_{SS}(r) = a \cdot e^{-br} \tag{5.7}$$

has been used. Additionally to other authors in these calculations there are not only calculations of surface phonons, but also of the surface structure (relaxation of surface atoms). In these calculations the slab-method has

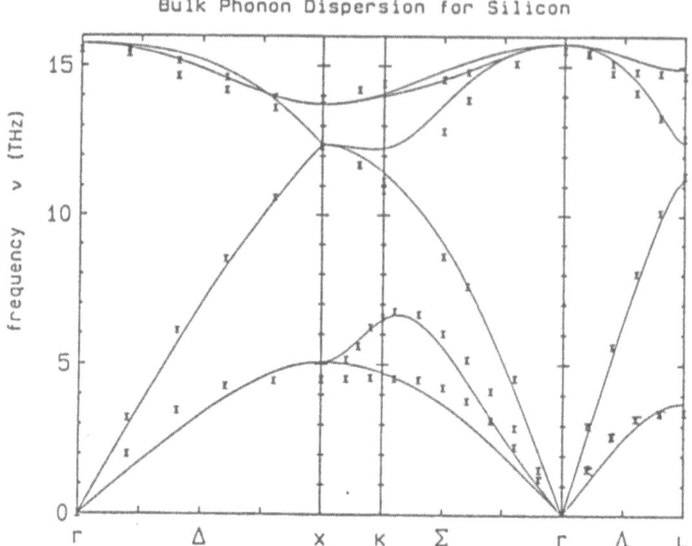

Fig. 4 Bulk phonon dispersion for Si in the pure force constant model. Force constants are determined by a least quare fit to the experimental values.

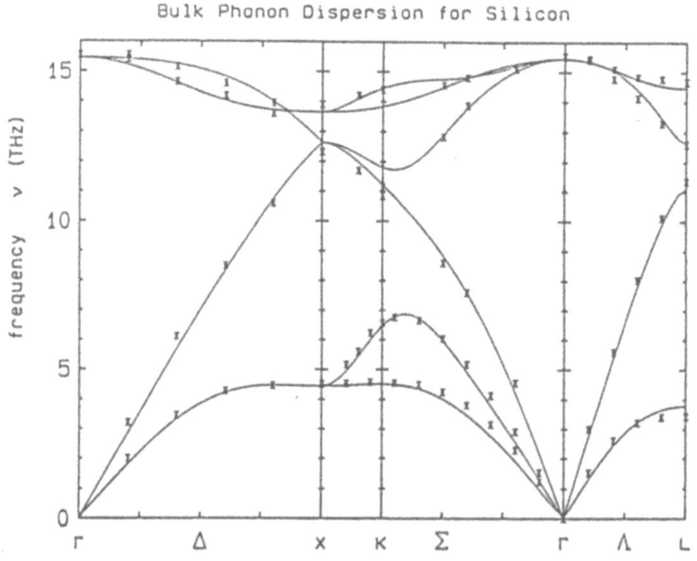

Fig. 5 Bulk phonon dispersion for Si in the dipole model. Parameters are determined by a least quare fit to the experimental values.

116

been used, whereas in the first surface phonon calculations of alkali halides at all by G. Benedek [4] the Green function method has been used.

Dynamical surface effects in metals have been investigated with the use of the rigid ion model in many cases; volume forces have been added sometimes. The force constants in this case have been derived from pair potentials and special angular dependent potentials. Discussions of this type are mainly due to Bortolani and his group [25–29].

But also in this case the arguments against force constant fitting remain valid [9,10]. The best way would be to predict surface modes without any fitting, that means, one had to do first principles calculations. One method has been recently proposed [30] which seems to be adequate for such discussions, the *embedded atom method*, being already applied in some cases [11,12].

In this model, the total energy of an arbitrary set of atoms is given by

$$\Phi = \sum_m F^m(\rho^m) + \frac{1}{2} \sum_{m \neq n} \varphi_{mn}(R^{mn}). \tag{5.8}$$

ρ^m is the electron density at atom (ion) m due to all other atoms (apart from m itself) and $F^m(\rho)$ is the energy to embed atom m into the electron density of all the other atoms. It represents the interaction of one atom with the local electron distribution of all other atoms. $\varphi_{mn}(R^{mn})$ is a short range interaction between ion m and n.

The main problem lies in the determination of ρ^m. Generally it is assumed to be a superposition of ionic electron densities

$$\rho^m = \sum_{n(\neq m)} \rho_I^n(R^{mn}). \tag{5.9}$$

$\rho_I^n(R^{mn})$ is the electronic density at ion m due to ion n; thus the $\rho_I^n(R^{mn})$ remain to be determined. This has been done via the quasi-atoms according to Stott et al. [31]. A more unique consistent way seems to be the use of the partial densities of Falter [32]. Some parameters are left in any case which then can be obtained by fitting to equilibrium distances, elastic constants, and other bulk quantities.

Once (5.8) has been determined, the force constants can be obtained from the second derivatives (Sect. 2) with respect to R_i^m. After doing this, it turns out that Φ_{ij}^{mn} has two contributions: the first one can be looked upon as arising from pair potentials, the second one contains effective many-body interactions, both are dependent on the structure of the system; thus they contain surface effects in the sense that the force constants in the vicinity of the surface are different from those of the bulk. Relaxation effects and force-constant changes could be calculated in a consistent procedure. However, the quantitative calculations are quite tedious. Fig. 6b shows the bulk phonons for Cu according to [11]; thus this model gives rather good results.

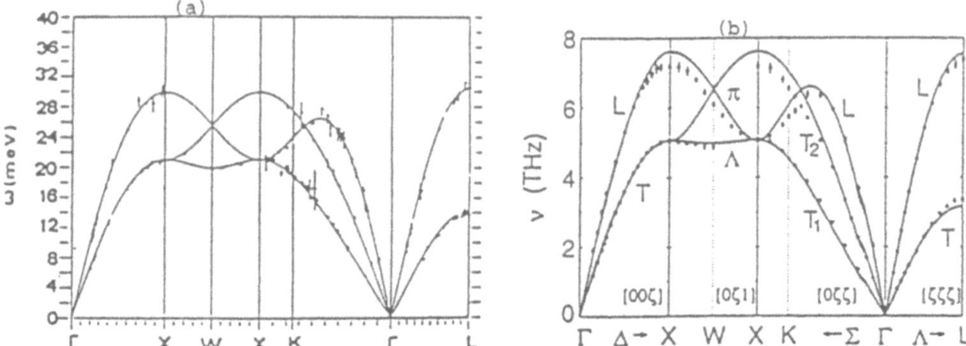

Fig. 6 Bulk phonon dispersion for Cu according to Bortolani et al. [26] (a) and to Nelson et al. [11] (b).

Apart from this method other first principle calculations make use of the *frozen-phonon* approximation [13,14]. Because of the difficulties of such calculations they can be done only for high-symmetry points of the Brillouin zone; the displacements (eigenvectors) of the phonons of these q-vectors are assumed to be frozen-in instead of being determined. This is particularly difficult for surface phonons where the amplitude decreases with an unknown attenuation constant into the bulk. If one wants to have complete dispersion curves for the surface phonons one has to fit the high symmetry phonons with force constants using these for the complete dispersion.

Other self-consistent pseudopotential theories of surface phonons use nearly free electron approximations and thus have limited applications, e.g. to Al [15,53].

6 Results

6.1 Surface Phonons

Surface states for the phonons do always occur, independent of the models and methods. This is due to the fact, that also the free surface of any elastic solid shows such states. The extension of these states to higher frequencies gives the localized surface states of the different lattices.

That means that in general surface states already occur only if there is a (plane) surface, without any changes of the force constants near the surface. These surface states are not very far in frequency from the bulk phonon states, lying below the bulk states. Further localized surface states lie in the gaps between the bulk bands. Figs. 7,8 show these states for a Si-(111)-surface according to the pure force constant model and according to the long range dipole model [33–37].

118

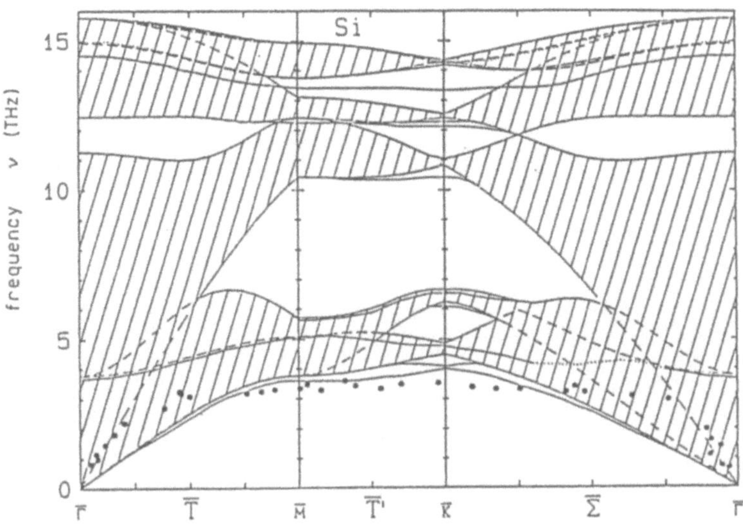

Fig. 7 Surface phonons at an ideal Si-(111) surface within the force constant model. Bulk phonon bands are hatched, band edges are indicated by broken lines. Full lines indicate surface phonons or resonances, weak resonances are indicated by dotted lines. Experimental points according to [38]. See also the 13.6 THz mode at the $\overline{M}$-point.

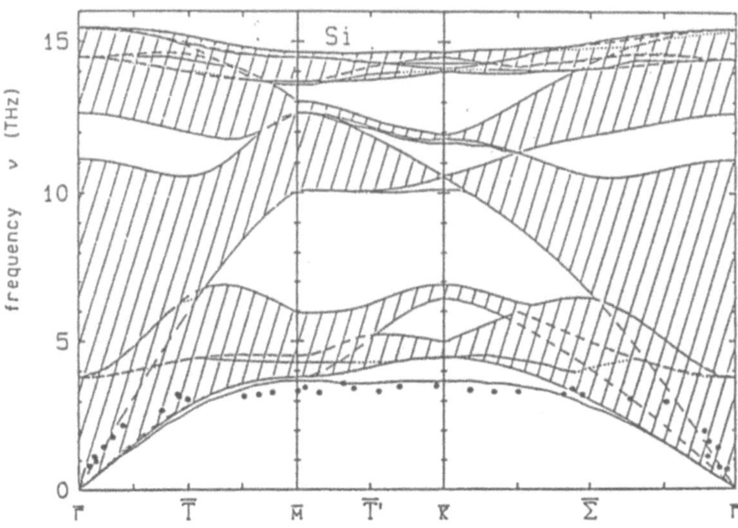

Fig. 8 Surface phonons at an ideal Si-(111) surface within the dipole model. Compare Fig. 7 for the further description.

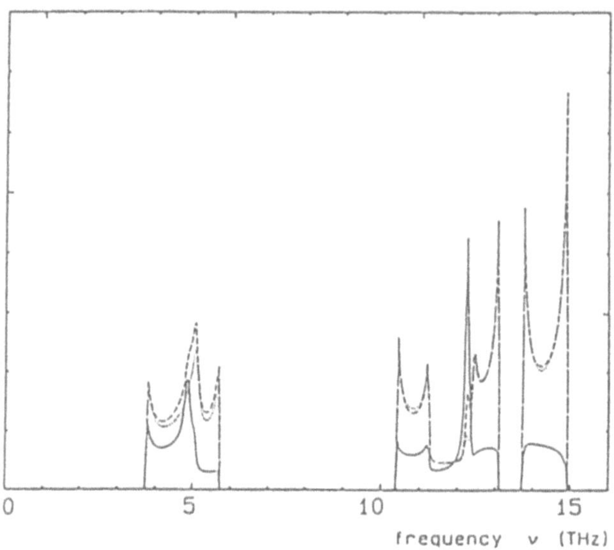

Fig. 9 Density of phonon states at the $\overline{M}$-point for an ideal Si-(111) surface in the force constant model (Fig. 7) for the three uppermost layers of atoms. First layer: full lines, second layer: broken lines, third layer: dotted lines. The figure shows that the resonances are mainly limited to the first layer. Localized states are not included.

Besides these true localized states there are resonance states within the bulk bands which have been calculated according to Eq. (4.8). Fig. 9 shows the contributions of different layers parallel to the surface indicating that the van Hove type singularities at the band edges as well as the resonances are mainly localized in the uppermost layer (see also Fig. 2).

I will refer to one single mode here, which occurs in nearly all the calculations of modes on Si-(111) surfaces independent of models. It is the 56-meV-surface mode ($\nu =13.6$ THz) at the $\overline{M}$-point. Probably this is the mode found experimentally by Ibach [6] for the first time. It occurs at (2×1)-reconstructed Si-(111)-surfaces. Now, the calculations have been done with unreconstructed surfaces. But introducing the 2 × 1-reconstruction means, that the $\overline{M}$-point of the ideal surface corresponds to the $\overline{\Gamma}$-point of the surface Brillouin zone of the reconstructed 2 × 1-surface [3,34]. EELS measurements are only sensitive to phonons at the $\overline{\Gamma}$-point. This explains why this mode is invisible in (1 × 1) and (7 × 7) surface geometries. Since this mode proved to be very stable against variations of the surface parameters one should not expect drastic changes for this mode upon reconstruction apart from this geometric effect. The polarization of this mode is in chain direction. After folding back the $\overline{M}$-point to the $\overline{\Gamma}$-point in a (2 × 1) structure this mode falls into the orthogonal LO band and is no longer localized, but only (strongly) quasi-localized.

120

Thus in this explanation the Ibach-phonon is related to TO-vibrations at
the $\overline{M}$-point. The (2×1)-structure is not the origin of the phonon, but gives
rather the possibility of measuring this phonon at the $\overline{\Gamma}$-point with EELS.
The statement that this phonon can only be explained in the π-bonded
chain model is not valid [44–46]. In the (7×7) structure the $\overline{M}$-point does
not coincide with the $\overline{\Gamma}$-point when folding back the surface Brillouin zone.

However, in realistic cases the force constants at the surface are different
from those in the interior and most calculations have been done with surface
force constants changed. The surface force constants are then used as free
parameters or calculated from models or first principles, if possible (Sect. 5).
Lowering surface force constants (or increasing the dipole parameters in the
dipole model) leads to a decrease of the acoustic surface phonon frequencies
and also to the occurrence of further surface states. This is shown again
with the example of the Si-(111) surface in the two models of Figs. 7,8. The
parameters have been changed in such a way, that just soft modes occur
(Figs. 10,11). It can be seen that there is a soft mode which corresponds to
a 7×7-reconstruction in the force-constant-model (short range) and a soft
mode leading to a 2×1-reconstruction in the dipole-model. The essential
point is that these soft modes always occur independent of the special choice
of the changed force constants. All changes leading to soft modes cause a
phonon softening at the same q-vector in these models.

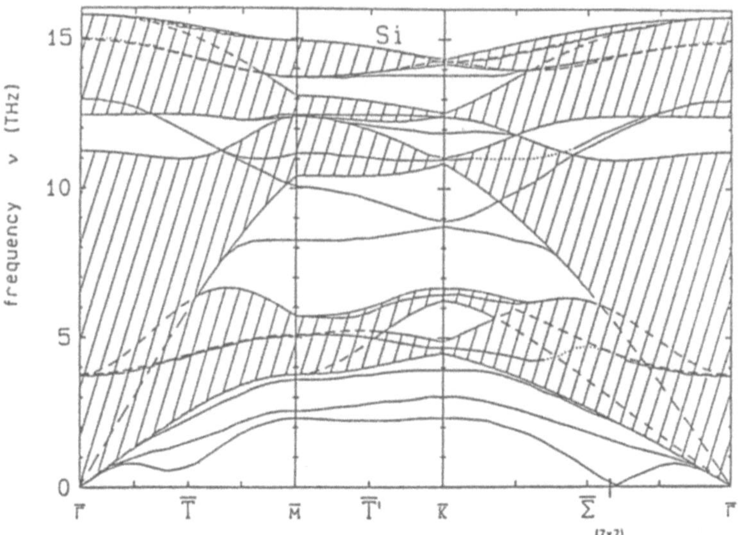

Fig. 10 Surface phonons at a Si-(111) surface with a (7×7)-soft mode in the force
constant model with surface force constants changed according to Table 1. The 13.6 THz
mode nearly coincides with the TO band edge.

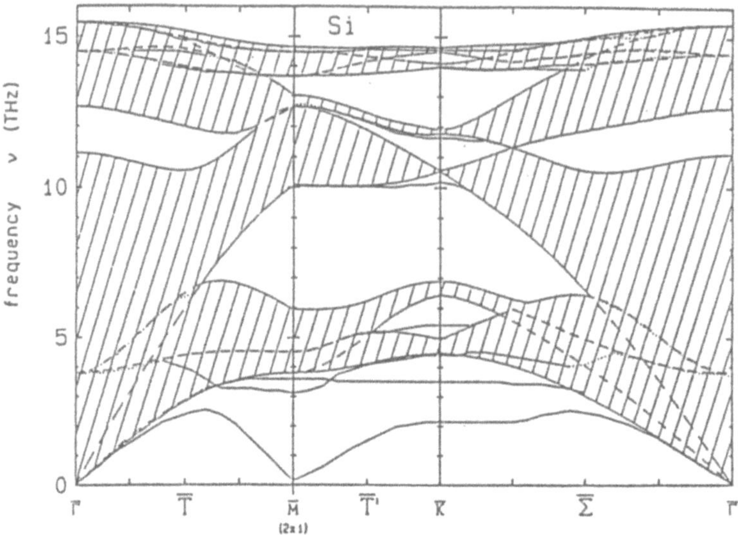

Fig. 11 Surface phonons at a Si-(111) surface with a (2 × 1)-soft mode in the dipole model with surface parameters changed.

The Ibach-mode is nearly invariant against these changes. The only difference in comparison to the ideal surface in Figs. 7,8 is that the mode is shifted upward and now lies slightly below the TO band edge. Despite the drastically changed surface parameters the agreement is not influenced very much. This shows that a unique determination of surface force constants from experiments is hardly possible.

The results for the (111)-Si surface using the bond charge model are also very similar. Such investigations have been done by the Benedek-group [21,22] for a hydrogen-covered surface as well as for a (2 × 1) reconstructed surface. Since the low frequency modes are not influenced by the hydrogen-covering one can use such calculations for a comparison (Fig. 12). Of course, the Ibach-mode is present, too. Other calculations have been done for C, Ge, α-Sn surfaces, (111)-surfaces as well as (110)-and (100)-surfaces [35–37]. The results are similar apart from the fact that in the short-range force constant model the soft modes occur at different positions. There is no soft mode for C in the force constant model.

GaAs as a zincblende-structure has been discussed, too. The results are shown in Fig. 13 and compared with experiments of the Toennies-group [39] and of Doake et al. [40]. The agreement again is rather good. The surface parameters in Fig. 13 are not changed with respect to the bulk. Since the experiments have been performed on a (1×1) reconstructed surface the good agreement again shows that the determination of surface force constants from experimental surface phonon frequencies is not unambiguous.

122

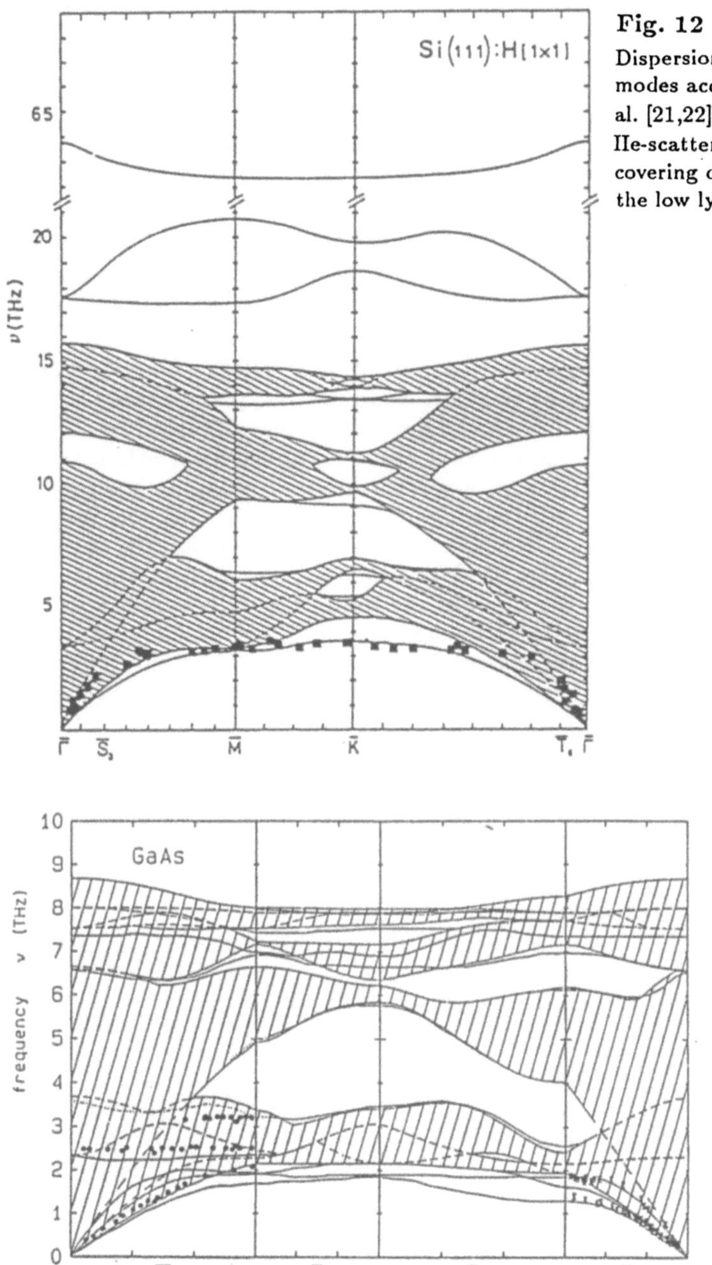

Fig. 12
Dispersion curves for surface modes according to Miglio et al. [21,22] compared with He-scattering data. Hydrogen covering does not influence the low lying modes.

Fig. 13 Surface phonons at a GaAs-(110) surface with the dipole model. For the description see legend of Fig. 7. Experimental points are due to Harten et al. [39] ($\overline{\Sigma}$-line) and Doak et al. [40] ($\overline{\Delta}$-line).

The figure shows that He-surface scattering does not only show surface or resonance modes but sometimes scattering mainly takes place at the band edges of the bulk phonon-bands, which can be understood quite well looking at the frequency density of states shown for example in Figs. 3 or 9, respectively. As in Figs. 7,8 one can see that for long wavelengths the scattering is apparently dominated by band edges and by resonances but not by the Rayleigh mode.

Alkali halogenides have been investigated at first by G. Benedek [4] using the Green function method and later to a large extent including MgO and PbS by Kress, de Wette et al. [23,24] using the slab method and the model described in Sect. 5. One example is given in Fig. 14.

The general features are similar to those of other crystals. A tendency for soft modes does not seem to be present. Perovskite type crystals ($KZnF_3$, $KMnF_3$) seem to be of more interest. They have been discussed by Reiger, Prade et al. [16–18]. Fig. 15 shows that there is, apart from other surface states, a soft mode at the $\overline{M}$-point, whereas the bulk phonons become soft at the R-point in $KMnF_3$. Without going into the details, we mention only one interesting feature. When a special force constant A_1 between K-F is changed (e.g. by temperature) to lower values, at first the surface phonon becomes soft (Fig. 16), whereas the bulk phonon becomes soft only with a force constant being lower by about 1%. This might lead to the conclusion, that the surface reconstruction by the soft $\overline{M}$-mode triggers the bulk phase transition (antiferro-distortive) which has to be confirmed by further investigations.

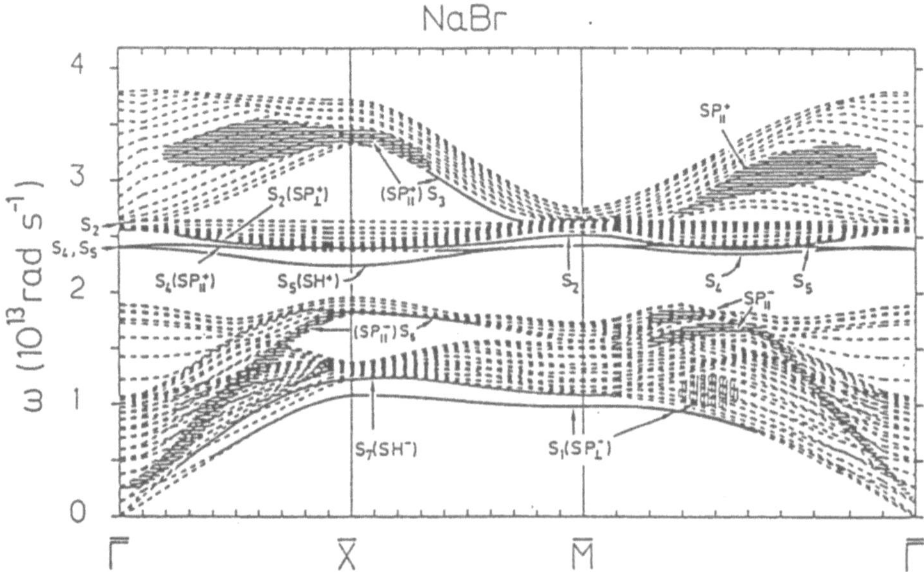

Fig. 14   Dispersion curves for a relaxed 15-layer slab of NaBr with free (001) surfaces.

124

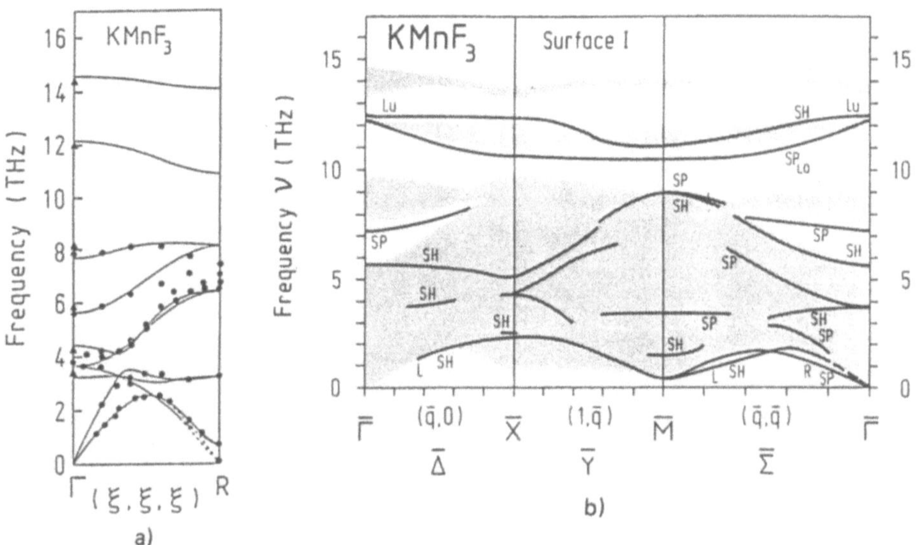

Fig. 15 Soft surface modes at a $KMnF_3$-(001) surface according to Prade et al. [18]. Bulk phonons (a) show a soft mode at the $\overline{R}$-point, surface phonons (b) at the $\overline{M}$-point.

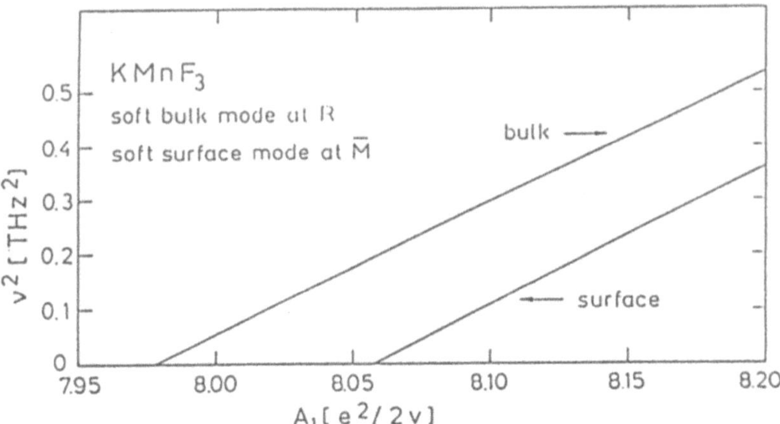

Fig. 16 Soft mode frequencies as a function of the A_1-force constant according to Prade et al. [18]. The surface soft mode occurs with smaller changes of the force constant.

The theory of surface phonons at metal surfaces has been put forward by Bortolani and the Modena group [25–29] using potential and force constant models and by Nelson et al. [11] with the embedded atom model. As a comparison of their results for Cu shows (Figs. 17,18) the dispersion curves for

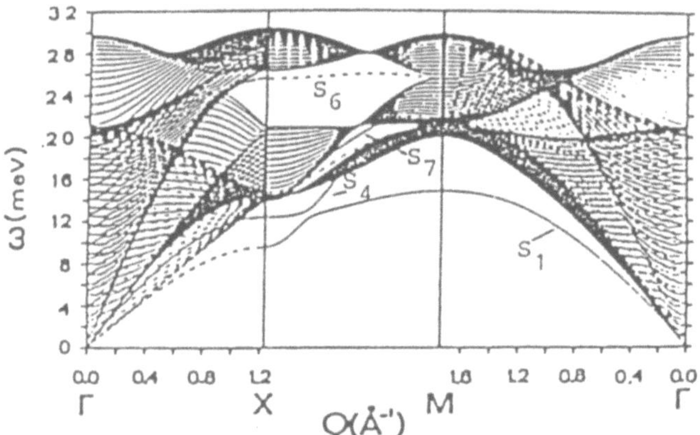

Fig. 17 Surface phonons for a Cu-(001) surface according to Bortolani et al. [26]. Compare with Fig. 18.

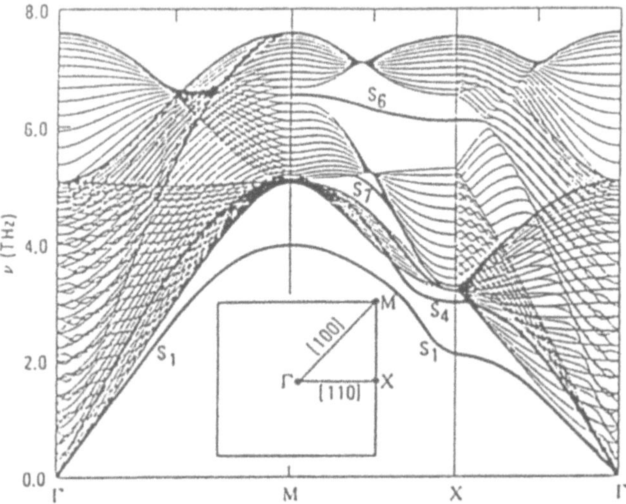

Fig. 18 Surface phonons for a Cu-(001) surface according to Nelson et al. [11]. Compare with Fig. 17; realize that Fig. 17 shows the directions Γ-M-X-Γ, whereas this figure has Γ-X-M-Γ. Though the models are different, the localized surface modes agree quite well. There is also a good agreement with experimental results of Toennies et al. [41] and Wuttig et al. [42].

126

the (001)-surface do agree qualitatively, though there are quantitative differences. Both the models also give reasonable agreement with experimental results [41,42].

Phonons have also been calculated at (100)- and (111)-surfaces of Pt, Ag, Au, Ni [26]. In all cases one finds reasonable agreement with experiments, but as has already been mentioned, such an agreement should not be taken too seriously, if the results depend to much on the special choice of the force constants. Rather one should concentrate on such statements which are independent of special assumptions. First principle calculations have been started by Ho and Bohnen [13,14]. But they give ab initio surface phonons for high symmetry points only, further discussions are necessary.

6.2 Reconstructions

As already mentioned, surface reconstructions are strongly related with the occurrence of soft surface phonon modes. Of course, a detailed discussion of surface phase transitions leading to a reconstruction needs anharmonic terms in the interaction and can thus not be discussed in this simple way. But in any case soft modes (zero frequencies) mean instabilities of those positions one has started from. In these modes the dynamical vibrational displacements of the phonons change to static displacements which would be infinite without anharmonic forces. However, the relative magnitude of the static displacements leads to some pattern which if locked in may be looked upon as the reconstructions connected with them.

We have mentioned already the interesting features of the $\overline{M}$-point soft mode in Perovskite-structures, especially in $KMnF_3$ [24]. It might be that this surface mode even stimulates the bulk phase transition in these structures.

A rather systematic investigation has been done for soft modes at (111)-, (110)-, (100)-surfaces of diamond-type crystals [33–37]. The results are compiled in Table 1. The essential point is that the soft modes occur independent of the special choices of the surface force constants provided the changes are of that magnitude given in the table, and the soft modes always occur at the same point in q-space.

Nearly all the reconstructions have been found experimentally. Only the (8 × 8)-Ge reconstruction is sometimes discussed as a (8 × 2) or (8 × 4) reconstruction, and (111)-α-Sn surfaces have not been investigated. However, experimental results show that there are more than these reconstructions, some of them correspond to q-space points lying outside symmetry directions. These have not been discussed until now. Others may be related to more complex interactions. Though the force constants in Si and Ge are not very different and fit in the general trend, the reconstructions for Si and Ge sometimes show a very different behavior (Fig. 19).

The soft mode in Si shows a very broad minimum, corresponding to (2×1)-, (9×1)-, (7×1)-, (5×1)-reconstructions, whereas the minimum in Ge is rather sharp corresponding to a (2×1)-reconstruction. This shows that even small differences in force constants for the same structure give very sensitive results for soft modes and reconstructions but not so much for the higher modes (see above).

Table 1 Reconstructions of semiconductor surfaces as a result of different models. $\Delta f/f$ stands for a mean relative change of surface parameters producing soft modes.

(111)-surface	force constant model (short range)		dipole model (long range)	
		$\Delta f/f$		$\Delta f/f$
C	none	even for 100%	2×1	150%
Si	7×7	45%	2×1	40%
Ge	8×8	40%	2×1	35%
α-Sn	3×3	30%	2×1	28%
(110)-surface				
Si	$2 \times 1;\ 5 \times 1$	50%		
	$7 \times 1;\ 9 \times 1$			
Ge	2×1	40%		
α-Sn	?	40%		
(100)-surface				
Si	2×2	30%		
Ge	2×2	20%		
α-Sn	2×2	20%		

In general, short range forces imply *higher order* reconstructions, long range forces lead to *lower order* reconstructions. In Figs. 20,21 we give the reconstructions of Si as calculated from the static displacements connected with the soft modes. Though these patterns do not completely agree with the experimental results, there are large similarities to them; the pattern in Fig. 21 is not so far from the pattern of the π-bonded chain model. For the (7×7) reconstruction there is no generally accepted model. The pattern in Fig. 20 has some similarities with a model proposed by Chadi et al. [51,52]. It should be remarked, too, that the patterns in Figs. 20,21 are not unique because of degeneracies. The (7×7)-soft mode is 6-fold, the (2×1)-mode is 3-fold degenerate. Therefore, one could build up different linear combinations of the corresponding static displacements. In Fig. 20

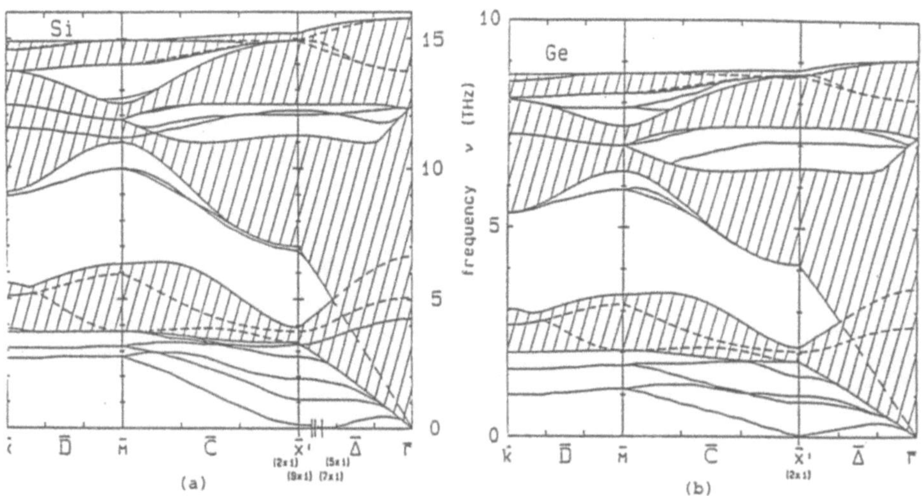

Fig. 19 Surface phonons with soft modes for a Si-(110) surface (a) and for a Ge-(110) surface (b) in the force constant model. Though the force constants and their changes are very similar, the behavior at a Si-or Ge-surface shows remarkable differences. According to [35,37].

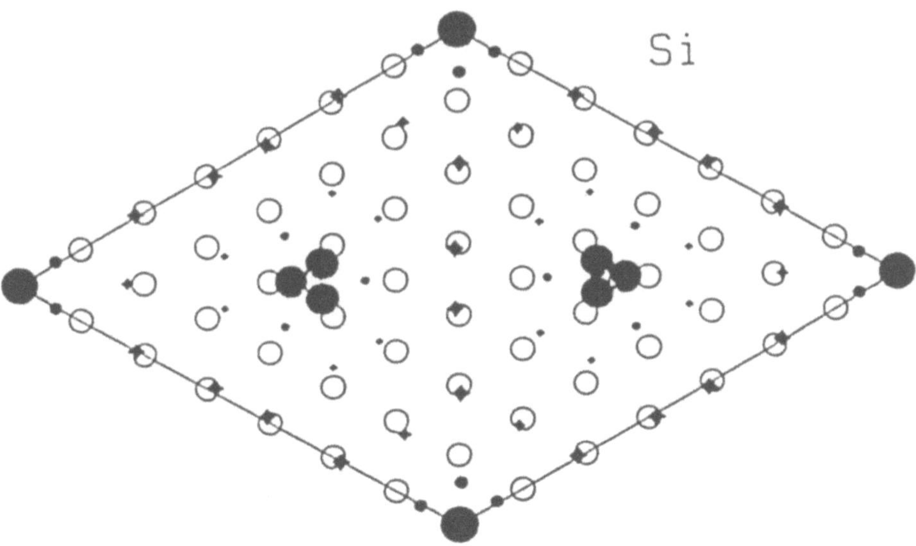

Fig. 20 Model for a (7 × 7)-reconstructed Si-(111) surface. The magnitude of circles and squares corresponds to the magnitude of displacements perpendicular to the surface, circles or squares indicate different signs of displacements. According to [33,35,37]

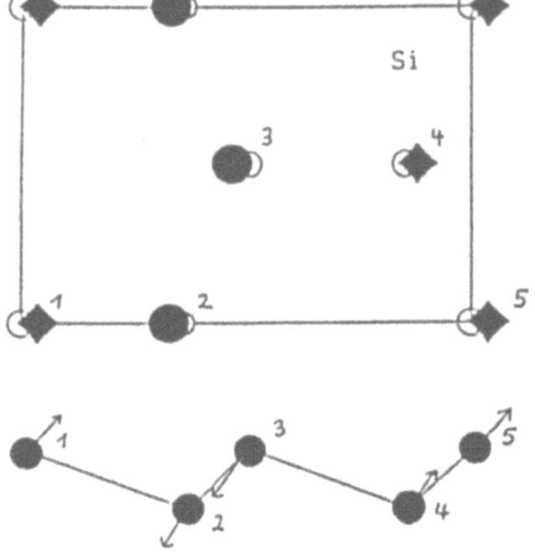

Fig. 21

Model for a (2×1)-reconstructed Si-(111) surface. According to [37].

the most symmetric combination has been taken for simplicity, in Fig. 21 the amplitudes of two of the three degenerate modes have been put to zero.

If in our model the long range dipole forces are diminished by a factor two and the short range forces are adjusted in such a way that the bulk phonons are represented in a least square fit (*intermediate model*) one finds, that lowering of short range forces in the surface leads to a soft mode at the (7×7)-point, whereas an increase of the dipole parameters in the surface leads to a soft mode at the (2×1)-point$(\overline{M})$. Both reconstructions can be discussed with one model, and soft modes occur always at (7×7) or (2×1), respectively and at no other position. Similar statements hold for the other crystals. This discussion shows that soft modes and reconstructions are inherent properties of these crystals and that long range and short range forces are competing with each other in a certain sense.

7 Conclusions

In this paper I have given a short review on the theory of surface modes, which by far is not complete. However, these examples already show that even simple models can explain many important features. This holds for a comparison with experimental surface phonon frequencies as well as for the discussion of soft modes and the reconstructions connected with these modes. Of course, complete quantitative agreement cannot be expected as long as the surface force constants are not exactly known. One could determine the surface parameters from measured frequencies, but such a procedure would not give unique results. So one should use microscopic calculations for getting statements on surface force constants. Methods for

130

this are now in progress. When satisfactory results are obtained, these surface force constants have to be used as the input data for a calculation of surface phonons. The methods for this are available and can be used as a standard procedure.

Acknowledgements

I wish to thank my collaborator W. Goldammer for his valuable contributions to this paper and all my colleagues for their hints to other work. However, the list of references is by far not complete, but further references can be found in those papers cited.

References

[1] *Rayleigh, Lord (J.W. Strutt)*, Proc. Math. Soc. London **17**, 4 (1885)

[2] *R. Zimmermann*, Applied Physics **3**, 235 (1974)

[3] *W. Ludwig*, Jpn. J. Appl. Phys. Suppl. **2**, pt. 2, 879 (1974)

[4] *G. Benedek*, Surf. Sci. **61**, 603 (1976)

[5] *R.E. Allen, G.P. Alldredge, F.W. de Wette*, Phys. Rev. B **4**, 1648; 1661; 1682 (1970)

[6] *H. Ibach*, Phys. Rev. Lett. **27**, 253 (1971)

[7] *J.P. Toennies*, Verhandl. DGP (VI) 23, DY-1.1 (1988)

[8] *K.C. Pandey*, Phys. Rev. Lett. **47**, 1913 (1981), **49**, 233 (1982)

[9] *B. Szigeti*, Proc. Int. Conf. Phonons, Flammarion, Paris, **43** (1971)

[10] *R.S. Leigh, B. Szigeti, V.K. Tewary*, Proc. Roy. Soc. London A **320**, 505 (1971)

[11] *J.S. Nelson, E.C. Sowa, M.S. Daw*, Phys. Rev. Lett. **61**, 1977 (1988)

[12] *J.L. Ningsheng, X. Wenlan, S.C. Shen*, Sol. State Comm. **67**, 837 (1988)

[13] *K.M. Ho, K.P. Bohnen*, Phys. Rev. Lett. **56**, 934 (1986)

[14] *K.M. Ho, K.P. Bohnen*, Phys. Rev. B **38**, 12897 (1988)

[15] *A.G. Eguiluz, A.A. Maradudin, R.F. Wallis*, Phys. Rev. Lett. **60**, 309 (1988)

[16] *R. Reiger*, Dissertation, Regensburg (1986)

[17] *R. Reiger, J. Prade, U. Schröder, F.W. de Wette, W. Kress*, J. Electr. Spectr. a. Related Phen. **44**, 403 (1987)

[18] *J. Prade, A.D. Kulkarni, F.W. de Wette, R. Reiger, U. Schröder, W. Kress*, Surf. Sci. (in print)

[19] *K.H. Wanser*, Dissertation, Irvine (1982)

[20] *W. Goldammer, W. Ludwig, W. Zierau, K.H. Wanser, R.F. Wallis*, Phys. Rev. B **36**, 4624 (1987)

[21] *L. Miglio, P. Ruggerone, G. Benedek, L. Colombo*, Physica Scripta **37**, 768 (1988)

[22] *L. Miglio et al.*, to be published

[23] *F.W. de Wette, W. Kress, U. Schröder*, Phys. Rev. B **32**, 4143 (1985)

[24] *W. Kress, F.W. de Wette, A.D. Kulkarni, U. Schröder*, Phys. Rev. B **35**, 5783 (1987)

[25] *V. Bortolani, A. Franchini, F. Nizzoli, G. Santoro*, Phys. Rev. Lett. **52**, 429 (1984)

[26] *V. Bortolani, A. Franchini, G. Santoro*, Electr. Structure, Dynamics and ..., Plenum Publ. Corp., **401** (1985)

[27] *G. Santoro, A. Franchini, V. Bortolani, U. Harten, J.P. Toennies, Ch. Wöll*, Surf. Sci. **183**, 180 (1987)

[28] *V. Bortolani, A. Franchini, G. Santoro, J.P. Toennies, Ch. Wöll, G. Zhang*, Phys. Rev., to be published

[29] *V. Bortolani, F. Ercolessi, E. Tosatti, A. Franchini, G. Santoro*, Phys. Rev. Lett., submitted

[30] *M.S. Daw, M.I. Baskes*, Phys. Rev. **B 29**, 6443 (1984)

[31] *M.J. Stott, E. Zaremba*, Phys. Rev. **B 22**, 1564 (1980)

[32] *C. Falter*, Physics Reports **164**, 1–117 (1988)

[33] *W. Goldammer, W. Ludwig, W. Zierau, C. Falter*, Surf. Sci. **141**, 139 (1984)

[34] *W. Goldammer, W. Ludwig*, Phys. Lett. **A 133**, 85 (1988)

[35] *W. Goldammer, W. Ludwig, W. Zierau*, Surf. Sci. **173**, 673 (1986)

[36] *W. Goldammer, J. Backhaus, W. Ludwig*, Physica Scripta **38**, 155 (1988)

[37] *W. Goldammer*, Dissertation, Münster (1988)

[38] *U. Harten, J.P. Toennies, Ch. Wöll, L. Miglio, O. Ruggerone, L. Colombo, G. Benedek*, Phys. Rev. **B 38**, 3305 (1988)

[39] *U. Harten, J.P. Toennies*, Europhys. Lett. **4**, 833 (1987)

[40] *R.B. Doak, D.B. Nguyen*, Journ. of Electron Spectroscopy and Related Phenonema **44**, 205 (1987)

[41] *J.P. Toennies, Ch. Wolf*, Phys. Rev. **B 36**, 4475 (1987)

[42] *M. Wuttig, R. Franchy, H. Ibach*, Z. Phys. **B 65**, 71 (1986)

[43] *M. Wuttig, R. Franchy, H. Ibach*, Solid State Comm. **57**, 445 (1986)

[44] *O.L. Alerhand, N.J. DiNardo, E.J. Mele*, Surf. Sci. Lett. **173**, L659 (1986)

[45] *O.L. Alerhand, D.C. Allan, E.J. Mele*, Phys. Rev. Lett. **55**, 2700 (1985)

[46] *O.L. Alerhand, E.J. Mele*, Phys. Rev. **B 37**, 2536 (1988)

[47] *E.O. Kane*, Phys. Rev. **B 31**, 7865 (1985)

[48] *M. Lax*, Phys. Rev. Lett. **1**, 133 (1958); Lattice Dynamics, Pergamon Press, Oxford **179** (1965)

[49] *J. Ginter*, J. Phys. **C 6**, 819 (1973)

[50] *S.E. Trullinger, S.L. Cunningham*, Phys. Rev. **B 8**, 2622 (1973)

[51] *D.J. Chadi et al.*, Phys. Rev. Lett. **44**, 799 (1980)

[52] *D.J. Chadi*, Surf. Sci. **99**, 1 (1980)

[53] *J.E. Black*, Dynamical Properties of Solids 6(1989), to be published

Scanning tunneling microscopy and spectroscopy on clean and metal-covered Si surfaces

Henning Neddermeyer and Stephan Tosch

Institut für Experimentalphysik, Ruhr-Universität Bochum, Postfach 10 21 48, D-4630 Bochum, Federal Republic of Germany

Summary: Main theoretical concepts and experimental details of scanning tunneling microscopy (STM) are summarized and methods for image processing of the data are described. The spectroscopical use of STM is explained. The application of STM to clean and metal-covered Si surfaces is demonstrated. In particular, results are presented for the spectroscopy of the local electronic structure of Si(111)7x7, for the initial stages of metal atom (Ag) condensation on room temperature Si(111)7x7 in the submonolayer and monolayer range and for an analysis of a metal-induced reconstruction (Si(111)$\sqrt{3}$x$\sqrt{3}R30°$-Ag). Results from metal islands on Si(111) (Ag) are compared with those of a bulk metal single-crystal (Au(111)).

1 Introduction

The investigation of clean and metal-covered semiconductor surfaces is one of the most attractive applications of scanning tunneling microscopy (STM). Already the second important publication by Binnig *et al.* has been devoted to the study of Si(111)7x7 [1], where in the STM images the adatom pattern of the 7x7 reconstructed unit cell is beautifully resolved in real space. Not much later the analysis of Si(111)7x7 by using electron transmission microscopy has led to the development of the dimer-adatom-stacking fault-model (DAS) [2], which provides a convincing explanation of the STM results of Ref. [1].

Subsequently, the surface of Si(111)7x7 played a considerable role as a model system for the general development of the STM techniques and in particular for a spectroscopy of the local electronic structure. By comparing images obtained with negative and positive sample bias voltage U (corresponding to tunneling from the filled states and into the empty states of the sample, respectively) characteristic differences in form of asymmetries in both halves of the 7x7 unit cell were interpreted in terms of the local electronic structure [3,4]. The relation of occupied and unoccupied states in the images obtained with opposite polarities of U was discussed for Si(111)2x1 by Stroscio *et al.* [5].

Since in these measurements the entire range of states between the Fermi level E_F and eU contributes to the tunneling current I the images reflect a kind of energy-average of accessible states. The idea to change the sample bias during scanning of the surface and to measure the I/U or dI/dU characteristics was first reported by Binnig *et al.* for various surfaces [6] and for Si(111)7x7 by Becker *et al.* [7]. Since in Refs. [6] and [7] the distance between sample and tip was not kept constant due to regulating of the tunneling current to a constant value, the dI/dU dependencies show a marked oscillatory behavior, which is not directly related to the genuine local electronic structure of the sample surface. This problem was solved by Hamers *et al.*, who at every lateral tip position regulated the tunneling current for only a small time interval and measuring the I/U's at a constant sample-tip distance [8]. By using the latter method a map of the electronic properties of the surface may be constructed on an atomic scale.

The condensation of metal atoms on a semiconductor substrate has several important aspects, which may be addressed by STM. Firstly, the initial stages of nucleation and film growth can be investigated. In case of deposition on Si surfaces the interaction between the deposited atoms and the substrate may be quite large and lead to well defined adsorption sites in the submonolayer coverage range even for low temperature (about room temperature) experiments [9,10]. Secondly, for overlayers following the island/interface growth mode, shape and morphology of the condensed metal islands may be analyzed [11,12]. Thirdly, after annealing of the films metal-induced reconstructions are observed frequently. Atomic models for these reconstructions may be elucidated on the basis of STM images, although the nature of the atomic species cannot simply be derived from the STM data. The analysis of the Si(111)$\sqrt{3}$x$\sqrt{3}R$30°-Ag structure [13-15] and the Cu-induced 5x5-like reconstruction of Si(111) [16-18] are important examples in this field.

In the present work an overview is given on the theoretical models used for interpretation of the STM images and the spectroscopical results (Chapter 2). A brief description of experimental equipment and image processing follows in Chapter 3. The main part of the paper (Chapter 4) concerns the presentation and discussion of STM results from Si(111)7x7 and the system Ag on Si(111) for room temperature deposition and in the annealed state. Our aim is a demonstration of the capabilities of STM for a measurement of clean and metal-covered semiconductor surfaces. A full description of the measured systems and an extensive relation to other published work is beyond the scope of the present article.

2 Theory

In the simplest form of the planar electrode model a rectangular potential barrier is assumed between sample and metallic tip, which is not dependent on the coordinates x and y parallel to the surface and extends into infinity [19]. By solving a one-dimensional Schrödinger equation the tunneling conductivity σ may be expressed by

$$\sigma = const_1 \cdot e^{-const_2 \cdot \Phi^{1/2} \cdot \Delta z},$$

where Δz is the width and Φ the height of the potential barrier. This model already explains the basic principle of the tunneling microscope [20], where the tunneling current I is electronically stabilized by means of a feed-back circuit, which regulates the sample-tip distance Δz to a constant value. If the tip is scanned over the sample surface under the condition of a constant tunneling current the actual z value supplied by the feed-back circuit may be used for the construction of an image of the surface. Such an image is usually called constant current topography (CCT).

While the observation of atomic steps on a metal surface [20] may adequately be described by this simple model, it breaks down for surfaces and tunneling conditions, where an atomically resolved corrugation is obtained. This is quite obvious, since the assumption of a potential barrier independent of the xy coordinates is no longer fulfilled. To include the lateral variation of the potential barrier, which is defined by the atomic charge density of the surface, Tersoff and Hamann [21] have used Bardeen's work [22] to derive an expression for the tunneling current

$$I = \frac{2\pi e}{\hbar} \sum_{\mu,\nu} f(E_\mu)[1 - f(E_\nu + eU)] \, |M_{\mu\nu}|^2 \, \delta(E_\mu - E_\nu) \,,$$

where $f(E)$ is the Fermi function, $M_{\mu\nu}$ the tunneling matrix element between states ψ_μ of the metal probe tip and ψ_ν of the sample surface, E_μ and E_ν the energies of the states in the absence of tunneling and U the sample bias voltage. The contribution of reverse tunneling has been neglected in this equation. For small sample bias voltages U, low temperature and in the limit where the tip is replaced by a point probe the tunneling current may be written as

$$I \propto \sum_\nu |\psi(\vec{r}_0)|^2 \, \delta(E_\nu - E_F) \,.$$

The right hand side of this equation corresponds to the charge density of states at E_F measured at the position $\vec{r}_0$ of the point probe. According to this expression the tip follows the contours of charge density of states at E_F during scanning of the surface. Note that the charge density extending into the vacuum region normally also contains a factor $\propto e^{-constant \cdot z}$, which means that the general dependency on the sample-tip distance is the same as in case of the planar electrode model.

In the more general theory by Feuchtwang *et al.* the tunneling current is derived by folding of spectral densities of both sample and tip [23]. If the spectral density is defined by

$$\rho(\vec{r}_\mu, \vec{r}_\nu, E) = \sum_i \psi_i(\vec{r}_\mu, E)\psi_i^*(\vec{r}_\nu, E) \, \delta(E - E_i),$$

the tunneling current for a planar sample is obtained by:

$$I = \frac{e\pi\hbar^3}{m} \int dE \, [f_\mu(E) - f_\nu(E + eU)]$$

$$\times \int_S \int_S d^2\vec{r}_\mu d^2\vec{r}_\nu \, \frac{\partial^2}{\partial z_\mu \partial z_\nu} \, [\rho_{tip}(\vec{r}_\mu - \vec{r}, \vec{r}_\nu - \vec{r}, E) \, \rho_{sample}(\vec{r}_\nu, \vec{r}_\mu, E)],$$

where the tip is located at the position $\vec{r}$. For small sample bias voltage and low temperature the tunneling conductivity is obtained by the expression:

$$\sigma = \frac{e^2 \pi \hbar^3}{m} \int_S \int_S d^2 \vec{r}_\mu d^2 \vec{r}_\nu \frac{\partial^2}{\partial z_\mu \partial z_\nu} \left[\rho_{tip}(\vec{r}_\mu - \vec{r}, \vec{r}_\nu - \vec{r}, E_F) \, \rho_{sample}(\vec{r}_\nu, \vec{r}_\mu, E_F) \right].$$

In this approximation the contribution of states with $l \neq 0$ may be computed. For example, the contribution of d states to the tunneling current is estimated to be only 10 % of that of the free-electron-like sp states.

In a recent theoretical work based on a Green's function formalism Noguera has given an exact expression of the tunneling current [24]. It is shown in Ref. [24] that the electronic structure of sample and tip does not enter the tunneling current through the local density of states directly. In particular, no tunneling current is transported by surface states of the sample, since the latter have no group velocity normal to the surface. This is at variance with the results of Tersoff and Hamann [21] and leads to the discrepancy that in experimental measurements such contributions are indeed observed (see below). Noguera suggested the possibility of contributions by surface states, which propagate parallel to the surface. The possibility of this effect will be discussed below on the basis of our experimental results.

For semiconductors and their smaller conductivity (compared to metallic samples) bias voltages of 1–2 V are utilized for a measurement of the CCTs. This situation has been considered theoretically by Bono and Good, who used a free-electron model for their computation of the tunneling current from a semiconducting sample [25]. Qualitatively, a whole range of energy levels of the sample contributes to the tunneling current and the corrugation reflects in a first approximation the spatial distribution of the charge density integrated over an energy interval (here we are using again the Tersoff and Hamann picture of the tunneling process). In a golden-rule formulation of the tunneling current one has to sum up the contributions corresponding to transitions from or into the various energy levels of the sample (depending on the polarity of the sample bias voltage), which have to be weighed by an appropriate tunneling probability. According to the planar electrode model the tunneling probability is proportional to $\exp(const \cdot U \cdot z/\Phi^{1/2})$. This means, e.g., that the tunneling current from a discrete occupied state of the semiconductor into an empty state of the tip will increase exponentially with the sample bias voltage.

A general theory for the use of STM for a spectroscopy of electronic surface states on an atomic scale has not yet been developed. In some of the previous spectroscopical applications one has assumed that the derivative of the local tunneling current over the sample bias voltage $\partial I(U, z)/\partial U$ under the condition $z = const$ is proportional to the local density of states [8,26]. To consider the influence of the tunneling probability in an approximate way, Stroscio et al. divide their experimental dI/dU values by I/U, which leads to a better agreement between experimental curves and theoretical density of states [5]. The reason for calculating this quantity is based on the following arguments [27]. If the tunneling current is expressed by $I \propto \int_0^U N(E)T(E,U)dE$ ($N(E)$ being the density of states of the sample and $T(E,U)$ the transmission or tunneling probability; the density of states of the tip is assumed to be constant here) and $T(E,U)$ is expressed in the limits of the planar electrode model and assuming a mean barrier height $\overline{\Phi} = (\Phi_{sample} + \Phi_{tip})/2 - E + U/2$, the quantity $d\ln I/d\ln U$ may be written as

$$\frac{d\ln I}{d\ln U} = \frac{N(E) + A(E,U))}{B(E,U)}.$$

Proportionality of $d\ln I/d\ln U$ to the local density of states $N(E)$ is only found for $A(E,U) < N(E)$ and absence of strong energy-dependencies in $B(E,U)$, which is not generally the case, however.

If the spatially resolved tunneling current I obtained at some sample bias U is plotted in a so called current image (CI) one may derive information on the spatial distribution of surface states [8].

3 Experiment

Most STM instruments follow the concept of fine positioning of the tip by a piezoelectric xyz system and coarse positioning of the sample by a piezoelectric walker [28-30]. A mechanical solution for the coarse approach between tip and sample has been reported by Demuth et al. [31]. To obtain atomic resolution the precision of the tip movement should be controllable within a few 0.1 Å lateral to the surface and within 0.1 Å (or better) normal to the surface. The electronic and mechanical noise must be correspondingly small. The achievement of these requirements is one of the main problems for realization of an STM. The decoupling from external mechanical vibrations is mostly accomplished by a system of damped metal or with rubber springs, which are used for suspension of the microscope. Since the sensitivity against external mechanical noise is decreasing with increasing main resonance frequencies of the microscope [32], its construction should be both rigid and light-weighed.

Formation and control of the metal tip, which is usually electrolytically etched from a tungsten wire, have not yet been solved satisfactorily. By scanning of the surface in the field emission regime it is in general not difficult to obtain atomic resolution in the CCT mode. For spectroscopical measurements, however, the requirements regarding stability and atomic structure of the tip are much higher and not easily to achieve.

For the study of semiconductor surfaces the microscope has to be operated in an ultra-high vacuum environment, where surface physical equipment for sample preparation and control is available. The experiments discussed here are mostly performed on polished p-type (B-doped, $\approx 2\Omega$cm) Si wafers.

Prior to mounting in the microscope the samples are cleaned in acetone, methanol and deionized water. Sometimes an etch in fluoric acid was also employed. The latter procedure did not lead to significant differences in the results. *In situ* treatment consisted of degassing and final heating up to 800°C, until the Si(111) samples showed a clear 7x7 low energy electron diffraction (LEED) pattern. For the deposition of Ag a quartz-controlled effusion cell was utilized. Even during evaporation the pressure was better than 2×10^{-10} mbar. Good vacuum conditions are necessary, since after heating of the samples the STM measurements can only be started about 1 h later, until thermal drift effects in the STM images become sufficiently small.

A possibility of adequate image processing of the data is a further important demand on the experimental set-up. Otherwise, the analysis of the measured STM results is extremely time-consuming and difficult. The main parts of the electronic equipment used for the present studies are a fast low-current amplifier (Ithaco 1211), a feed-back circuit for regulation of the tunneling current to a constant value, low-noise high-voltage amplifiers for the xyz piezoelectric drives of the microscope and a microprocessor system (Commodore CBM 8032) for data acquisition and measurement control, which is connected to a second system (Hewlett-Packard 9000/320 SRX) for general data handling and image processing.

In Fig. 1 various ways of data representation are compared. During scanning of the surface for measurement of a CCT, at every lateral tip position in the xy plane the z value needed for stabilization of I to a constant value (more precisely, the specifically applied high-voltage for z positioning of the tip) is stored in the memory of the computer system and displayed consecutively on the monitor (Fig. 1 (a)). An illustrative overview on the quality of the measurement (noise, resolution, atomic order and general topological properties of the surface, cleanliness) may already be derived from such data. For comparison with atomic models top view images have to be computed from such data. In Fig. 1 (b) the same measurement is shown as top view image with lines of constant height. Whether a certain atomic structure is high or low may better be visualized by a grey tone plot (Fig. 1 (c)), where the protrusions are shown by bright and depressions by dark parts of the image. The quadratic raster slightly visible in Fig. 1 (c) corresponds to the x and y step size during the measurements. By a projection technique "three-dimensional" (3D) images (Figs. 1 (d) and (e)) may be computed from the ordinary grey tone top view image of the surface. The 3D representations provide a very illustrative impression of the surface. A general difficulty of presenting grey tone images is the limitation in grey tone scale of the reproduction, where small but real details are easily lost. In particular the representation of stepped surfaces is quite difficult in this respect. The use of "light sources" for computation of the 3D image may be advantageous in this case (Fig. 1(e)).

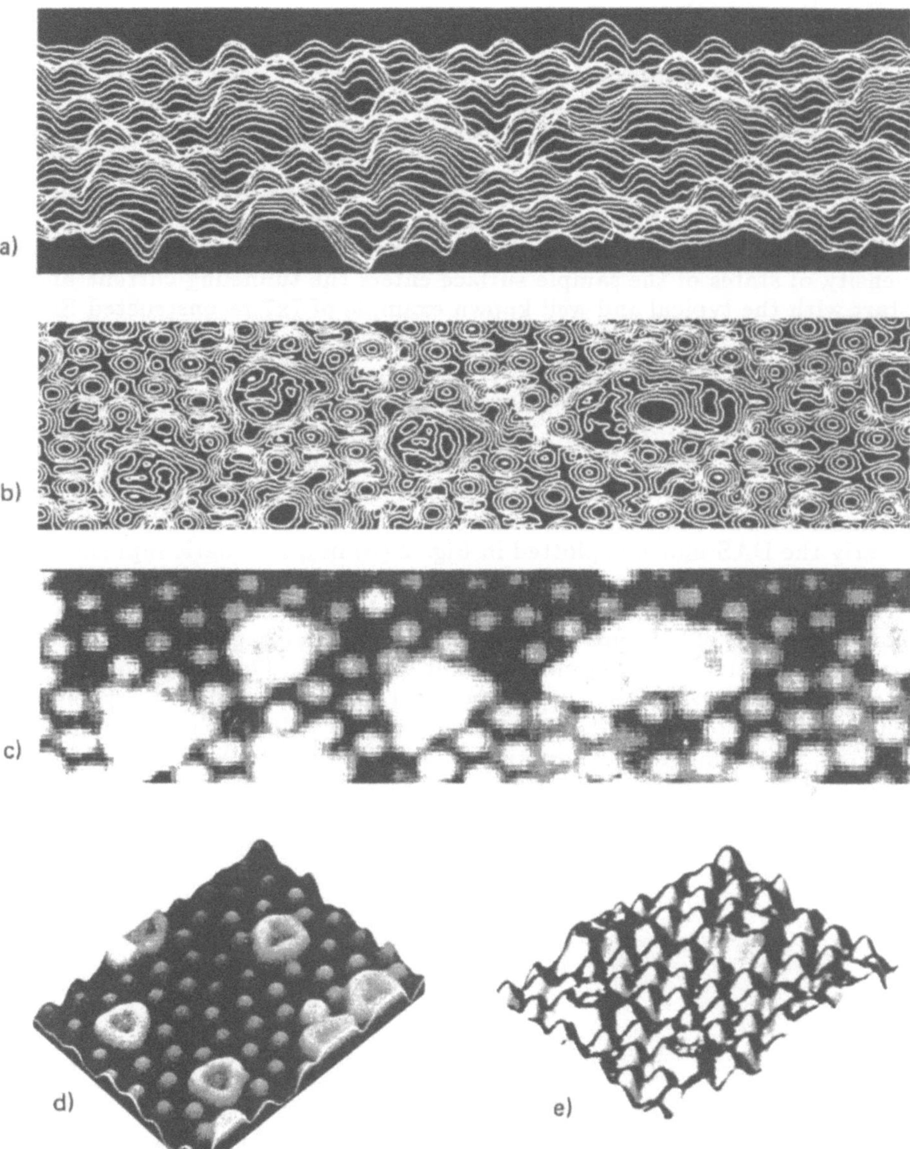

Fig. 1 Various representations of a CCT from Si(111), where 1/3 monolayer of Ag has been deposited at 90°C. (a) Line scan along the x direction. The maximum corrugation of the lines is 0.23 nm. (b) Lines of constant height in a distance of 0.02 nm. (c) Top view grey tone image. (d) Three-dimensional representation in form of a grey tone image. The smaller protrusions correspond to Si adatoms of clean Si(111)7x7. Ag is visible by the bright ring-like structures. (e) Three-dimensional representation with additional use of "light sources". The area of the CCT's shown in (a)–(b) is $\approx 4 \times 16$ nm^2, and in (d) and (e) $\approx 6 \times 8$ nm^2.

4 Results and Discussion

Although the first applications of STM were mainly dedicated to the study of the atomic structure of surfaces it became clear quite soon that in the above summarized simplifying picture STM probes the electronic states of a surface, which as a matter of fact are strongly related to the atomic structure. The main question is now, how can the effects of atomic and electronic structure be unravelled in the STM images, which are normally obtained as CCT's. In Chapter 2 an overview has been given, how the local density of states of the sample surface enters the tunneling current and we start with the typical and well known example of 7x7 reconstructed Si(111) to demonstrate the electronic effects in CCT's.

In Fig. 2 (a) a CCT from Si(111)7x7 obtained at a sample bias $U = -2$ V (i.e., for tunneling from the filled states of the sample into the empty states of the tip) [4] is reproduced. The marked black regions (corresponding to depressions) in the image are the so called corner holes of the 7x7 unit cell and the bright dots (corresponding to protrusions) the Si adatoms of the DAS structure of Ref. [2]. To understand the observed STM pattern more clearly the DAS model is plotted in Fig. 2 (b) in a schematic representation using only the three uppermost atomic layers of Si. The outermost (i.e., first) layer consists of 12 adatoms (open circles), which either saturate three dangling bonds of Si atoms in the second layer. Six dangling bonds of second layer atoms (small solid circles) are not saturated. The dimers of the third atomic layer are depicted by heavy solid lines. The model exhibits two mirror planes, one containing the long and the other the short diagonal of the 7x7 unit cell. A stacking fault is necessary in one half of the unit cell (denoted faulted half (FH) in contrast to the unfaulted half (UH)) to fit this structure to an ideally terminated Si(111) surface. We note that for bulk Si(111) the mirror plane containing the short diagonal of the 7x7 unit cell is absent.

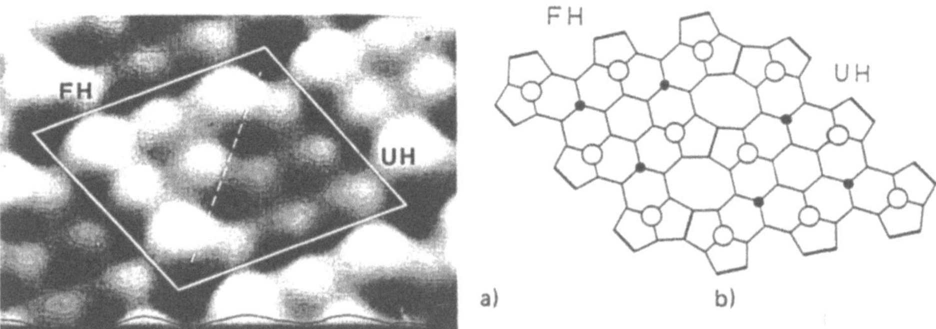

Fig. 2 (a) CCT obtained on Si(111)7x7 at $U = -2$ V and $I = 2$ nA. The 7x7 unit cell (and its short diagonal) is indicated by the white solid (dashed) lines. The area of the CCT is $\approx 4.5 \times 6$ nm^2. (b) Schematic top view of the 7x7 unit cell according to the DAS model [2].

Surprisingly, the CCT from Si(111)7x7 obtained at $U = -2$ V shows a marked height difference between both halves of the unit cell of about 0.05 nm, which leads to a distinct threefold symmetry of the entire pattern (Fig. 2 (a)). We note that Keating-type calculations of atomic displacements in the DAS model and a comparison with He ion scattering results have shown a height difference in both halves of less than 10^{-3} nm [33], which indicates that the marked experimental height differences in the STM images are caused by the presence of strong electronic effects in the CCT. On the other hand, for tunneling into the empty states of Si(111)7x7 (i.e., for positive sample bias U) such a strong asymmetry is absent and the height of the adatoms practically appear at the same level [3,4].

The first attempt to solve this problem has been reported by Hamers et al., who measured the local I/U characteristics of Si(111)7x7 at fixed sample-tip distance [8]. A detailed inspection of these I/U dependencies shows characteristic differences on the various locations of the 7x7 unit cell, which become even more apparent in the dI/dU curves computed from the I/U's.

In particular, the adatom positions have to be associated with metallic-like (ohmic) behavior near the Fermi level E_F (i.e., for small sample bias U allowing only tunneling within the band gap region of Si) in contrast to the dangling bond positions of the second atomic layer (see Fig. 2 (b)). The latter locations show a distinct maximum in dI/dU at -0.8 eV (referred to $E_F = 0$ V). This maximum is partly responsible for an appreciable increase of the tunneling current for $U < -1$ V. As is discussed more thoroughly in Refs. [4] and [34], the second reason for the observed increase of I on these locations (which for $U < -1$ V is becoming larger than I from the adatom positions) is the increase of the tunneling or transmission probability with increasing sample bias U. A comparison with photoemission results (see, for example, Ref. [35]) strongly supports the identification of density of states features in the I/U or dI/dU curves.

A graphical representation of the local tunneling current provides information on the spatial distribution of the density of states [8]. In Fig. 3 two

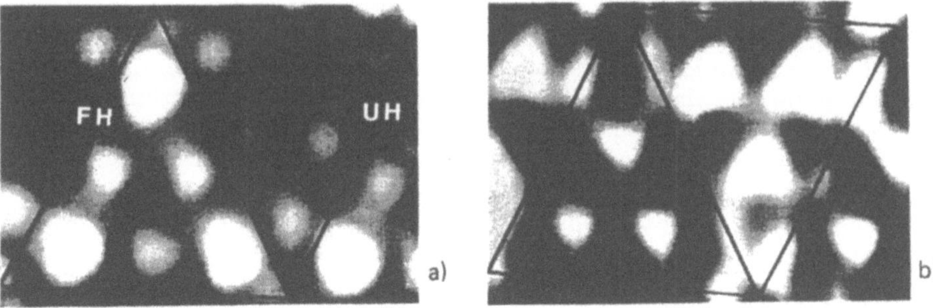

Fig. 3 CI's from Si(111)7x7 obtained at a stabilization voltage $U_0 = 2$ V for $U = -1.3$ V (a) and $U = -2.0$ V (b).

CI's from Si(111)7x7 are reproduced for $U = -1.3$ V (Fig. 3 (a)) and $U = -2$ V (Fig. 3 (b)) [4], which correspond to tunneling from the filled states of the sample into the empty states of the tip and which show distinct differences in both halves of the unit cell. In these images bright and dark parts indicate high and low tunneling current, respectively. As has been discussed above, I is obtained with disabled feed-back circuit. For stabilization of the sample-tip distance $U_0 = 2$ V has been employed. This means that during measurement of the I/U's the tip follows the corrugation as defined by $U = 2$ V (tunneling into empty states of the sample) and $I = const$ (2 nA in this case). The CI's shown in Fig. 3 consistently explain the differences of the CCT's measured at negative (Fig. 2 (a)) and positive sample bias (not shown here; see, for example Ref. [4]). Since the adatoms of the faulted half (FH) provide a larger current than those of the unfaulted half (UH), the tip retracts from the surface in the FH (when measuring a CCT at $U = -2$ V) to reach the same tunneling current as in the UH. On the other hand, for $U_0 = -2$ V and $U = -1.3$ V the local tunneling current on the adatom sites of both halves of the 7x7 unit cell is practically the same (not shown here, see Ref. [4]).

For a stabilization voltage $U_0 = 2$ V and $U = -2$ V the tunneling current from the dangling bond sites of the second monolayer exceeds that of the adatom sites. This means that the corresponding CI essentially images the dangling bonds of the second atomic layer (Fig. 3 (b)).

It has to be emphasized that the general shape of the I/U characteristics is influenced by the sample-tip distance through a change of the tunneling probability. For example, opposite polarities of the stabilization voltage U_0 give rise to differences in the contours of constant tunneling current, which are followed by the tip upon scanning of the surface, and therefore to differences in the I/U characteristics, which are not simply related to density of states effects. The influence of the sample-tip distance on the tunneling current has the consequence that the magnitude of the tunneling current can generally not be used as a measure of the local density of states.

The differentiation between density of states effects and influences of the tunneling probability is less difficult in dI/dU's. The previous work of Refs. [4,8,34] has shown that dominant features of the local density of states are convincingly established by using the spectroscopical mode of STM. The metallic-like behavior of the adatoms and the presence of a density of states maximum on the dangling bond positions have to be mentioned in this respect. Other results will be presented below, where the general suitability of STM for a local spectroscopy of electronic structure will be illustrated.

The previous paragraphs have shown that the measured CCT's reflect details of the density of occupied and unoccupied states. We note that for Si(111)7x7 the typical adatom pattern is visible for a wide range and both polarities of U. Differences exist in the measured height of the adatoms but not in their lateral location in the xy plane of the surface, which means that the xy coordinates of the adatoms may be directly determined from the

CCT in contrast to the z position, which may be influenced by electronic effects. In general, the position of atomic-like structures in CCT's does not always correspond to that of the related atomic nuclei. An important example for this fact was discussed by Stroscio *et al.* [5], who interpreted the measured corrugation on Si(111)2x1 as due to the wavefunction behavior of the π-bonded chains. Whether an atomic structure may directly be related to atomic nuclei has to be ascertained by a measurement of the same surface with opposite polarities of the sample bias voltage. This check is particularly important for semiconductors, where the occupied and unoccupied states may have a completely different spatial charge distribution. Another illustrative example has been reported by Hamers *et al.* for Si(100)2x1 [36]. States near the valence band edge (using $U = -2$ V) led to "bean-shaped" dimer structures in the CCT's, whereas the unoccupied antibonding states of the Si dimer became visible for $U = 1.2$ V in form of two distinct protrusions.

After the CCT's have been measured with both polarities of the sample bias voltage (reflecting the charge density of the occupied and unoccupied states) and the influence of electronic effects on the CCT's is determined qualitatively, the experimental atomic pattern has to be interpreted by an atomic model of the surface. Clean Si surfaces are always reconstructed and the STM images provide information on the more protruding features of the surface and on the location of atomic subunits within the unit cell. Since a number of atomic layers take part in the formation of the reconstruction, the evaluation of an atomic model is always a difficult task and STM images are very useful to obtain the general arrangement of atomic subunits in the surface region.

Typical example for this procedure have, e.g., been published for Si(112) [37] and Si(110) [38]. The latter surface is particularly interesting since it develops several reconstructions with quite different unit cells. Differences in reconstruction behavior (in particular for Si(110)) is often related to metallic impurities, which already at small concentrations (a few % of a monolayer) may change completely size and nature of the observed atomic pattern [39]. Metallic impurities may also give rise to defect-like structures on reconstructed surfaces. Niehus *et al.* have observed that Si(100) develops a 2x8 reconstruction by the presence of a small Ni concentration (typically 1 at. % in the surface region) [40]. The defects appear as missing Si dimers on 2x1 reconstructed Si(100) and show the tendency of lateral ordering. In the STM image these defect structures produce quasi "ordered" channels, which at the same time give rise to 1/8 order beams in the LEED pattern. The observed reconstructions are further influenced by thermal treatment of the sample. For example, a non-uniform heating of an initially flat and clean Si(111) sample produced terraces on the surface with a width of less than the size of the 7x7 unit cell. On the terraces the development of a $\sqrt{3}x\sqrt{3}R30°$ reconstruction was observed [41]. The same kind of reconstruction and additional ones have been found by Becker *et al.*, if a Si(111) surface has been exposed to laser-annealing [42]. Finally, the analysis of steps on Si surfaces has to be mentioned as important application of STM [43-45]. Steps may play a significant role in growth processes of hetero-systems, which will be discussed in the next part of the paper.

The evaluation of the atomic nature of growth processes on semiconductor surfaces belongs to a major application of STM. Such studies are not only interesting for our basic understanding of the phenomena in the initial stages of film growth but also with regard to technical applications, where the metal/semiconductor interface plays a considerable role. Atomic effects on such systems have so far mostly been studied by other surface physical methods as electron microscopy, LEED, and Auger electron spectroscopy (AES) [46], which allow general conclusions on the growth process. STM may deliver information on such systems directly on an atomic scale. In the following paragraphs our STM studies of the condensation of Ag on Si(111)7x7 will be described in a representative way, results from a Ag-induced reconstruction on Si(111) will be discussed, and STM work on metal surfaces will briefly be mentioned in connection with the growth of three-dimensional metal islands on the Si substrates.

We start with a description of Ag condensation on a low-temperature Si(111) substrate. Low means in this context that the substrate temperature is not yet sufficient for the $\sqrt{3}x\sqrt{3}$ phase transition, which is observed for temperatures above 500°C and an Ag coverage (Θ) of at least 1/3 monolayer (ML). The number of atoms in 1 ML is as usual referred to the number of atoms in the outermost atomic layer of the ideally terminated Si substrate. Previous studies have already established the *general* nature of the growth process of Ag on Si(111)7x7. Room-temperature deposition leads essentially to a layer-by-layer growth mode up to Θ of a few ML according to the nearly exponential dependency of the Si $L_{2,3}$ and Ag M_4VV Auger intensities [47]. For $\Theta = 10$ ML, angle-resolved photoemission could be explained by the growth of an ordered overlayer structure in form of two Ag(111) domains. Additional information on this system may be inferred from electron microscopic work [46].

The experiments described here in the submonolayer coverage range have been performed by deposition of 1/3 ML Ag on a Si(111)7x7 substrate at 90°C [10]. In Fig. 4 a CCT from this surface (reflecting the empty states of the sample) is displayed. Ag is identified in form of ring-like and more triangular-shaped atomic structures on top of the otherwise undistorted Si(111)7x7 pattern. The ring-like structures constitute the smallest Ag-induced features on the surface and we interpret them as being the critical nuclei for further Ag condensation. This view is supported by observation of various stages of formation of larger islands, where additional Ag atoms are added to the initial nuclei. The initial nuclei are located on the inner adatoms of the 7x7 unit cell halves, preferentially on the faulted halves. This observation may only be explained by a definite interaction between Ag and the Si substrate by formation of bonding states with the Si surface and subsurface atoms (see below). The largest Ag-induced structures do mostly not extend over the edges of 7x7 unit cell halves. Since the edges

144

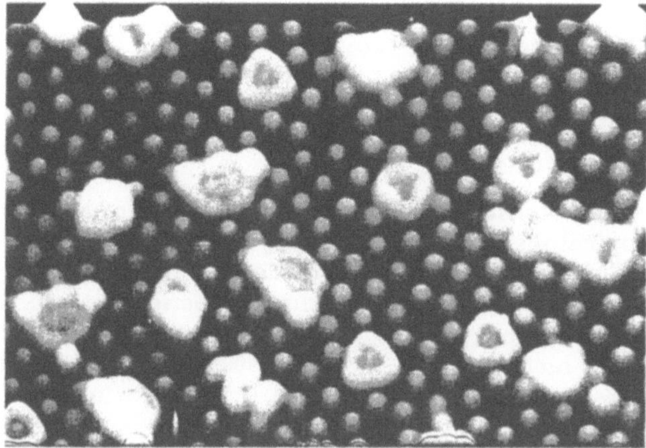

Fig. 4 CCT from 1/3 ML Ag on Si(111)7x7 obtained at $U = 2$ V and $I = 3$ nA. The area is $\approx 10 \times 15$ nm^2.

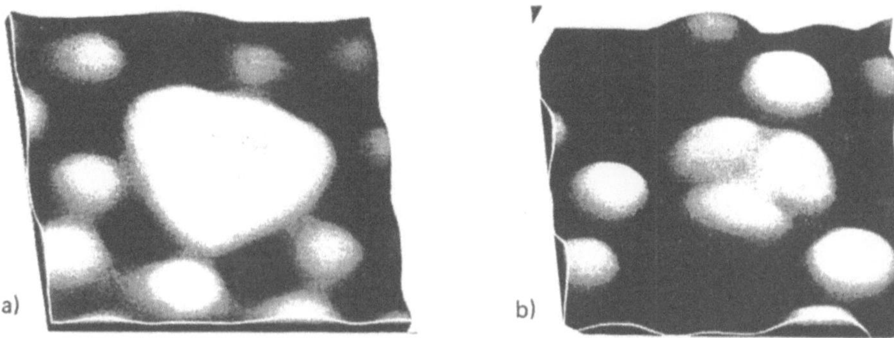

Fig. 5 CCT's from the critical Ag nuclei on Si(111)7x7 obtained with $U = 2$ V (a) and $U = -2$ V (b). $I = 3$ nA and the size is $\approx 4 \times 4$ nm^2 in both cases.

correspond to the positions of dimers in the DAS structure [2], a very weak or even repulsive interaction between Ag and the dimer sites is suggestive.

To derive an atomic model for the critical nuclei, we have measured the Ag clusters using both polarities of U (Fig. 5). The results for tunneling into the unoccupied states (Fig. 5 (a)) are clearly different from those obtained by tunneling from the occupied states (Fig. 5 (b)), where the subunit appears as a three-lobe structure pointing into the direction of the inner adatoms of the unit cell halves. Note that the three single protrusions adjacent to the Ag-induced features are Si adatoms in corner positions of the unit cell half. These differences already indicate that the Ag 5s valence electrons occupy covalent bonds with Si atoms with distinct differences of the bonding (Fig. 5 (b)) and antibonding (Fig. 5 (a)) states. The results shown in Fig. 5 again demonstrate that atomically resolved protrusions in STM images may not directly be assigned to atoms.

Before an attempt is made to construct an atomic model of the Ag clusters we present spectroscopical results obtained on Ag covered parts of the surface and on Si adatoms of the same sample. The density of states information has been measured in form of local I/U characteristics with fixed sample-tip distance as described above. The data are presented in four different ways. In Fig. 6 (a) the measured I/U characteristics are reproduced. From the data dI/dU (Fig. 6 (b)), $d\ln I/dU$ (Fig. 6 (c)), and $d\ln I/d\ln U$ (Fig. 6 (d)) have been computed. The results from the Ag covered parts

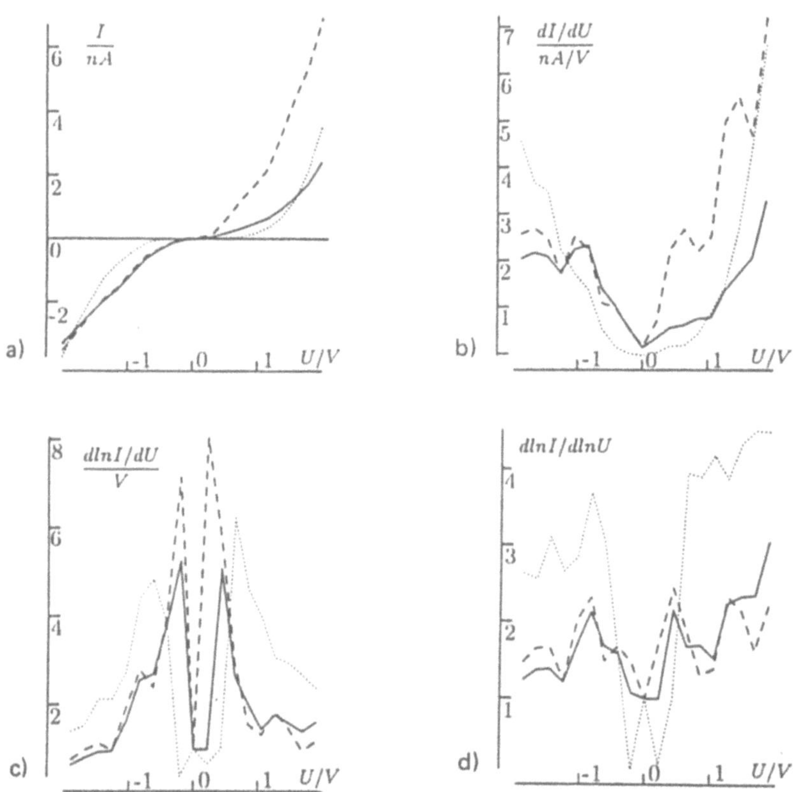

Fig. 6 (a) I/U, (b) dI/dU, (c) $d\ln I/U$, and (d) $d\ln I/d\ln U$ from 1/3 ML Ag on Si(111)7x7. Solid lines: Si adatom positions in the FH. Dashed lines: Si adatom positions in the UH. Dotted lines: Ag covered part of the surface.

146

of the surface are displayed by dotted lines, those from Si adatoms by solid (dashed) lines for the FH (UH) of the 7x7 unit cells of the same surface. As may be seen in Fig. 6 (a) the main difference between the I/U's is an extremely small current on the Ag islands in the approximate range -0.6 V$< U < 0.6$ V. On the adatom positions I increases almost linearly already in the limit if small U (≈ -0.2 V and ≈ 0.2 V, which were the smallest values of U used in this case) corresponding to the above mentioned ohmic behavior of the tunneling resistance in the vicinity of the Fermi level E_F on the adatom positions of Si(111)7x7. Accordingly, dI/dU is nearly zero on the Ag clusters, whereas the Si positions are characterized by nonnegligible values of dI/dU. The latter curves show distinct maxima at $U \approx -1$ V and some more structures for positive U. The feature at -1 V is probably related to the already mentioned density of states maximum on the positions of the dangling bonds of the second atomic layer of clean Si(111)7x7. We believe that this structure is seen in the curves because of insufficient resolution of the tip in this case, which may be deduced from the CCT measured simultaneously with the CI's (the data are reproduced in Ref. [10]).

We interpret the differences in the current characteristics between the Ag covered and uncovered parts of the surface by a reduction of available states for tunneling in the vicinity of E_F upon adsorption of Ag. The same effect is found in the photoemission results, where the adsorption of 1 ML of Ag on Si(111)7x7 is connected with a noticeable reduction of emission below E_F (see Ref. [10]). A reduction of the local density of states at E_F is expected, if the dangling bonds of clean Si(111)7x7 are saturated by Ag atoms and if bonding (distinctly below E_F) and antibonding states are formed in the Ag clusters. The formation of such states has already been proposed qualitatively on the basis of the shape of the initial nuclei for tunneling with positive or negative U.

In Fig. 6 (d) the quantity $d\ln I/d\ln U$ has been evaluated following the proposal of Stroscio *et al.* [5]. The formation of a distinct band gap by Ag adsorption is also seen in these results. In Fig. 6 (c) we show in addition $d\ln I/dU$, which, as we believe, compensates for the with U exponentially increasing tunneling probability. It is difficult to decide at present, which of the chosen quantities is particularly suitable for derivation of density of states information. We prefer the discussion of I/U's or dI/dU's, were influences of the energy-dependent tunneling probability remain still visible.

For derivation of an atomic model for the Ag nuclei additional information should be considered. Firstly, if we compare the surface area of the visible islands with the Ag coverage (in the ML scale) we have to conclude that Ag is condensed in a close-packed arrangement of Ag atoms (provided the sticking coefficient is one and assuming a radius of the Ag atoms in their metallic state). In particular the assumption of a "metallic" radius for the Ag nuclei should be investigated further, since the formation of Si-Ag bonds could lead to somewhat different Ag-Ag distances. Secondly, previous AES results for the submonolayer range at moderate substrate temperature can only be explained by an atop condensation of Ag atoms [47]. It is useful in this context to analyze the corrugation of an Ag nucleus and of a neighboring more complete triangular Ag island as shown in Fig. 4. The corrugation is plotted in Fig. 7 (a)). It shows that the Ag nuclei appear ≈ 0.04 nm higher than an uncovered corner adatom, which is much smaller than expected on the atomic (i.e., metallic) radius of Ag (0.14 nm [48]) and an atop position. On the other hand, for an Ag covered corner adatom we find a height difference of 0.12 nm, which is almost in agreement with the expected value of an Ag atom in atop position. The smallness of the height difference for the initial Ag nuclei is consistent with the above discussed reduction of the local density of states. The tip has to approach the surface more closely on these positions in order to reach the same tunneling current as on the uncovered part of the surface.

For the construction of an atomic model of the Ag deposit we may therefore assume that the Ag atoms in the initial nuclei and the triangular islands are located at the same height and form nearly two-dimensional structures on Si(111)7x7. In Fig. 7 (b) we have plotted the shape of the observed occupied features of an Ag nucleus (Fig. 5 (b)) including the Si corner atoms (indicated by the hatched area) on the three uppermost atomic layers of a 7x7 unit cell half. The three lobes point to the inner adatoms and fill the space within these adatoms and the dangling bonds of the second atomic layer. In principle, six electrons would saturate these dangling bonds, which places a lower limit of six Ag atoms in the critical Ag nucleus. We suggest that each lobe corresponds to a pair of Ag atoms.

The AES results of Ref. [47] indicated layer-by-layer growth mode of Ag on Si(111)7x7 (at room temperature) up to a few monolayers. Deviations are observed for higher coverage and for deposition at higher substrate temperatures. In case of annealing and Θ in the 1 ML range the surface develops

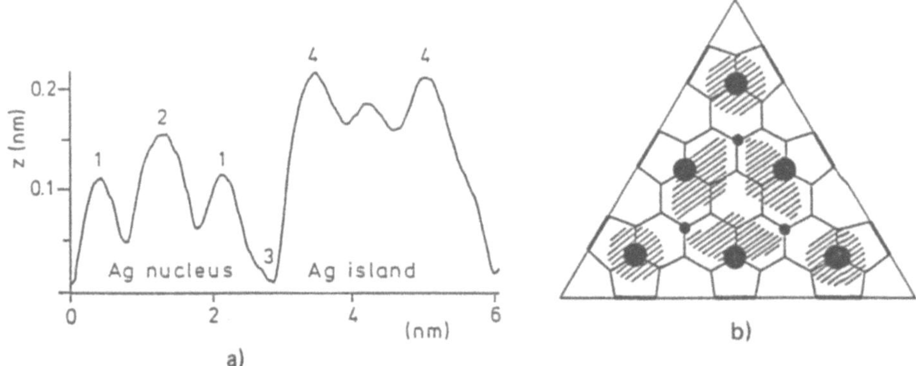

Fig. 7 (a) Corrugation on Ag nucleus and island obtained from the CCT shown in Fig. 4. 1: corner adatom, 2: inner adatom covered with Ag, 3: corner hole, and 4: corner adatom covered with Ag. (b) Schematic arrangement of structures from Ag nucleus (occupied states) on top of the DAS model.

148

a $\sqrt{3}\text{x}\sqrt{3}R30°$-Ag reconstruction (see below), which acts as interface for an interface+3D islands growth mode [46]. We have measured CCT's and spectroscopical information on the system 3 ML Ag on room temperature Si(111)7x7. In the CCT's (Fig. 8 (a)) we find flat islands of different heights corresponding to integral number of dense-packed Ag layers (Fig. 8 (b)). The edges of the islands are fairly straight and inclined by angles, which are expected from the growth of ordered 3D islands of Ag(111). The orientation of the edges is consistent with the orientation of the initial triangular two-dimensional islands, which in the submonolayer range cover halves of the 7x7 unit cell (see Fig. 4). The image clearly shows strong deviations from the layer-by-layer growth mode already at moderate coverage, which has also been deduced from ion scattering data [49] but in the AES results is only seen for larger coverage. We note that the same kind of islands have been observed for $\Theta = 4.8$ ML on a room temperature Si(111)$\sqrt{3}\text{x}\sqrt{3}R30°$ -Ag substrate [11]. The possibility of observing an atomic corrugation on the Ag islands is discussed below.

Local I/U characteristics on this surface have been measured at the base (labeled A on Fig. 8 (a), solid lines in Fig. 8 (c)), on an Ag islands with a height of 4 ML (labeled B on Fig. 8 (a), dashed lines in Fig. 8 (c)), and on an Ag island with a height of 7 ML Ag (labeled C on Fig. 8 (a), dotted lines in Fig 8 (c)). Noticeable differences exist between the results from the base, which corresponds to a coverage of approximately 1 ML Ag, and the Ag islands with a height of several ML. The initial slope at $U = 0$ V (corresponding to tunneling near E_F) is much steeper for positions on the islands than on the base. Qualitatively, the results from the 1 ML structure resemble those of the Ag clusters (see Figs. 6 (a) and 6 (b)). The differences in shape indicate metallic character for the I/U's from the islands and more semiconducting properties for the first ML. It seems plausible to explain this behavior by a particular strong bonding and charge transfer of the first

Ag layer and, as a consequence, a reduction of the density of states near E_F. The dI/dU curves on the islands (right hand side of Fig. 8 (c)) show distinct maxima at $U = -1$ V and 0.5 V. We cannot explain these structures in terms of the density states of bulk Ag and do not believe that they are caused by the interaction between Ag and the Si substrate. I/U and dI/dU curves from Au(111) and Cu(111) do not show this effect [50].

As may be seen in Fig. 8 (b), the noise in the corrugation is in the order of 0.01 nm and within these limits a regular corrugation due to a dense-packed metal structure on top of the islands could only be resolved in a fragmentary way. If, on the other hand, we measured a bulk dense-packed clean Au(111) single-crystal, an atomic corrugation of about 0.005...0.01 nm could clearly be detected (Fig. 9). We note that an atomic corrugation on a dense-packed metal atom surface (evaporated Au on mica) was first observed by Hallmark *et al.* [51] and later for Al(111) by Wintterlin *et al.* [52]. Sometimes an atomic corrugation of nearly 0.1 nm is found for Al(111), which in Ref. [52] is explained as due to an enlargement effect by the existence of an atomic cluster on the tip apex. It has to be mentioned

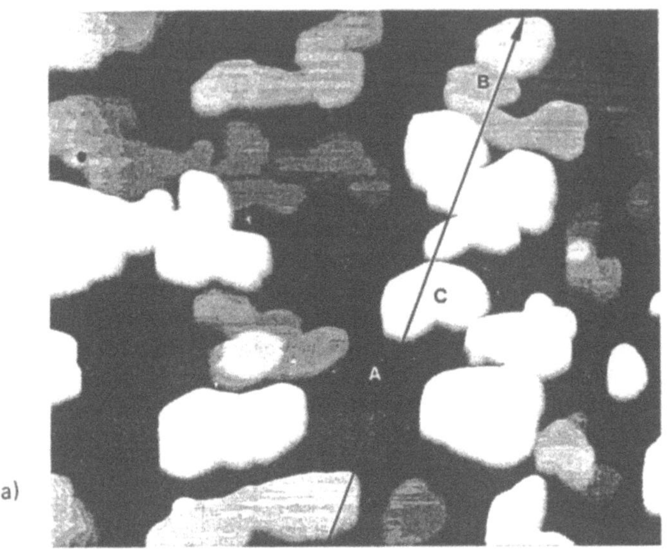

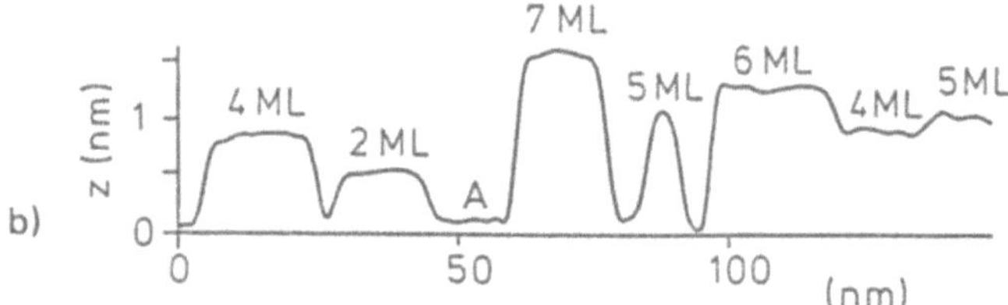

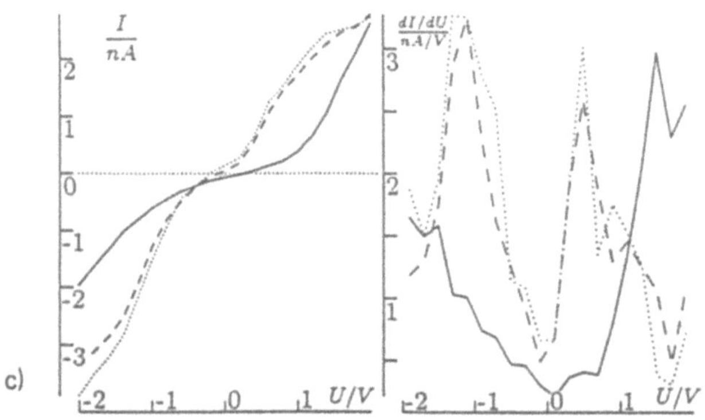

Fig. 8 (a) CCT from 3 ML Ag on room temperature Si(111)7x7. The area is 150×150 nm^2. (b) Corrugation of the CCT along the arrow shown on (a). "A" denotes the base of the islands.

150

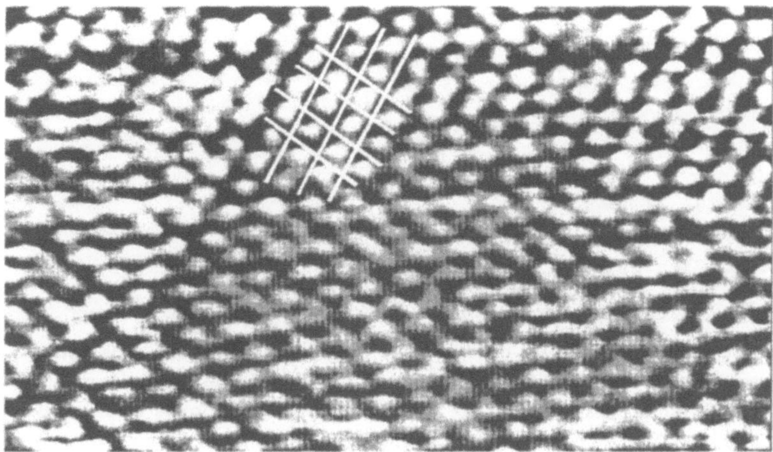

Fig. 9 Atomic corrugation on bulk Au(111) obtained at $U = 40$ mV and $I = 4$ nA. The size of the CCT is $\approx 4.5 \times 7$ nm^2.

that He scattering (which is sensitive to the charge density of the surface at about the same distance as the tip is located during tunneling) on dense-packed metal surfaces provides evidence for a corrugation of a few 0.001 nm [53]. Therefore in STM images a corrugation of such surfaces can hardly be detected without enlargement effects.

Finally, the possibility of analyzing metal-induced reconstructions on Si surfaces will be described briefly for Si(111)$\sqrt{3}$x$\sqrt{3}R30°$-Ag. Numerous publications exist for this surface, which is generally used as model system to demonstrate the capabilities of the various measurement techniques [54]. When the first STM results were published on this surface [13,14], a generally accepted atomic model of this reconstruction did not yet exist. Many groups favored a model, where the Ag atoms are embedded in the uppermost double layer of Si and others located the Ag atoms in atop positions of the substrate. It was very striking that the STM results of Refs. [13] and [14] although quite similar were discussed either in favor of the "embedded Ag trimer" model [14] or in terms of the "Ag atop" structure [13]. This discrepancy is not surprising since, as a matter of fact, the identity of atomic species cannot yet be distinguished by using STM. Additional surface analytical techniques have normally to be applied for this purpose.

Under suitable experimental conditions the topological appearance of surface structures may be used for an identification of atomic arrangements. Since the STM images from Si(111)$\sqrt{3}$x$\sqrt{3}R30°$-Ag are characterized by a honeycomb pattern [13,14], the main question is, whether the observed features are the Si atoms of the original Si(111) lattice (favoring then the "embedded Ag trimer" model) or the Ag atoms, which would explain the

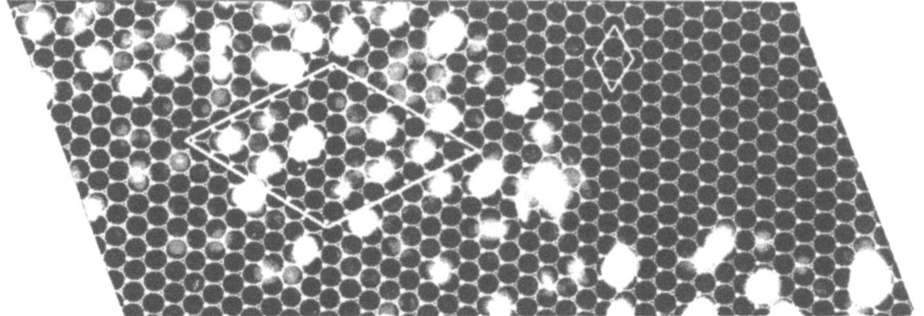

Fig. 10 Registration of the lattice of ideal Si to neighboring 7x7 unit cells and $\sqrt{3} \times \sqrt{3}R30°$-Ag reconstructed parts of Si(111) obtained at $\Theta \approx 0.5$ ML and $U = -2$V. A double layer of ideal Si(111) is shown. The little rhomboid corresponds to the $\sqrt{3}$ unit cell, the large one is plotted slightly larger than the 7x7 unit cell.

reconstruction in terms of the "Ag atop" structure. The atomic position relative to the ideal Si lattice is different in both models by ≈ 0.2 nm. This fact can be utilized to distinguish between both models, if the position of the original lattice is known within an accuracy of ≈ 0.1 nm. Wilson and Chiang have shown [15] that the registration of the Si lattice relative to the measured STM image is possible for this surface structure because in the submonolayer coverage range ($\Theta \approx 0.5$ ML) neighboring Ag-induced $\sqrt{3}$ reconstructed parts of the surface are found adjacent to 7x7 unit cells the latter one being used for alignment of the ideal Si lattice.

In Fig. 10 such a fit of an ideal Si double layer to a measured CCT is shown. In the left part of the Fig. the 7x7 pattern is distinctly visible in form of the white dots and the right part of the Fig. corresponds to the $\sqrt{3} \times \sqrt{3}$ reconstruction. The reproduction of this result in Fig. 10 may not be as convincing as the processed image on the monitor of the computer system, which shows a larger range of grey tones.

4 Concluding Remarks

It was the main aim of the present work to provide an overview on theoretical and experimental details of scanning tunneling microscopy and spectroscopy and illustrate the use of these methods for a study of clean and metal-covered Si surfaces. Important applications of STM in this field have not been discussed, neither an attempt has been made to extend presentation and discussion to every clean and metal-covered Si surface studied so far. It has also to be noted that similar studies have been performed on other semiconductor surfaces, where the principal use of STM to particular problems have been developed. In order to facilitate the access to some of the more innovative experimental work on Si surfaces, we summarize recent progress in this field.

The investigation of chemical reactions on semiconductor surfaces certainly belongs to the more spectacular applications of STM. The interaction of NH_3 with Si(111)7x7 has recently been studied by Wolkow and Avouris [55]. The differences in the energy-level structure of the clean and reacted Si(111) surfaces and the possibility of a local spectroscopy with STM leads to direct information on the distribution of the reacted atoms on the surface. The dangling bonds of the second atomic layer were found to be more reactive than those of the adatoms and the center adatoms more reactive than the corner adatoms. This observation is consistent with our findings of the above described Ag clusters, which are formed on the inner part of the 7x7 unit cell half. Another interesting application in this respect is the study on the initial stage of oxygen adsorption or oxidation of Si(111)7x7 [56,57]. The observed effects appear here in form of missing adatoms, which, to our opinion, have to be explained by reacted adatoms with reduced density of states accessible by the tunneling electrons. The position of adsorbed O atoms would therefore be those of the seemingly missing adatoms.

The initial growth morphology of other semiconducting materials deposited on Si surfaces are further important applications of STM. Here we briefly mention results obtained on Si(111)-As [58] and on the epitaxial growth of Si on Si(100) and Si(111) [59].

The beautiful STM results published by many authors and recent progress does not mean that future developments are unimportant or unnecessary in STM. This concerns both experimental and theoretical problems on which further effort have to concentrated. The STM experiments on clean semiconductor materials are generally quite difficult and time-consuming, even if the experimental equipment is basically suitable for measurements with atomic resolution. Difficulties arise mostly due to tip artifacts, instabilities, and improper macroscopic shape. To our experience, not yet fully cleaned semiconductor surfaces or those containing oxide and other insulating islands tend to increase the mean tip radius during the measurements probably due to mechanic contact with the nonconducting parts of the surface. A few day's use of the tip may lead to an increase of its radius from typically 10–100 nm to 10 μm. Tip artifacts are mostly recognized by a doubling of the observed atomic structures (see, for example, Ref. [60]). Sometimes they are only visible for one polarity of the sample bias, while the opposite polarity delivers correct results. These effects are particularly disturbing for spectroscopic measurements. Concerning future theoretical work, the computation of voltage-dependent STM images and of I/U characteristics of model systems would be highly desirable. We note that previous work in this field has been published by Selloni *et al.* for graphite [61].

Acknowledgement

This work has been financially supported by the Deutsche Forschungsgemeinschaft. The Si wafers have been provided by Wacker-Chemitronic (Burghausen). We are indebted to G. Hackel for her careful photographical work. Discussions with Dr. Th. Berghaus, D. Badt, A. Brodde, H. Wengelnik and their technical help are gratefully acknowledged.

References

[1] G. Binnig, H. Rohrer, Ch. Gerber, and E. Weibel, Phys. Rev. Lett. **50**, 120 (1983)

[2] K. Takayanagi, Y. Tanishiro, M. Takahashi, and S. Takahashi, J. Vacuum Sci. Technol. **A3**, 1502 (1985)

[3] R.M. Tromp, R.J. Hamers, and J.E. Demuth, Phys. Rev. **B34**, 1388 (1986)

[4] Th. Berghaus, A. Brodde, H. Neddermeyer, and St. Tosch, Surface Sci. **193**, 235 (1988)

[5] J.A. Stroscio, R.M. Feenstra, and A.P. Fein, Phys. Rev. Lett **57**, 2579 (1986)

[6] G. Binnig, K.H. Frank, H. Fuchs, N. Garcia, B. Reihl, H. Rohrer, F. Salvan, and A.R. Williams, Phys. Rev. Lett. **55**, 991 (1985)

[7] R.S. Becker, J. Golovchenko, D.R. Hamann, and B.S. Swartzentruber, Phys. Rev. Lett. **55**, 2032 (1985)

[8] R.J. Hamers, R.M. Tromp, and J.E. Demuth, Phys. Rev. Lett. **56**, 1972 (1986)

[9] U.K. Köhler, J.E. Demuth, and R.J. Hamers, Phys. Rev. Lett. **60**, 2499 (1988)

[10] St. Tosch and H. Neddermeyer, Phys. Rev. Lett. **61**, 349 (1988)

[11] J.E. Demuth, E.J. van Loenen, R.M. Tromp, and R.J. Hamers, J. Vac. Sci. Technol. **B6**, 18 (1988)

[12] St. Tosch and H. Neddermeyer, to be published in J. Microscopy

[13] R.J. Wilson and S. Chiang, Phys. Rev. Lett. **58**, 369 (1987)

[14] E.J. van Loenen, J.E. Demuth, R.M. Tromp, and R.J. Hamers, Phys. Rev. Lett. **58**, 373 (1987)

[15] R.J. Wilson and S. Chiang, Phys. Rev. Lett. **59**, 2329 (1987)

[16] R.J. Wilson, S. Chiang, and F. Salvan, Phys. Rev. **B38**, 12 696 (1988)

[17] J.E. Demuth, U.K. Köhler, R.J. Hamers, and P. Kaplan Phys. Rev. Lett. **62**, 641 (1989)

[18] St. Tosch and H. Neddermeyer, to be published in Surface Sci.

[19] G. Binnig, N. Garcia, H. Rohrer, J.M. Soler, and F. Flores, Phys. Rev. **B30**, 4816 (1984)

[20] G. Binnig, H. Rohrer, Ch. Gerber, and E. Weibel, Phys. Rev. Lett. **49**, 57 (1982)

[21] J. Tersoff and D.R. Hamann, Phys. Rev. **B31**, 805 (1985)

[22] J. Bardeen Phys. Rev. Lett. **6**, 57 (1961)

[23] M.S. Chung, T.E. Feuchtwang, and P.H. Cutler, Surface Sci. **187**, 559 (1987); T.E. Feuchtwang, to be published in J. Microscopy

[24] C. Noguera, to be published in J. Microscopy

[25] J. Bono and R.H. Good, Surface Sci. **175**, 415 (1986)

[26] A. Baratoff, G. Binnig, H. Fuchs, F. Salvan, and E. Stoll, Surface Sci. **168**, 734 (1986)

154

[27] *R.M. Feenstra, J.A. Stroscio, and A.P. Fein*, Surface Sci. **181**, 295 (1987)

[28] *Ch. Gerber, G. Binnig, H. Fuchs, O. Marti, and H. Rohrer*, Rev. Sci. Instrum. **57**, 221 (1986)

[29] *B. Drake, R. Sonnenfeld, J. Schneir, P.K. Hansma, G. Slough, and R.V. Coleman*, Rev. Sci. Instrum. **57**, 441 (1986)

[30] *G.F.A. van de Walle, J.W. Gerritsen, H. van Kempen, and P. Wyder*, Rev. Sci. Instrum. **56**, 1573 (1985)

[31] *J.E. Demuth, R.J. Hamers, R.M. Tromp, and M.E. Welland*, IBM J. Res. Develop. **30**, 396 (1986)

[32] *D.W. Pohl*, IBM J. Res. Develop. **30**, 417 (1986)

[33] *T. Yamaguchi*, Phys. Rev. **B32**, 2356 (1985)

[34] *Th. Berghaus, A. Brodde, H. Neddermeyer, and St. Tosch*, J. Vac. Sci. Technol. **A6**, 483 (1988)

[35] *H. Neddermeyer, U. Misse, and P. Rupieper*, Surface Sci. **117**, 405 (1982)

[36] *R.J. Hamers, Ph. Avouris, and F. Boszo*, Phys. Rev. Lett. **59**, 2071 (1987)

[37] *Th. Berghaus, A. Brodde, H. Neddermeyer, and St. Tosch*, Surface Sci. **184**, 273 (1987)

[38] *H. Neddermeyer and St. Tosch*, Phys. Rev. **B38**, 5784 (1988)

[39] *E.J. van Loenen, A.J. Hoeven, and D. Dijkkamp*, to be published in J. Microscopy

[40] *H. Niehus, U.K. Köhler, M. Copel, and J.E. Demuth*, to be published in J. Microscopy

[41] *Th. Berghaus, A. Brodde, H. Neddermeyer, and St. Tosch*, Surface Sci. **181**, 340 (1987)

[42] *R.S. Becker, J.A. Golovchenko, G.S. Higashi, and B.S. Swartzentruber*, Phys. Rev. Lett. **57**, 1020 (1986)

[43] *Th. Berghaus, A. Brodde, H. Neddermeyer, and St. Tosch*, J. Vac. Sci. Technol. **A6**, 478 (1988)

[44] *P.E. Wierenga, J.A. Kubby, and J.E. Griffith*, Phys. Rev. Lett. **59**, 2169 (1987)

[45] *R.M. Feenstra and J.A. Stroscio*, Phys. Rev. Lett. **59**, 2173 (1987)

[46] *J.A. Venables, G.D.T. Spiller, and M. Hanbücken*, Rep. Prog. Phys. **47**, 345 (1984)

[47] *M. Hanbücken, H. Neddermeyer, and P. Rupieper*, Thin Solid Films **90**, 37 (1982)

[48] *C. Kittel*, Introduction to Solid State Physics, 5th Edition (John Wiley, New York, 1976)

[49] *E.J. van Loenen, M. Iwami, R.M. Tromp, and J.F. van der Veen*, Surfac Sci. **137**, 1 (1984)

[50] *A. Brodde, St. Tosch, and H. Neddermeyer*, to be published in J. Microscopy

[51] *V.M. Hallmark, S. Chiang, J.F. Rabolt, J.D. Swalen, and R.J. Wilson*, Phys. Rev. Lett. **59**, 2879 (1987)

[52] *J. Wintterlin, J. Wiechers, H. Brune, T. Gritsch, H. Höfer, and R.J. Behm*, Phys. Rev. Lett. **62**, 59 (1989)

[53] *K.-H. Rieder*, private communication

[54] Relevant literature on Si(111)$\sqrt{3}$x$\sqrt{3}R30°$-Ag is cited in Refs. [13-15]

[55] *R. Wolkow and Ph. Avouris*, Phys. Rev. **60**, 1049 (1988)

[56] *F.M. Leibsle, A. Samsavar, and T.-C. Chiang*, Phys. Rev. **B38**, 5780 (1988)

[57] *D. Badt and H. Neddermeyer*, unpublished results

[58] *R.S. Becker, B.S. Swartzentruber, J.S. Vickers, and M.S. Hybertsen*, Phys. Rev. Lett. **60**, 116 (1988)

[59] *R.J. Hamers, U.K. Köhler, and J.E. Demuth*, to be published in Ultramicroscopy

[60] *H.A. Mizes, Sang-il Park, and W.A. Harrison*, Phys. Rev. **B36**, 4491 (1987)

[61] *A. Selloni, P. Carnevali, E. Tosatti, and C.D. Chen*, Phys. Rev. **B31**, 2602 (1985)

Optical Dephasing and Orientational Relaxation of Wannier-Excitons and Free Carriers in GaAs and GaAs/Al$_x$Ga$_{1-x}$As Quantum Wells

Jürgen Kuhl
Max-Planck-Institut für Festkörperforschung
Heisenbergstr. 1, D-7000 Stuttgart 80, FRG

Alfred Honold
NTT Basic Research Laboratories
Musashino-shi, Tokyo 180, Japan

Lothar Schultheis
ASEA Brown Boveri Corp. Research
CH-5405 Baden, Switzerland

Charles W. Tu
University of California at San Diego, La Jolla, CA 92093, USA

Summary: The dynamics and mechanisms of optical dephasing and orientational relaxation of photoexcited excitons have been investigated in the $GaAs/Al_xGa_{1-x}As$ system by means of time-resolved Degenerate-Four-Wave-Mixing (DFWM) experiments with picosecond optical pulses. We interpret experimental results measured on two different samples with $GaAs$ layer thicknesses L_z which are large and small, respectively, compared to the 3D exciton Bohr radius in order to reveal the influence of confinement on to the exciton kinetics. The influence of exciton/exciton, exciton/free-carrier and exciton/acoustic-phonon scattering on the phase relaxation rate is discussed. In addition, the present knowledge about the corresponding relaxation phenomena for free-electrons and holes in the $GaAs/Al_xGa_{1-x}As$ system is reviewed.

1 Introduction

The nonlinear optical properties and the ultrafast dynamics of excitons in semiconductors are a major field of present semiconductor research [1,2]. This interest is explained by the potential applications of excitonic nonlinearities as ultrafast optical switching devices [3,4,5] in future optical communication systems. The design and development of such optoelectronic devices require detailed information about the intrinsic relaxation and recombination times of photoexcited excitons, which set an ultimate limit for the attainable operation speed. Simultaneously, the exploration of the relaxation phenomena can provide deep insight into the fundamental interaction

processes of excitons with their crystal surrounding, in particular, collisions
of excitons with crystal defects, as well as with other quasi-particles like
phonons, excitons or free electrons and holes.

The present knowledge about exciton dynamics [6] is still relatively fragmentary. Whereas several papers have been published which report on measurements of exciton lifetimes [7,8,9], very little information is available about
the initial relaxation steps of photoexcited excitons, like the decay of the
photo-induced phase coherence or the scattering of the electron and hole,
which form the exciton, out of the optically coupled states. These rapid relaxation phenomena, which occur on a time-scale of a few picoseconds are of
fundamental importance for conceivable optoelectronic devices utilizing e.g.
the ac Stark effect as the switching mechanism [10,12].

In addition, measurements of the dephasing time can permit quantitative
proof of the theoretically predicted relations between the radiative lifetime
[9], as well as the 3rd order nonlinear susceptibility [13] and the homogeneous
linewidth of the optical transition.

In this paper we will review our recent investigations on optical dephasing
and orientational relaxation of excitons in $GaAs$ and $GaAs/Al_xGa_{1-x}As$ quantum wells (QW)[14,15,16,17]. The rapid kinetics have been directly studied
in the time domain by means of time-resolved Degenerate-Four-Wave-Mixing
(DFWM) with picosecond light pulses. Special attention has been addressed
to the influence of the dimensionality of the excitonic system on the relaxation processes. We present and discuss experiments performed on two
samples with $GaAs$ layer thicknesses L_z of 194 nm and 12 nm, which are large
and small, respectively, as compared to the Bohr diameter ($2a_B$= 27nm) of
the 3D exciton. Comparison of the results, therefore, directly reveals the
influence of the increased confinement of the exciton for the direction perpendicular to the layer in the second sample on the relaxation kinetics.

The paper is organized as follows: After briefly explaining the various relaxation steps of excitons in a semiconductor (Sec.2), we shall discuss the
potential of DFWM-experiments with picosecond light pulses to study the
ultrafast dynamics of excitons (Sec.3). Then we describe the experimental
set-up and the samples (Sec.4). The bulk of the paper (Sec.5-8) is devoted to
a discussion of experimental results and their interpretation. Besides data for
the intrinsic relaxtion rates determined at low excitation levels, we present
a detailed study of the dependence of the polarization dephasing rate on ex-
citon/exciton, exciton/free-carrier and excition/acoustic-phonon scattering
which provides important information concerning the respective interaction
mechanisms. In Sec. 9 we summarize the present state of knowledge about
the corresponding relaxation phenomena of free electrons and holes in 3D
and 2D-$GaAs$. The paper will conclude with a summary.

2 Relaxation Phenomena of Excitons

In both our samples the exciton mobility is restricted perpendicular to the
layer (see Sec.5). For photoexcitation, momentum conservation has thus to
be fulfilled only for the components k_x and k_y parallel to the layer. Opti-

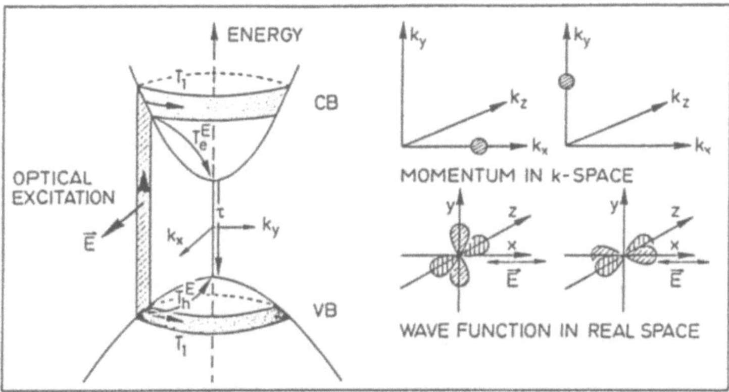

Fig. 1 Anisotropic phase-space filling and subsequent relaxation in a zinc-blende type semiconductor produced by optical excitation with polarization parallel to the x-direction.

cal excitation of carriers from the heavy-hole valence band in such a 2D-semiconductor and the subsequent relaxation steps are schematically depicted in Fig. 1. The valence and conduction band states at the center of the Brillouin zone in a material with zinc-blende symmetry are composed by predominantly p-type and s-type electronic wavefunctions, respectively.

Irradiation with linearly polarized light induces transitions between valence and conduction band states with well-defined energies (determined by the photon energy) and well-defined momentum (depending upon the polarization direction of the field). The electric field couples more strongly to electrons at the top of the heavy-hole valence band with a momentum perpendicular to the field than to those parallel to the field. Electrons in the valence band, with a momentum $\vec{k} = (k_x, 0, 0)$ do not couple to a field polarized parallel to x since their wavefunction is a linear superposition of the p_y and p_z-function and has negligible small dipole moment in field direction. In contrary, electrons with $\vec{k} = (0, k_y, 0)$ couple efficiently to the same field owing to the dipole moment of the p_x-part of their wavefunction.

The same polarization-dependent anisotropic population of electron and hole states in phase space will be created if the generated electron and hole are bound via Coulomb interaction and form a Wannier exciton.

The excitation of excitons can be discussed to a first approximation in the frame of a simple two-level model. Optical excitation of such a two-level system, which is theoretically formulated in a density matrix formalism [18], results in a coherent superposition of the two states. This coherence decays due to scattering processes with the phase coherence time T_2 if the driving field is switched off. The phase coherence time T_2 corresponds to the homogeneous linewidth Γ_h of an optical transition ($\Gamma_h = 1/\pi \cdot T_2$) which is very often covered in the frequency domain by inhomogeneous broadening. In

159

the case of inhomogeneous broadening Γ_h can be determined either by spectral hole burning experiments in the frequency domain [19] or by DFWM or photon-echo experiments in the time domain [18,20].

The next relaxation step is the decay of the population of the optically coupled states. Whereas for the ideal two-level system recombination of the electron and the hole represents the unique decay channel of the excited system, a variety of relaxation channels is available for the depopulation of the optically coupled states in a real semiconductor. First, both carriers can scatter out of their initial momentum state where the corresponding momentum randomization in k-space may or may not involve energy relaxation. This momentum randomization associated with a reorientation of the excitonic dipole moment governs the decay of the population of the optically coupled states. We call the respective time constant orientational relaxation time T_1.

The next step involves energy relaxation of electrons and holes via carrier-carrier or carrier-phonon interaction. In the present paper we do not treat this energy relaxation in detail, since the studied excitons were generated resonantly (vanishing excess energy) and the sample temperature was kept at 1.85 K. Relaxation of the exciton due to acoustic phonon emission or absorption leads to changes of the exciton energy, which are small compared to the applied bandwidth of the picosecond pulse.

The final relaxation step is the recombination of the electron and the hole described by the exciton lifetime τ which amounts typically to a few 100 ps up to 1 ns.

3 Transient Four-Wave Mixing

A comprehensive theoretical analysis of transient grating effects observed in time-resolved DFWM-experiments with pico or subpicosecond optical pulses in two or three beam configurations as well as of the application of these dynamical gratings to study relaxation phenomena in solids or molecules, has been presented in several review articles [20,21,22,23,24].

In a typical DFWM-experiment the interaction of the electric fields of 3 pulses $E_i(\omega, k_i, t_i)$ with i= 1,..,3 which may propagate at different times $t_1 \leq t_2 \leq t_3$ through a thin slab of material creates a nonlinear polarization. For a (hypothetic) material with an infinitely fast response (no polarization memory) the 3rd order nonlinear polarization is given by the following expression:

$$\vec{P}^{(3)}(\omega, \vec{k}_S, t_1, T_{21}, T_{23}) \;=\; \chi^{(3)} \cdot \vec{E}_1(\omega, \vec{k}_1, t - t_1) \cdot \tag{1}$$
$$\vec{E}_2(\omega, \vec{k}_2, t - t_2) \cdot \vec{E}_3(\omega, \vec{k}_3, t - t_3)$$

with

$$T_{12} = t_2 - t_1 \text{ and } T_{23} = t_3 - t_2.$$

This nonlinear polarization in turn leads to the emission of two signal beams with electric field $\vec{E}_S(\omega, \vec{k}_S, t - t_1, T_{12}, T_{23})$ (see Fig. 2). The emission direction is determined by the phase matching condition $\Delta \vec{k} = \vec{k}_3 \pm (\vec{k}_2 - \vec{k}_1) - \vec{k}_S$.

160

Maximum signal would be observed for $\Delta \vec{k} = 0$. It should be noted, however, that for noncollinearly propagating beams a small phase-mismatch is unavoidable (see Sec. 4). In a real material the excitation decays with a finite time constant. The magnitude of the overall polarization induced by three successive pulses, therefore, depends on the relaxation of the excitation created by a single pulse before the arrival of the next pulse and consequently will vary with the delay between the pulses. Thus, measurements of the time-averaged signal $I_S(\omega, \vec{k}_S, T_{12}, T_{23}) \sim \int_{-\infty}^{+\infty} |P^{(3)}(\omega, \vec{k}_S, t - t_1, T_{12}, T_{23})|^2 dt$ as a function of T_{12} and/or T_{23} can provide detailed information about the various ultrafast relaxation steps of the excited state.

Two fundamentally different categories of gratings may be classified: population and orientational gratings.

3.1 Population Gratings

Temporal and spatial superposition of two pulses ($t_1 = t_2, T_{12} = 0, T_{23} = T$) with parallel polarization (perpendicular to the plane of incidence) on the sample leads to a spatially periodic modulation of the resulting intensity due to the interference of the two coherent optical fields (see Fig. 3). The absorption of the pulses thus generates a spatially modulated density of the excited states. It is important to note however, that even for the case of parallel polarization of the two grating forming pulses the excitation is associated with a preferential orientation of the exciton wavevector parallel to the layer as well as of the exciton dipole moment. Thus, the grating can partially decay by momentum reorientation at the beginning (see Fig.8 and Sec.3.2). The incoherent density grating being left after randomization of the excitonic states creates a slight spatially periodic modulation of the linear susceptibility $\chi^{(1)}$ of the material due to the excitation density dependence of the eigenenergy E_T, the oscillator strength $f_x \propto E_L - E_T$ and the homogeneous

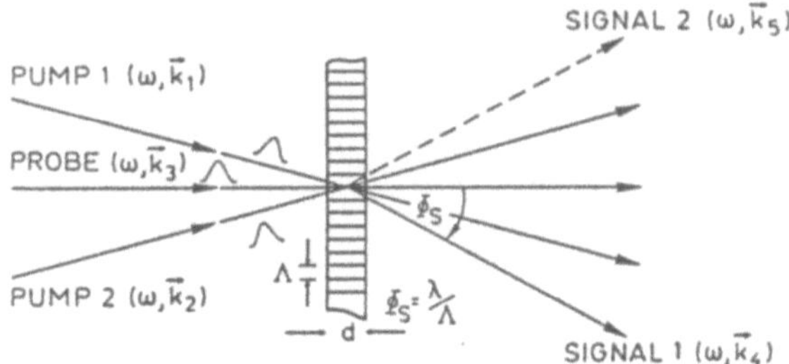

Fig. 2 Transient grating generation by superposition of two light pulses and its detection by diffraction of a third pulse.

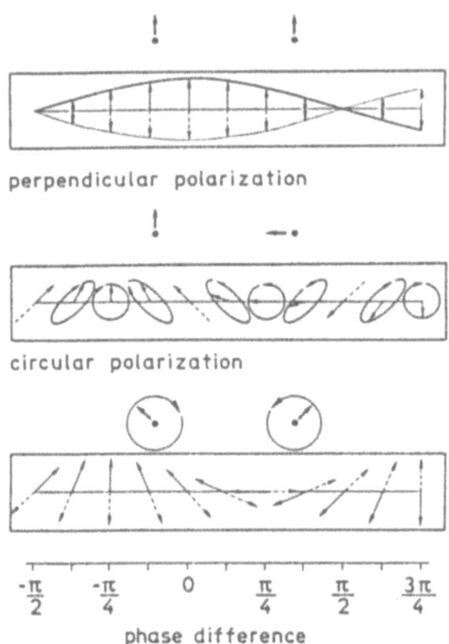

Fig. 3 Superposition of two beams with parallel, perpendicular and opposite circular polarization. The first configuration results in an intensity modulation, the second and third in a periodic variation of the polarization of the incident light with spatial coordinates.

linewidth Γ_h of the exciton:

$$\chi^{(1)} \propto \frac{E_L(N_x) - E_T(N_x)}{(E_T(N_x) - E) + \frac{i}{2}\Gamma_h(N_x)} \tag{2}$$

The spatial modulation of $\chi^{(1)}$ results in a corresponding variation of the optical constants n and κ ($\tilde{n} = n + i\kappa/2|\vec{k}|$ is the complex refractive index). In a linear approximation the density dependence of n and κ can be expressed as:

$$\Delta\kappa = \kappa_o - \kappa = \sigma \cdot N \tag{3}$$
$$\Delta n = n_o - n = \epsilon \cdot N. \tag{4}$$

Here, N is the electron-hole pair density, and σ and ϵ are the changes of the absorption coefficient and refractive index, respectively, induced by the excitation of one electron-hole pair. The nonlinearities can be extremely large in the excitonic region. For a $GaAs/Al_xGa_{1-x}As$ MQW structure with $L_z = 96\text{Å}$ Miller et.al.[25] measured $\sigma = 7 \cdot 10^{-14}cm^2$ and $\epsilon = 3.7 \cdot 10^{-19}cm^3$.

The light-induced amplitude and phase-grating diffracts some of the intensity of the third pulse into directions 4 and 5 (see Fig. 2) when this probe pulse traverses the grating area. The diffraction efficiency is

$$\eta = \left(\frac{\pi \cdot \Delta n \cdot d}{\lambda}\right)^2 + \left(\frac{\Delta\kappa \cdot d}{4}\right)^2 \tag{5}$$

with d the thickness of the grating.

162

The diffracted intensity as a function of the delay T of the probe pulse with respect to the two grating forming pump pulses is a measure of the decay of the grating which can be completely destroyed only by recombination and/or spatial diffusion. The latter contribution can be eliminated on a picosecond time scale by choice of a sufficiently large grating constant.

The initial reorientation of the excitonic dipoles changes the tensor character of the nonlinear polarization [21,22,23]. Before orientational relaxation the polarization is determined by the 3rd order nonlinear susceptibility $\chi^{(3)}$. For times long compared to the orientational relaxation the polarization is given by the product of two $\chi^{(1)}$-processes. The first step is the generation of a population change by the two initial pulses. After orientational relaxation this population is probed by a second $\chi^{(1)}$-process.

3.2 Orientational Gratings

Superposition of two perpendicularly polarized coherent pulses on the sample leads to a periodic modulation of the polarization state of the resulting field with spatial coordinates whereas the intensity remains uniform (see Fig.3). This excitation results in a spatially anisotropic state filling in $\vec{k}$-space and a corresponding spatially periodic orientation of the excitonic dipole moments, since in zinc blende-type semiconductors linearly polarized light excites preferentially electrons and holes in conduction and valence band states with a $\vec{k}$-direction perpendicular to the polarization vector. In spite of the uniform density of excited species, a linearly polarized probe pulse will observe a periodic variation of populated $\vec{k}$-states (anisotropic state filling) with spatial coordinates and will thus be diffracted off this orientational grating. This orientational grating is destroyed if both the electrons and holes are scattered out of those states which are coupled directly by the radiation field. Therefore, its decay measures the momentum randomization in $\vec{k}$-space and the associated orientational relaxation of the excitonic dipole moments.

3.3 Two-Pulse Self-Diffraction

DFWM with two noncollinear pulses leads to the phenomenon of self-diffraction by an optically-induced grating. In this case the first pulse sets up a macroscopic polarization of excitonic dipoles all oscillating in phase. The second (delayed) pulse probes the coherent part of the polarization left from the first pulse via interference of the two electric fields and formation of a grating. Subsequent self-diffraction of some intensity of the second pulse by this grating results in a signal with $\vec{k}_3 = 2\vec{k}_2 - \vec{k}_1$ emitted from the phased oscillator array. The decrease of this signal with increasing delay between the two pulses can be used to measure the excitonic phase coherence time T_2.

In the small signal region the two-pulse DFWM experiment is equivalent to the commonly employed photon echo experiment. One should note, however, that the DFWM signal of a homogeneously broadened line leaves the sample together with the second pulse ($T_{signal} = T_{12}$), in contrast to the photon echo of an inhomogeneously broadened line for which the delay with respect to pulse 1 is two times the delay between pulses 1 and 2 ($T_{signal} = 2T_{12}$).

For pulses much shorter than the dephasing time the intensity of the signal decay is known to vary exponentially as a function of the delay T_{12} between the two pulses as

$$I(T_{12}) \propto exp(-T_{12}/T_D) \tag{6}$$

with $T_D = T_2/2$ and $T_D = T_2/4$ for a homogeneously and inhomogeneously broadened transition, respectively.

Difficulties in the evaluation of the experimental data arise if the dephasing occuring during excitation can no longer be neglected. In this case, one has to solve the optical Bloch equations of a two-level system taking into account the finite pulse duration [18]. Fortunately, we are interested only in the small density region where bleaching effects (Bloch saturation) are not important. In this limit the Bloch equations can be iteratively solved for an arbitrary pulse shape. The third order density matrix theory of Yajima and Taira [18] yields the following expression for the polarization in the $2\vec{k}_2 - \vec{k}_1$ direction produced by the pulses $E_1(t)$ and $E_2(t - T_{12})$ with corresponding wave vectors $\vec{k}_1$ and $\vec{k}_2$ and frequency ω:

$$
\begin{aligned}
P(t, T_{12}) \quad \propto \quad & \int_0^\infty g(\omega_0) \int_{-\infty}^t \int_{-\infty}^{t'} \int_{-\infty}^{t''} dt''' dt'' dt' d\omega_0 \\
& \cdot \{ E_2(t' - T_{12}) E_2(t'' - T_{12}) E_1^*(t''') \exp(i(\omega_0 - \omega)(t' + t'' - t''' - t)) \\
& + E_2(t' - T_{12}) E_1^*(t'') E_2(t''' - T_{12}) \exp(i(\omega_0 - \omega)(t' - t'' + t''' - t)) \} \\
& \cdot \exp[(1/T_2 - 1/T_1)(t' - t'') + (t''' - t)/T_2],
\end{aligned}
\tag{7}
$$

where $g(\omega_o)$ is the exciton distribution excited by the pulses. All one has to know is the <u>electric field</u> of the pulse and the shape and width of a possible inhomogeneous distribution of the exciton eigenfrequencies. Since the electric field amplitude of the pulses cannot be measured directly we assume a Gaussian pulse shape with a FWHM determined from a pulse autocorrelation measurement. As the power spectrum of our nearly transform limited pulses is broader than the exciton absorption lines of our samples, the exciton distribution $g(\omega_o)$ is determined by the inhomogeneous linewidth of the excitonic transition.

Integration of Eq.7 for various values of T_1 and T_2 yields

$$I(T_{12}) = \propto \int_{-\infty}^{+\infty} |P(t, T_{12})|^2 dt \tag{8}$$

the diffracted intensity as a function of the delay between the two pulses, which can be compared to experimental traces. For a detailed discussion of the dynamics of the nonlinear signal, in particular for finite pulse lengths, see Ref. 26.

4 Experimental

The experimental arrangement is shown in Fig. 4. The pulses for excitation
of the transient gratings and probing of the subsequent decay are provided by
a standard synchronously pumped, tunable dye laser with Styryl 9 operating
at a repetition rate of 76.55 MHz. The pulse autocorrelation width is 3.7 ps,
the width of the power spectrum 0.9 meV. The beams are focused to a 250 μm
spot on the sample and the total average power is typically in the range of
0.1-5 mW. The second dye laser provides a highly synchronized (width of
the cross-correlation between the two laser outputs 6.4 ps) independently
tunable pulse which is employed for the injection of excitons or free electrons
and holes into the excitation volume during the pump-probe experiments
described in Sec. 8.

We use for our experiments high purity $GaAs$ layers with a thickness L_z of
194 nm (3D-sample) and 12 nm (2D-sample) cladded by $Al_{0.3}Ga_{0.7}As$ and
grown on n^+-substrates. During the experiments the samples are immersed
in superfluid He. For the conventional DFWM in a forward diffraction geometry, the substrates were polished down to wedges of about 30 μm average
thickness. The residual substrate layer is transparent below 1.540 eV, owing
to the Burstein-Moss shift and has proven to be sufficiently thick to avoid
strain. The increased confinement energy of the exciton in the 12 nm QW
requires a considerably thinner substrate thickness or even a complete remo-
val of the substrate in order to get rid of the disturbing absorption at the
transition frequency of the exciton in the QW.

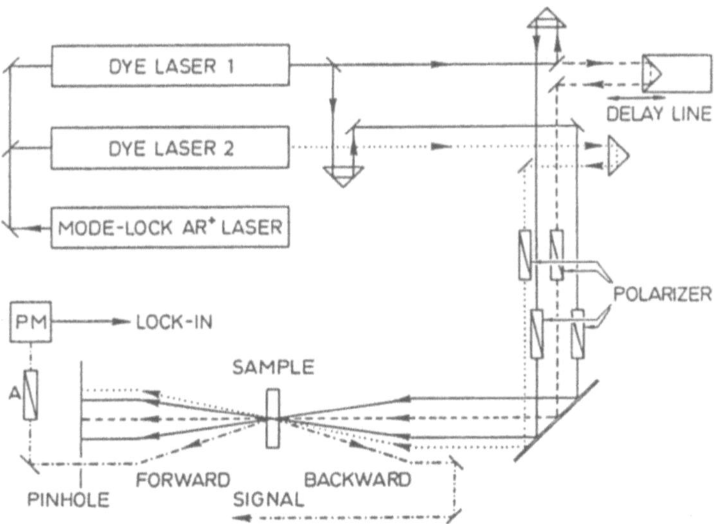

Fig. 4 Experimental set-up (A = polarization analyzer, PM =
photomultiplier).

Therefore, all time-resolved DFWM experiments on the 12 nm QW sample are performed in a new backward (reflection) geometry [27] where all optical pulses are incident from one side of the sample and the nonlinear signal is detected in a backward direction. This technique permits to perform the measurements on the original sample without the need of polishing and etching of the substrate which very often result in the generation of mechanical stress and a modification of the optical properties. The generation of the backwardly diffracted signal is due to the partial breakdown of momentum conservation in thin layers: only the wavevector components parallel to the layer have to be conserved ($k_{n\parallel} = 2k_{2\parallel} - k_{1\parallel}$). The phase matching geometry is depicted in Fig. 5a for the case of a two-pulse self-diffraction experiment. This figure illustrates that DFWM results in the excitation of two diffracted signals, both conserving the component of the polarization wavevector parallel to the layer but differing in the component perpendicular to the layer and the resulting phase mismatch between signal and polarization. The well-known signal in forward direction is close to the direction of the nonlinear polarization and has only a small phase mismatch. The backward signal propagates with the opposite direction for the wavevector component perpendicular to the layer and, consequently, exhibits a considerably larger phase mismatch. For small angles ϕ_1 and ϕ_2 of the exciting beams, the phase mismatches of both signals in the direction perpendicular to the layer have the value

$$forward: \quad \Delta k_3 = k(\phi_1 + \phi_2)^2 \tag{9}$$

$$backward: \quad \Delta k_4 = k(2 - 3\phi_2^2 - 2\phi_1\phi_2) \quad . \tag{10}$$

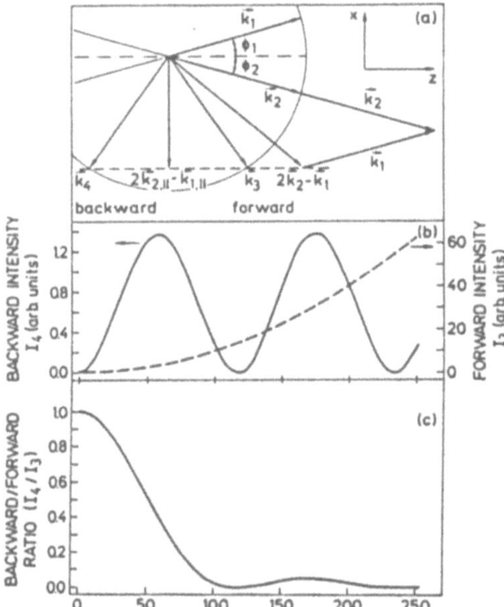

Fig. 5 (a) Geometry of the exciting beams and generated signals for DFWM. Calculated intensities in forward (dashed line) and backward (solid line) direction (b) and intensity ratio of the two signals versus L_z (c).

166

Normally, it is expected that large phase mismatch between signal and polarization in nonlinear spectroscopy drastically reduces the signal intensity. The dependence of the nonlinear signal intensity I_n on the phase mismatch Δk_n in the z-direction between signal and nonlinear polarization is described by the following equation [28]:

$$I_n \propto L_z^2 sin^2(\Delta k_n L_z/2)/(\Delta k_n L_z/2)^2 \qquad (11)$$

with n=3 and 4 for the forward and backward direction, respectively. In thin layers ($L_z \leq 15nm$) the reduction of the backward signal intensity is negligible in spite of the large phase mismatch since the interaction length is extremely small. Figure 5c presents the ratio of the two signals in backward and forward direction which proves that for thin quantum wells ($L_z \leq 50nm$), the intensity diffracted in backward direction is almost the same as in forward direction. The detected backward signal provides the same information as the forward signal and allows the study of excitations in thin layers without being influenced by the optical properties of absorbing substrates.

5 Characterization in the Frequency Domain

Sample characterization is accomplished by conventional reflectance, transmission, photoluminescence (PL) and photoluminescence excitation (PLE) spectroscopy. Figure 6 depicts typical transmission and photoluminescence spectra of the 194 nm thick layer. The transmission exhibits a strong absorption peak at 1.51508 eV, accompanied by considerably weaker peaks at 1.5158 eV and 1.5165 eV. These absorption lines can be attributed to the ex-

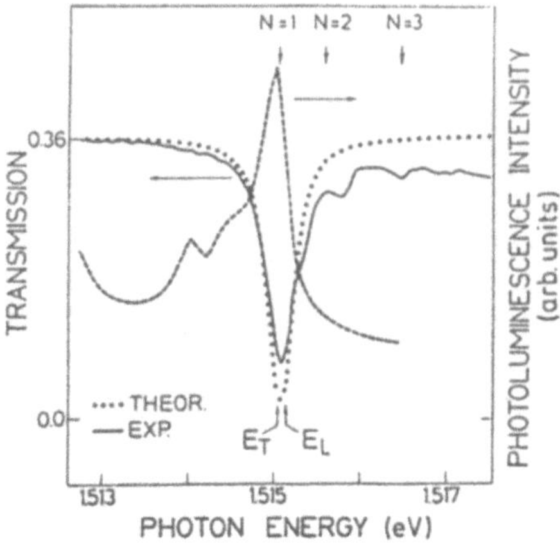

Fig. 6   Experimental (solid line) and theoretical (dotted line) transmission and photoluminescence (dashed line) spectra of the 194nm $GaAs$ layer near the 1s exciton resonance at 2K.

citon transitions originating from the electronic subbands with index N=1, 2 and 3, respectively[14]. For an electron mass of 0.067 m_o, a heavy-hole mass of 0.45 m_o and a slightly increased (by 0.17 meV) binding energy as compared to the bulk binding energy the heavy-hole exciton transition energies E_N are calculated within the model of a QW with infinitely high barriers

$$E_N = 1.5149eV + 0.17meV \cdot N^2. \tag{12}$$

The excellent agreement between the theoretical values indicated by the arrows in Fig. 6 and the experimentally observed absorption peaks confirms that the excitonic states can be appropriately described by the quantized (single particle) electron and hole states.

The dashed line represents a theoretical transmission spectrum calculated within a simple one-layer dielectric oscillator model using a longitudinal-transverse splitting $E_L - E_T = 0.08$ meV [29] and exciton-free surface layers with an effective thickness $L_{eff} = 140$ nm. This lineshape analysis yields the excitonic eigenenergy $E_T = 1.51502$ eV and a homogeneous linewidth of $\Gamma_h = 0.2$ meV. As the PL-spectrum exhibits no shift with respect to the eigenenergy of the exciton determined from the transmission spectrum, the exciton of our sample can be well described within the model of a two-level system.

The PL-spectrum of the 12 nm QW sample shows two peaks with a linewidth of 0.59 ± 0.02 meV at 1.53932 and 1.53818 eV. Transmission, PL, and PLE-experiments reveal no Stokes shift between the high energy luminescence and the absorption line of the excitons and a linear dependence of the photoluminescence intensity on the excitation intensity over three orders of magnitude. Thus, we conclude that the studied excitons are free excitons and their lifetime is not shortened remarkably by impurity-related recombination. The low energy luminescence line is due to excitons emitting from regions of the QW one monolayer of $GaAs$ thicker than the major part of the sample. This assignment is substantiated by the linear dependence of the luminescence intensity on the excitation density. From the analysis of reflection and PLE-spectra we can estimate that the relative area of these thicker QW islands amounts to less than 10% of the focal spot size.

6 Relaxation Times for Low-Excitation

Figures 7 and 8 show typical diffraction curves for the 3 different DFWM-configurations schematically depicted as insets in Fig. 8 for the 194 nm and the 12 nm layer, respectively. The excitation density is kept below $2 \cdot 10^{14} cm^{-3}$ and $2 \cdot 10^9 cm^{-2}$ for the 3D and 2D sample, respectively. The observed relaxation times do not depend on the excitation intensity at such low levels and reflect the intrinsic exciton dynamics of our samples. The initial decay of the population grating (Fig. 8(left), upper curve) reflects the rapid randomization of the orientation of the excitonic dipoles as well as of the momentum parallel to the layer which transform the initially coherent grating into an incoherent density grating. The slow decay of the latter is determined by the exciton lifetime.

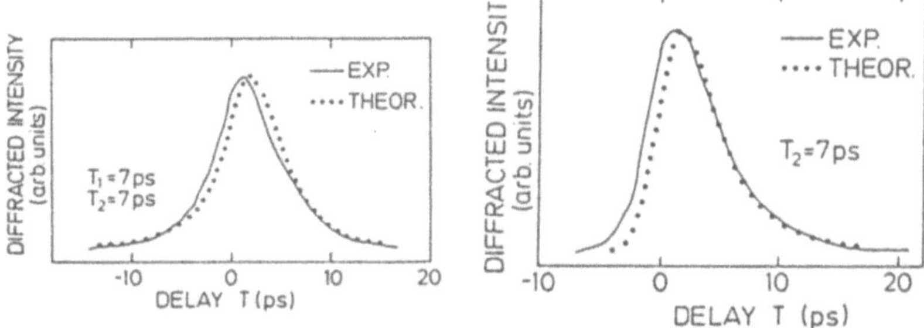

Fig. 7 Experimental diffraction curves (solid lines) for a transient orientational grating (left) and a two-pulse self-diffraction experiment (right) on the exciton transition of the 194nm $GaAs$ layer. The dotted lines are best fits calculated from the optical Bloch equations.

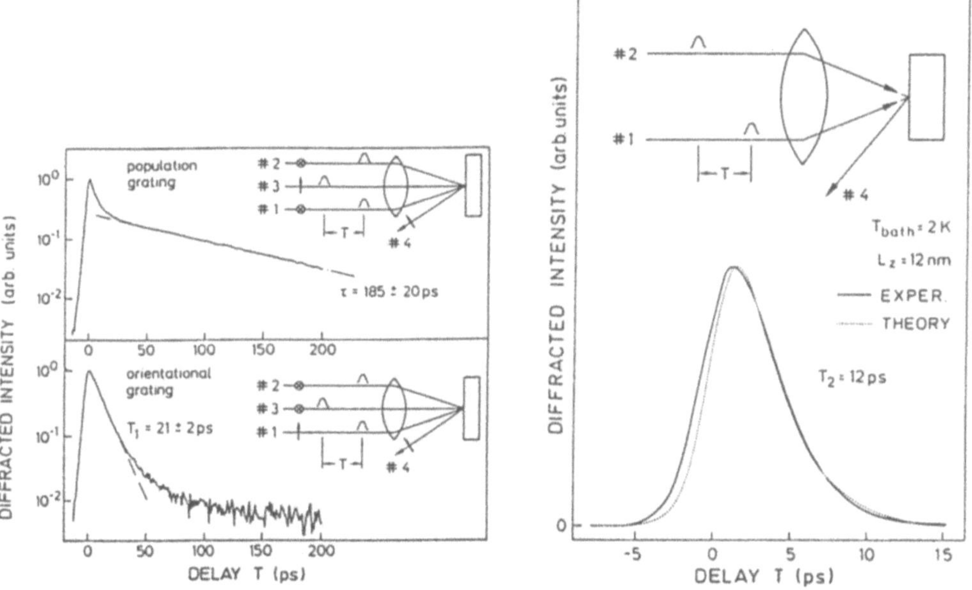

Fig. 8 Diffracted intensity versus delay for different time-resolved DFWM experiments on the 1s heavy-hole exciton transition of the 12nm single QW at 2K. (Left) upper curve: population grating, lower curve orientational grating. (Right) Experimental (solid line) and theoretical (dashed line) diffraction signal for an optical dephasing experiment.

169

The curves in the left part of Fig. 7 and 8 (lower curve) display the diffracted signal of a pure orientational grating. Two mechanisms contribute to the diffraction, in this case: 1) For positive delays, pulse 3 is diffracted from the orientational grating formed by the two orthogonally polarized pulses 1 and 2. The decay of the signal is governed by the orientational relaxation time T_1. 2) The rise of the signal is determined by diffraction of pulse 2 off a grating set up by early pulse 3 and later pulse 1 if their delay is comparable or smaller than the dephasing time T_2. Thus, T_2 determines the slope of the leading edge of the signal.

The solid lines of Fig. 7 (right) and 8 (right) present diffraction curves for two-pulse DFWM. The line shape analysis of these signals yields the phase coherence time T_2. The dotted and dashed lines depict numerical solutions of the optical Bloch equations providing the best fit to the experimental traces. Analysis of these signals under the assumption of homogeneous broadening would yield $T_2 = 7$ ps (corresponding to a homogeneous linewidth $\Gamma_h = 0.18$ meV) and $T_2 = 6$ps ($\Gamma_h = 0.22$ meV) for the 194 nm and 12 nm sample, respectively. Comparison of these values with the spectral widths observed in the frequency domain reveals the high quality of the 194 nm layer but substantial inhomogeneous broadening of the excitonic transition for the 12 nm sample. Model calculations using the 3rd order density matrix theory [18] predict a decay of the DFWM signal with time delay T between the two pulses as $I(T) \sim exp(-4T/T_2)$ if the inhomogeneous width $\Gamma_{inh} \geq \Gamma_h$. Taking into account the inhomogeneous broadening [30] the best fit of the optical Bloch equations to the experimental diffraction curve in Fig.8 (right) results in $T_2 = 12 \pm 1ps$ corresponding to $\Gamma_h = 0.11 \pm 0.01$ meV. Deconvolution of Γ_h from the spectral line profile yields $\Gamma_{inh} = 0.55 \pm 0.02$ meV for the 12nm QW sample.

The experimentally determined relaxation times are summarized in Table 1. The decrease of τ with decreasing layer thickness can be explained by the increased overlap of the electron and hole wavefunctions in the thinner layer resulting in an increased oscillator strength. The lifetime data are distinctly smaller than values measured on similar samples by time-resolved luminescence spectroscopy [7,8,9]. The discrepancy may be either explained by the dependence of τ on the sample quality or by the influence of exciton/exciton scattering at the higher excitation densities usually employed in luminescence experiments. Exciton/exciton scattering leads to a reduction of the dephasing time which implies an enhancement of the radiative lifetime according to the predictions of Ref. 9.

Table 1   Lifetime τ, orientational relaxation time T_1 and dephasing time T_2 for the two samples. The inhomogeneous linewidth Γ_{inh} has been calculated by deconvolution of the homogeneous linewidth $\Gamma_h = 1/\pi \cdot T_2$ from the linewidth Γ measured in the frequency domain.

L_z[nm]	τ[ps]	T_1[ps]	T_2[ps]	Γ_h[meV]	Γ[meV]	Γ_{inh}[meV]
194	500 ± 50	7 ± 0.5	7 ± 0.5	0.18 ± 0.02	0.20 ± 0.02	≤ 0.02
12	185 ± 20	21 ± 2	12 ± 1	0.11 ± 0.01	0.59 ± 0.02	0.55 ± 0.02

The increase of T_1 and T_2 with decreasing L_z can be attributed to the reduced interaction of the excitons with their environment owing to the restriction of the exciton mobility for the direction perpendicular to the layer. This explanation is substantiated by the previously observed increase of T_2 with increasing localization [31].

The experimentally observed rapid loss of the phase coherence within an excitonic ensemble raises the question about the nature of the phase-destroying events. In principle, two different groups of dephasing processes can be classified:

1. Interactions of excitons with the imperfections of the sample and

2. Collisions of excitons with other quasi-particles like optical or acoustic phonons, excitons and free electrons and holes.

The first group includes scattering of excitons by impurities, the roughness of the $GaAs/Al_xGa_{1-x}As$ interfaces or by fluctuations of the Al concentration within the $Al_xGa_{1-x}As$ barrier material. The latter contribution is expected to increase with decreasing well width because of the increasing extension of the exciton wavefunction into the barriers. The contributions originating from the crystal defects may vary considerably from sample to sample, depending on its quality. A well-defined quantitative variation of the magnitude and distribution of these crystal imperfections is still well beyond the potentials of MBE technology. Thus, a systematic study of the dependence of the dephasing rate on deviations of the crystal structure from the ideal order is presently impossible.

The interaction of excitons with optical phonons results in a rapid ionization of the excitons within 300 fs at room temperature [32]. This time scale is far beyond the time-resolution of our present experimental set-up. On the other hand, the modification of T_2 by the presence of acoustic phonons and free carriers or excitons can be easily explored on one and the same sample by temperature and excitation density dependent dephasing measurements, respectively.

7 Interaction of Excitons With Acoustic Phonons

Exciton/acoustic-phonon interaction can be readily studied by measuring the temperature dependence of T_2 [16,33]. Figure 9 displays diffraction curves measured directly in the time domain by two-pulse DFWM and reveals a strong decrease of T_2 for the 194 nm layer with increasing temperature. Figure 10 depicts plots of Γ_h-values versus temperature for the 3D and 2D sample. The low temperature variation of Γ_h versus sample temperature T_S can be fairly well described by a linear function

$$\Gamma_h(T_S) = \Gamma(0) + \gamma_{ph} \cdot T_S. \tag{13}$$

This linear temperature dependence is attributed to anti-Stokes scattering with acoustic phonons. The temperature independent term $\Gamma(0)$ summarizes

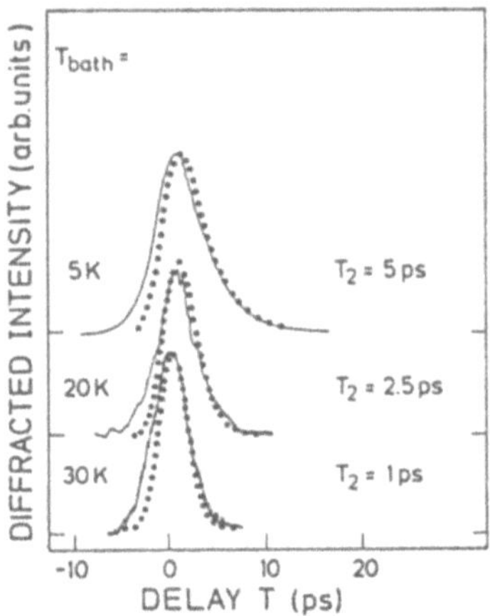

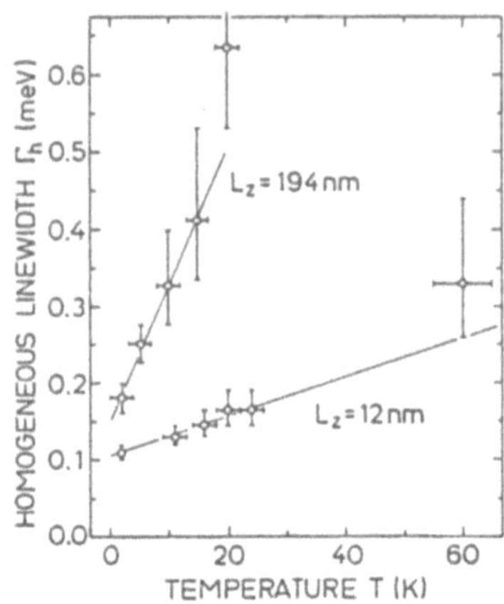

Fig. 9 Experimental (solid line) and theoretical (dotted line) dephasing curves for the exciton transition of the 194nm *GaAs* layer at 3 different temperatures.

Fig. 10 Homogeneous linewidth corresponding to the measured dephasing time versus temperature for the 194 nm *GaAs* layer and the 12nm single QW.

Table 2   Homogeneous linewidth and temperature coefficient for the line broadening.

L_z(nm)	194	27.7	13.5	12
$\gamma_{ph}(\mu$eV/K)	17 ± 1	9 ± 1	5 ± 1	2.5 ± 0.5
$\Gamma(0)$(meV)	0.18 ± 0.02	0.40 ± 0.02	0.48 ± 0.02	0.11 ± 0.01

the effects due to collisions with impurities, interface roughness and *AlAs* Mol fraction fluctuations in the barriers. Deviations from the linear temperature dependence observed above 15 K and 30 K for the 3D and 2D sample, respectively, may be explained by the onset of exciton ionization at these temperatures and a concommitant, highly effective dephasing of excitons originating from scattering with the free electrons and holes.

The parameters $\Gamma(0)$ and γ_{ph} determined from the linear fit are presented in Table 2. Evidently the dephasing efficiency of exciton/acoustic-phonon scattering in the 12 nm QW is distinctly smaller than in the 194 nm thick layer. This trend is substantiated by the additional results measured on 13.5 nm and 27.7 nm thick QWs [16]. From these results we have to conclude that the exciton/acoustic-phonon coupling is reduced with decreasing well width in contrast to recent theoretical predictions [34]. The origin of this discrepancy between experiment and theory is not yet clear. Possibly it can be explained by the lack of detailed knowledge about the phonon wavefunctions in a 2D system as well as the mass of the 2D excitons.

172

8 Enhanced Phase Relaxation of Excitons by Exciton/Exciton and Exciton/Electron-Hole Collisions

The faster loss of phase coherence for a low density coherent ensemble of excitons subjected to collisions with free carriers or incoherent excitons can be analyzed in a pump-probe type experiment which ultilizes the two-pulse self-diffraction as a probe to study the exciton/exciton and exciton/free-carrier scattering [15].

For this purpose free carriers or heavy-hole excitons are additionally created within the excitation volume by a third stronger optical pulse from the second dye laser the wavelength of which is tuned either to the heavy-hole exciton, or to the band-band transition. The heavy-hole excitons are created 20 ps before pulse #1 of the self-diffraction experiment arrives in the sample in order to ensure that all excitons created by pulse #3 have lost their phase coherence and act only as a background of scattering particles. The free carriers are created in temporal overlap with pulse #1 to prevent the condensation of these free carriers to excitons. In both configurations the phase coherence time T_2 is determined in dependence on the density either of the additionally created incoherent heavy-hole excitons or the free carriers.

Diffraction curves for the 12 nm QW are depicted as solid lines in Fig. 11. Figure 11(b) shows the curve recorded without pumping by pulse #3, which

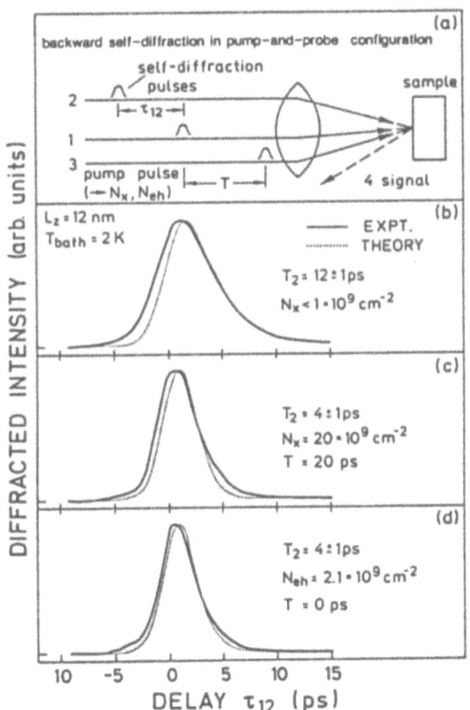

Fig. 11 Experimental diffraction curves (solid lines) and best fit calculated from the optical Bloch equations (dotted lines) for the 12nm *GaAs* single QW at 2K for low excitation (b) and additional pumping of either 1s heavy-hole excitons (c) or free carriers (d). (a) depicts the experimental arrangement.

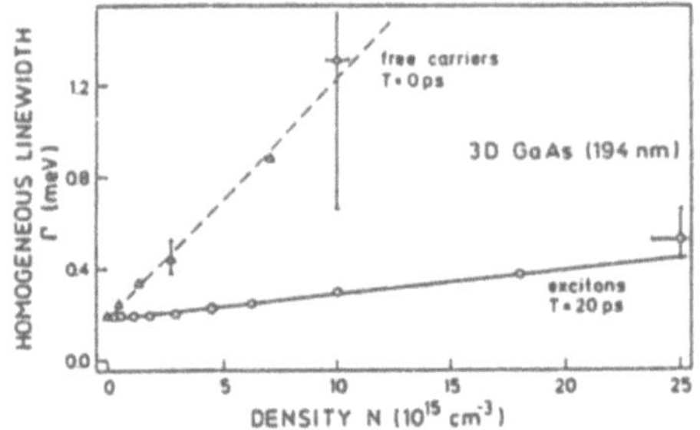

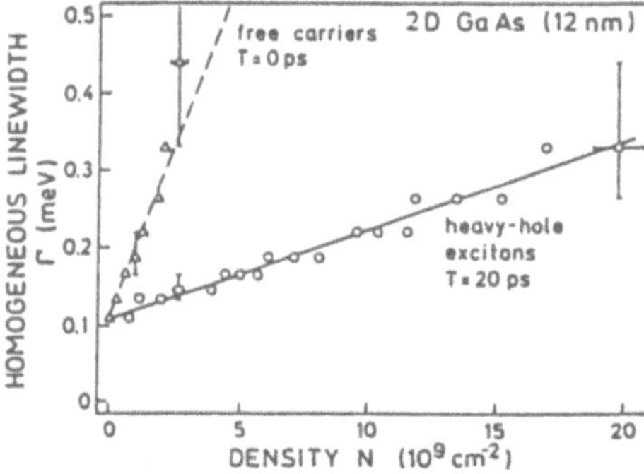

Fig. 12 Homogeneous linewidth corresponding to the measured dephasing time as a function of density of additionally created excitons and free carriers for the 194 nm $GaAs$ layer (upper part) and the 12nm single QW (lower part).

yields the low density limit of the phase coherence time T_2. Excitation of heavy-hole excitons at a density of $N_x = 2 \cdot 10^{10} cm^{-2}$ changes the diffraction curve remarkably, as demonstrated in Fig.11(c) and reduces the phase coherence time to $T_2 = 4 \pm 1 ps$. Similarly, the presence of free carriers at a density of $N_{eh} = 2.1 \cdot 10^9 cm^{-2}$ yields $T_2 = 4 \pm 1 ps$ (Fig.11(d)).

The homogeneous linewidths corresponding to the experimentally determined phase coherence times T_2 are depicted in Fig.12 for both samples, as a function of the density of additionally created scatterers. In case of the 12nm QW the homogeneous linewidth is influenced by collisions of the heavy-hole excitons with free carriers at densities as small as $N_{eh} = 2 \cdot 10^8 cm^{-2}$ and with incoherent excitons around $N_x = 2 \cdot 10^9 cm^{-2}$. For the 3D sample T_2 starts to drop for exciton densities of $N_x = 2 \cdot 10^{14} cm^{-3}$ and free carrier densities as small as $N_{eh} = 2 \cdot 10^{13} cm^{-3}$. Thus, free carriers destroy the excitonic phase coherence by a factor of about 10 more efficiently than excitons for both 2D as well as 3D systems. For both samples and both types of scatterers, the broadening of the homogeneous linewidth can be well described by a linear density dependence as expected in the low density region [35]. The function

$$\Gamma_h(N) = \Gamma_h(0) + \gamma \cdot a_B^n \cdot E_B \cdot N \qquad (14)$$

with $n = 2$ and $n = 3$ for the 2D and 3D exciton, respectively, is fitted to the experimental data and depicted in Fig.12 as straight lines. In this relation Γ_h is the homogeneous linewidth, N the density of the scatterers, a_B the exciton Bohr radius, and E_B the exciton binding energy.

174

Table 3 Broadening parameters for the homogeneous linewidth for exciton/exciton γ_{xx} and exciton/free-carrier scattering $\gamma_{xe,h}$

L_z[nm]	a_B[nm]	E_B[meV]	$\Gamma(0)$[meV]	γ_{xx}	$\gamma_{xe,h}$
194	13.5	4.2	0.18 ± 0.02	10 ± 1	100 ± 5
12	9.5	8.5	0.11 ± 0.01	1.5 ± 0.3	11.5 ± 2.0

γ is a dimensionless parameter for the line broadening and gives a measure of the interaction strength of the excitons with either free electrons and holes or other excitons. The best fit to the experimental data points using the a_B and E_B values given in the first and second column of Table 3 yields the line broadening parameters γ_{xx} and $\gamma_{xe,h}$ summarized in the last two columns of that table. For the 12 nm QW, E_B is determined to be 8.5 meV by excitation spectroscopy in good agreement with published values [36]. The corresponding value of $a_B = 9.5$ nm is taken from the variational calculations of Shinozuka and Matsuura [37].

The higher dephasing efficiency of exciton/free-carrier collisions is explained by the long-range Coulomb interaction. The surprisingly strong exciton/exciton interaction, which is detectable at exciton densities as small as $2 \cdot 10^9 cm^{-2}$, or $2 \cdot 10^{14} cm^{-3}$ for the 2D and 3D sample, respectively, is a consequence of the composite nature of excitons. This implies, in addition to the attractive part of the exciton-exciton interaction potential, due to screening of the Coulomb interaction, a repulsive contribution arising from the Pauli exclusion principle for identical fermions, (i.e. for the electrons and holes which form the exciton) and comprising phase- space filling and exchange effects [38].

The density independent contribution $\Gamma(0)$ to the homogeneous linewidth includes all residual interactions of the excitons with acoustic phonons, impurities, the interfaces of the quantum well and Al concentration fluctuations within the $Al_x Ga_{1-x} As$ barrier material. Collisions of excitons with acoustic phonons or impurities would result in a purely homogeneous broadening of the excitonic transition. Interface roughness and alloy disorder scattering contribute to homogeneous as well as inhomogeneous broadening, depending on the lateral extension (mean diameter Λ) of the irregularities. Well-width or alloy composition fluctuations with $\Lambda \ll 2a_B$ and $\Lambda \gg 2a_B$ will cause homogeneous and inhomogeneous broadening, respectively. The substantial interface roughness on a 5nm scale, which has been detected by transport measurements [39] or sophisticated imaging and pattern recognition techniques [40], may be the origin of $\Gamma(0)$. The experimentally observed inhomogeneous linewidth of our 12nm sample $\Gamma_{inh} = 0.55 \pm 0.02$ meV can be explained, e.g. by variations of the $AlAs$-Mol-content by $\pm$ 3% in the barrier layers on a length scale larger than 20nm.

The experimentally determined collision efficiencies can be compared with theoretically calculated values. A simple scattering model calculation yields for the 194nm layer $\gamma_{xx} = 10$ and $\gamma_{xe} = 100$ [15] in reasonable agreement with the experimental data. The line broadening due to free carriers is assumed to be mainly caused by exciton/electron collisions [15]. Manzke et al.[35] calculated the broadening constant for a 2D system in the case of exciton/exciton interaction to be $\gamma_{xx} = 0.42$, which is smaller by a factor of 4 than the experimental one. The calculations of Feng and Spector [41] yield a broadening constant for exciton/free-carrier interaction of $\gamma_{xe,h} = 0.16$ at a temperature of 10 K, which is nearly two orders of magnitude smaller than the experimental value. The large discrepancies between experiment and theory may be explained by the difficulties associated with the quantitative theoretical treatment of many body effects [38].

A direct comparison of the collisional broadening of the 2D and 3D exciton transition caused by the interaction with additional free carriers or incoherent excitons becomes available if the particle distance is used as the parameter for the excitation density. The inter-particle distance r_b normalized to the respective exciton Bohr radius, can be easily calculated from the particle density [42]

$$3D : \quad r_b = (4\pi a_B^3 N/3)^{-1/3} \tag{15}$$

$$2D : \quad r_b = (\pi a_B^2 N)^{-1/2}. \tag{16}$$

Figure 13 depicts the line broadening of 2D and 3D excitons due to exciton/exciton (X-X) and exciton/free-carrier (X-eh) collisions versus r_b. It is evident that 2D excitons are much more efficient in their interactions with free carriers or other excitons as compared with their 3D counterparts at the same normalized inter-particle distance. The line broadening of the 2D exciton transition measured at an inter-particle distance of 5-10 Bohr radii amounts to about 2-4 times the value found at the same distance for the 3D exciton for both exciton/exciton as well as exciton/free-carrier collisions.

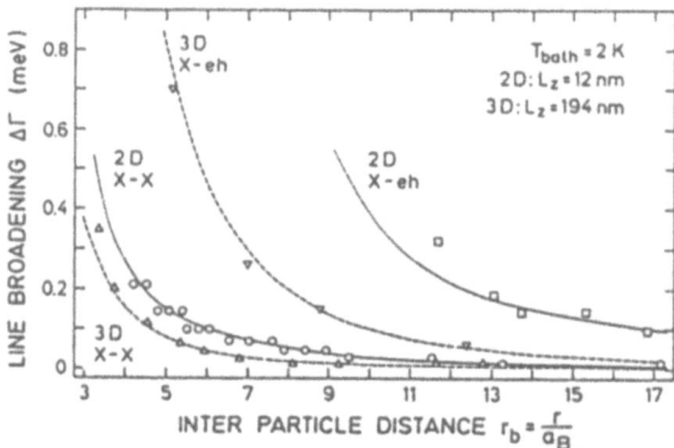

Fig. 13 Line broadening of 2D (circles and squares) and 3D-excitons (trigangles) subjected to collisions with excitons (X-X) or free carriers (X-e,h) as a function of the narmalized inter-particle distance.

This result can be understood as a consequence of the considerably weaker screening of the Coulomb interaction in 2D as compared with 3D [38]. The relative unimportance of screening in 2D systems leads to an increasing interaction of the colliding excitons via repulsive forces which originate from the Pauli exclusion principle for identical fermions [38].

In addition to the optical dephasing rate, the eigenfrequency and the oscillator strengths of excitonic transitions reveal distinct changes by exciton/exciton and exciton/free-carrier collisions, owing to screening of the Coulomb interaction as well as fermion-correlation and exchange effects. The 3D exciton exhibits a small blue shift of the eigenenergy [15] for exciton/exciton collisions and bleaching of the absorption for both collision processes [43]. In 2D *GaAs* exciton/exciton scattering leads to a distinctly stronger blue shift of the eigenenergy and reduction of the oscillator strength than in 3D-samples, whereas exciton/free-carrier interaction only causes a bleaching of the transition [44].

9 Orientational Relaxation and Optical Dephasing of Free Electrons and Holes

Free electrons and holes will lose their phase coherence on a much shorter time scale than Wannier excitons because of the efficient scattering associated with the long-range Coulomb potential of charged particles. The recent advances in short-pulse generation techniques, which have led to optical pulses as short as 6 fs [45] now provide however, the necessary time resolution for direct observations of polarization dephasing for a band to band transition in bulk *GaAs*. Becker et al.[46] measured a dephasing time as short as $T_2 = 14$ fs at 300 K for a carrier density of $n = 7 \cdot 10^{18} cm^{-3}$ and a rapid increase of T_2 with decreasing density to 44 fs at $n = 1.5 \cdot 10^{17} cm^{-3}$ according to $T_2(N) = 27.2 \cdot N^{-0.3} fs$ ($N = n/10^{18} cm^{-3}$) for the studied density region. Concerning the sublinear increase of Γ_h with N it is worth mentioning that Huang et al. [42] observed for electron densities $N_e \geq 2 \cdot 10^{10} cm^{-2}$ also a distinctly smaller broadening of the homogeneous linewidth for the exciton transition in a modulation-doped multiple QW than expected from our low-excitation experiments.

The strong density dependence of T_2 suggests that carrier/carrier interactions via the screened Coulomb potential are the dominant dephasing process. The determination of T_2 at lower densities has been prevented by the weakness of the echo signal which is explained by the much smaller 3rd order nonlinear susceptibility of the band to band transition compared to the excitonic resonance.

A remarkably longer dephasing time of $T_2 = 300$ fs at carrier densities of $6 \cdot 10^{17} cm^{-3}$ has been obtained by Oudar et al.[48] for near-band edge (excess energy ~ 20 meV) electrons and holes in *GaAs* at 15K by a hole-burning experiment with narrow-bandwidth subpicosecond pulses. The hole-burning effect results from a selective saturation of the optically coupled states. In principle, the large difference in the dephasing rates found in the two experiments could be due to the contribution of efficient alternative decay channels, like electron (hole)-optical phonon interaction or intervalley scattering for carriers with an initial excess energy of roughly 0.5 eV in the experiment of Becker et al. [46]. This explanation seems to be very unlikely, however, since carrier/carrier scattering has proven to be the dominant mechanism in this experiment. The initial energy spread of the photoexcited carrier distributions in the two experiments is extremely different and amounts to only a few meV for the near band edge populations of Ref.[48]. In contrast, the 6 fs pulses used in [46] have an energy width of 220 meV. Moreover, in the latter experiment electron wave packets with different excess energy are generated in the conduction band since the 2 eV photons permit simultaneous excitation of electrons from the split-off band and both the heavy and light-hole band. Thus, a strong variation of the dephasing efficiency of carrier/carrier collision with both the energy as well as the energy difference of the colliding particles may possibly explain the ultrarapid dephasing reported by Becker et al.[46]. In particular, the increased density of states at higher energies, i.e. the larger number of available (free) states for scattering events, would provide a reasonable explanation for the increase of the scattering rate with energy.

The spectral hole-burning technique has been also applied to dephasing studies of band/band transitions in $GaAs/Al_xGa_{1-x}As$ multiple quantum well structures (well width = 11.4 nm) [49]. For a near-band edge population the dephasing time is $T_2 = 150$ fs. Comparison of the T_2-values for QW and bulk samples indicates a significant effect of the reduced dimensionality on the dephasing efficiency of carrier/carrier collisions. As mentioned above, this result can be attributed to the reduced screening efficiency of electron/hole plasmas in 2D systems.

The orientational relaxation time of photoexcited electrons and holes has been studied by measuring the anisotropy of optical absorption induced by absorption saturation with an intense linearly polarized fs-pulse in a polarization sensitive pump-probe configuration [50]. The value $T_1 = 190$ fs observed for a near-band edge population in bulk $GaAs$ reveals a distinctly faster randomization of the initially anisotropic momentum distribution in $\vec{k}$-space compared to our experiments with Wannier excitons.

It should be noted that all data for T_2 and T_1 reported so far for free carriers in $GaAs$ concern the medium and high excitation region where the relaxation times may be considerably shortened by carrier/carrier collisions. The determination of the intrinsic (low density limit) values of T_2 and T_1 will be the subject of future research and requires an improved sensitivity of DFWM-experiments.

10 Conclusion

In this paper we have summarized the present stage of knowledge on optical dephasing and orientational relaxation of Wannier excitons and free carriers in the $GaAs/Al_xGa_{1-x}As$ system. Time-resolved DFWM experiments with picosecond optical pulses in various two and three-beam configurations are shown to be a powerful technique which provides fundamental understanding of the initial relaxation steps of photoexcited excitons or free electrons and holes. The influence of exciton/exciton, exciton/free-carrier and exciton/acoustic-phonon scattering on the relaxation dynamics of excitons is discussed. The rapid increase of the exciton dephasing rate by exciton/exciton collisions at exciton densities as low as $2 \cdot 10^{14} cm^{-3}$ and $2 \cdot 10^9 cm^{-2}$ for the 3D and 2D excitonic system, respectively, can be attributed to the composite nature of the exciton, which is a bound state of two fermions. Therefore, fermion exchange and phase-space filling effects contribute to the interaction potential in addition to screening of the Coulomb potential.

In this article we have particularly emphasized the influence of exciton confinement on the relaxation phenomena by comparing experimental results measured on a 194nm $GaAs$ layer with data for a 12nm QW. The dephasing as well as the orientational relaxation rate decreases for the transition from 3D to 2D. The dephasing efficiency of exciton/exciton, as well as exciton/free-carrier collisions, is higher in the 2D than in the 3D sample, as explained by the reduced importance of screening in 2D systems. In contrary, the dephasing efficiency of exciton/acoustic-phonon scattering is larger in the 194nm layer.

It will be a goal of future research to achieve a more comprehensive understanding of the influence of increasing confinement on the exciton dynamics by extending these studies to excitons in quantum wires and quantum dots.

References

[1]   *D.S. Chemla, S. Schmitt-Rink, and D.A.B. Miller* in "Nonlinear Optical Properties of Semiconductors", Ed. H. Haug (Academic, New York, 1987).

[2]   *S. Schmitt-Rink, D.S. Chemla, and D.A.B. Miller* to be published in Advances in Physics 1989.

[3]   *D.A.B. Miller, J.S. Weiner, and D.S. Chemla*, IEEE J. Quantum Electronics **QE-22**, 1816 (1986).

[4]   *N. Peyghambarian and H.M. Gibbs*, J. Opt. Soc. Am. **B2**, 1215 (1985).

[5]   *S. Tarucha and H. Okamoto*, Appl.Phys.Lett. **49**, 543 (1986).

[6]   *J. Singh*, "The Dynamics of Excitons", in Solid State Physics **38**, ed. *H. Ehrenreich and D. Turnbull* (Academic, New York 1984), p. 295.

[7]   *E.O. Göbel, H. Jung, J. Kuhl, and K. Ploog*, Phys. Rev. Lett. **51**, 1588 (1983).

[8]   *R. Höger, E.O. Göbel, J. Kuhl, and K. Ploog* in Proc. of 17th Int. Conf. on the Physics of Semiconductors, San Francisco 1984, Eds. *D.J. Chadi and W. Harrison* (Springer Verlag, New York, 1985), p.575.

[9]   *J. Feldmann, G. Peter, E.O. Göbel, P. Dawson, K. Moore, C. Foxon, and R.J. Elliot*, Phys. Rev. Lett. **59**, 2337 (1987).

[10] *A. Mysyrowicz, D. Hulin, A. Antonetti, A. Migus, W.T. Masselink, and H. Morkoc*, Phys. Rev. Lett. **56**, 2748 (1986).

[11] *A. von Lehmen, D.S. Chemla, J.E. Zucker, and J.P. Heritage*, Opt. Lett. **11**, 609 (1986).

[12] *B. Flügel, N. Peyghambarian, G. Olbright, M. Lindberg, S.W. Koch, M. Joffré, D. Hulin, A. Migus, and A. Antonetti*, Phys. Rev. Lett. **59**, 2588 (1987);
M. Joffré, D. Hulin, A. Migus, A. Antonetti, C. Benoit a la Guillaume, N. Peyghambarian, M. Lindberg, and S.W. Koch, Opt.Lett. **13**, 276 (1988).

[13]   E. Hanamura, Phys.Rev.B. **38**, 1228 (1988).

[14]   L. Schultheis, J. Kuhl, A. Honold, and C.W. Tu, Phys. Rev. Lett. **57**, 1797 (1986).

[15]   L. Schultheis, J. Kuhl, A. Honold, and C.W. Tu, Phys. Rev. Lett. **57**, 1635 (1986).

[16]   L. Schultheis, A. Honold, J. Kuhl, K. Köhler, and C.W. Tu, Phys. Rev. **B34**, 9027 (1986).

[17]   A. Honold, L. Schultheis, J. Kuhl, and C.W. Tu, to be published in Phys. Rev. B.

[18]   T. Yajima and Y. Taira, J.Phys.Soc.Jpn.**47**, 1620 (1979).

[19]   see e.g. J. Hegarty and M.D. Sturge, J. Opt. Soc. Am. **B2**, 1143 (1985).

[20]   I.D. Abella, N.A. Kurnit, and R.S. Hartmann, Phys. Rev. **57**, 391 (1966).

[21]   B.S. Wherrett, A.L. Smirl, and T.F. Boggess, IEEE Journ. of Quantum Electronics **QE-19**, 680 (1983).

[22]   A.L. Smirl, T.F. Boggess, B.S. Wherrett, G.P. Perryman, and A. Miller, IEEE Journ. of Quantum Electronics **QE-19**, 690 (1983).

[23]   A.L. Smirl in "Semiconductors Probed by Ultrafast Laser spectroscopy", Vol. I, ed. R. R. Alfano, (Academic Press, New York, 1984).

[24]   H.J. Eichler, P. Günther, and D.W. Pohl: "Laser-Induced Dynamical Gratings", (Springer Verlag, Berlin, Heidelberg, New York 1980).

[25]   D.A.B. Miller, D.S. Chemla, D.J. Eilenberger, P.W. Smith, A.C. Gossard, and W. Wiegmann, Appl. Phys. Lett. **42**, 925 (1983).

[26]   L. Schultheis and J. Hegarty, J. Physique, Colloque C7, Tome 46, 167 (1985).

[27]   A. Honold, L. Schultheis, J. Kuhl, and C.W. Tu, Appl. Phys. Lett. **52**, 2105 (1988).

[28]   see for example: F. Zernike and J.E. Midwinter, "Applied Nonlinear Optics" (Wiley, New York, 1973) p. 41.

[29]   R. Sooryakumar and P.E. Simmonds, Solid State Commun. **42**, 287 (1982).

[30]   The inhomogeneous broadening has been modeled by a Gaussian distribution function $g(\omega_o)$ for the exciton frequency ω_o with full-width half-maximum of 0.55 meV.

[31]   L. Schultheis, M.D. Sturge, and J. Hegarty, Appl. Phys. Lett. **47**, 995 (1985).

[32]   W.H. Knox, R.L. Fork, M.C. Downer, D.A.B. Miller, D.S. Chemla, C.V. Shank, A.C. Gossard, and W. Wiegmann, Phys. Rev. Lett. **54**, 1306 (1985).

[33]   L. Schultheis, A. Honold, J. Kuhl, K. Köhler, and C.W. Tu, Superlattices and Microstructures 2, 441 (1986).

[34]   T. Takagahara, Phys. Rev. **B31**, 6552 (1985) and Phys. Rev. **B32**, 7013 (1985).

[35]   G. Manzke, K. Henneberger, and V. May, phys. stat. sol.(b) **139**, 233 (1987).

[36]   R.C. Miller, D.A. Kleinman, W.T. Tsang, and A.C. Gossard, Phys. Rev. **B24**, 1134 (1981); U. Ehrenberg and M. Altarelli, Phys. Rev. **B35**, 7585 (1987); E.S. Koteles and J.Y. Chi, Phys. Rev. **B37**, 6332 (1988).

[37]   Y. Shinozuka and M. Matsuura, Phys. Rev. **B28**, 4878 (1983), and Phys. Rev. **B29**, 3717 (1984).

[38]   S. Schmitt-Rink, D.S. Chemla, and D.A.B. Miller, Phys. Rev. **B32**, 6601 (1985).

[39]   H. Sakaki, T. Noda, K. Hirakawa, M. Tanaka, and T. Matsusue, Appl. Phys. Lett. **51**, 1934 (1987) and R. Gottinger, A. Gold, G. Abstreiter, G. Weimann, and W. Schlapp, Europhys. Lett. **6**, 183 (1988).

[40]   A. Ourmazd, D.W. Taylor, J. Cunningham, and C.W. Tu, Phys. Rev. Lett. **62**, 933 (1989).

[41]   Y. Feng and H.N. Spector, Superlattices and Microstructures 3, 459 (1987).

[42]   G. Tränkle, H. Leier, A. Forchel, H. Haug, C. Ell, and G. Weimann, Phys. Rev. Lett. **58**, 419 (1987).

[43]   G.W. Fehrenbach, W. Schäfer, J. Treusch, and R.G. Ulbrich, Phys. Rev. Lett. **49**, 1281 (1982).

[44]   N. Peyghambarian, H.M. Gibbs, J.L. Jewell, A. Antonetti, A. Migus, D. Hulin, and
       A. Mysyrowicz, Phys. Rev. Lett. **53**, 2433 (1984) and D. Hulin, A. Mysyrowicz, A.
       Antonetti, A. Migus, W.T. Masselink, H. Morkoc, H.M. Gibbs, and N. Peyghamba-
       rian, Phys. Rev. **B33**, 4389 (1986).

[45]   R.L. Fork, C.H. Brito Cruz, P.C. Becker, and C.V. Shank, Opt. Lett. **12**, 483 (1986).

[46]   P.C. Becker, H.L. Fraquito, C.H. Brito Cruz, R.L. Fork, J.E. Cunningham, J.E.
       Henry, and C.V. Shank, Phys. Rev. Lett. **61**, 1647 (1988).

[47]   D. Huang, H.Y. Chu, Y.C. Chang, R. Houdré, and H. Morkoc, Phys. Rev. **B38**,
       1246 (1988).

[48]   J.L. Oudar, D. Hulin, A. Migus, A. Antonetti and F. Alexandre, Phys. Rev. Lett.
       55, 2074 (1985).

[49]   J.L. Oudar, J. Dubard, F. Alexandre, D. Hulin, A. Migus, and A. Antonetti,
       J. Physique, Colloque C5, Tome **48**, 511 (1987).

[50]   J.L. Oudar, A. Migus, D. Hulin, G. Grillon, J. Etchepare, and A. Antonetti, Phys.
       Rev. Lett. **53**, 384 (1984).

The Spectroscopic Evidence for the Identity of EL2 and the As_{Ga} Antisite in As-Grown GaAs

Ulrich Kaufmann

Fraunhofer-Institut für Angewandte Festkörperphysik
D-7800 Freiburg, Eckerstrasse 4, W. Germany

Summary: EL2, the dominant deep defect in undoped as-grown GaAs has attracted tremendous interest because of its peculiar optical properties and because of its technological importance for the growth of undoped semi-insulating GaAs. This paper first outlines the photoelectronic and optical properties of EL2. The second part describes the optical properties of the As_{Ga} antisite defect as inferred from magnetic resonance combined with optical techniques. A comparison of the data demonstrates that EL2 and the As_{Ga} antisite as defined by its electron-spin-resonance (ESR) behavior in undoped as-grown GaAs have the same optical properties. At present this fact is the most direct and convincing evidence that EL2 is the As_{Ga} antisite seen by ESR and related techniques. Whether this antisite is an isolated defect or a complex with an arsenic interstitial is a highly controversial question.

1 Introduction

Undoped bulk GaAs contains a dominant deep defect in concentrations near 10^{16} cm^{-3}. This defect manifested itself in various measurements already in the early sixties [1] but its ubiquitous nature and its great importance for the properties of bulk GaAs were not fully recognized until the second half of the seventies. During that period the defect in question was assessed mainly by space charge capacitance techniques and received its label, EL2, commonly used nowadays. At the beginning of this decade it had been established that EL2 is an intrinsic defect not related to impurities [2] and it was considered to be a single mid-gap donor with unusual optical properties. The latter had been inferred from photocapacitance measurements. On the basis of these data it had been suggested that the defect constituents can undergo an optically induced structural rearrangement leading to an excited metastable defect configuration at low temperatures [3, 4]. This property and the fact that EL2 plays a major role in the compensation of undoped semi-insulating (SI) GaAs [5] have made EL2 the most extensively studied deep defect in semiconductors to date.

Nothing positive was known about the chemical identity of EL2 at the beginning of this decade. The author was first confronted with this question at the "First Conference on Semi-Insulating III-V Materials" where he was asked [6]: "Can the As_{Ga} antisite defect be <u>the</u> deep donor in undoped SI GaAs?". No meaningful answer could be given at that time. Since then, however, the knowledge about the chemistry and structure of EL2 has advanced significantly. It can be safely stated today that As_{Ga} is the salient constituent of the EL2 defect. The exact microscopic nature of EL2 nevertheless remains controversial. However, among the many specific models that have been proposed, only two, namely that of an isolated As_{Ga} defect and that of an As_{Ga}-As_{int} complex are supported directly by spectroscopic evidence.

This article summarizes the spectroscopic data for EL2 and for As_{Ga} in as-grown GaAs which provide the basis for the correlation of these two species [7]. To establish this proved difficult since EL2 is a defect, originally detected and defined by electrical techniques, while As_{Ga} is a defect originally detected and defined by magnetic resonance. Fortunately, both techniques, when suitably combined with optical methods, provide also information about the optical properties of the respective species. This optical link is the basis for the identification of EL2 with As_{Ga}.

In the following the EL2 photocapacitance data will be presented first. Other optical properties of GaAs will then be shown to be EL2 induced by comparison with the photocapacitance results. Section 4 presents magnetic resonance data for the As_{Ga} antisite [8] and it will be shown that its optical properties are identical with those of EL2. Finally the magnetic-circular-dichroism absorption of As_{Ga} will be discussed and shown to be fully consistent with the optical behavior of EL2.

2 Optical Properties of EL2 from Photocapacitance Measurements

Apart from internal excitations the optical properties of a deep level defect are completely characterized by an electron ionization cross-section $\sigma_n°(h\nu)$ and a hole ionization cross-section $\sigma_p°(h\nu)$. These quantities describe the probability of electron and hole excitation into the conduction - and valence band, respectively, as a function of photon energy $h\nu$. They can be measured by a transient photocapacitance technique called "deep level optical spectroscopy" (DLOS) [9, 10].

For the mid-gap level of EL2 both cross-sections were reported in the early eighties [9, 10]. However, low temperature (80 K) data for $\sigma_{n1}°(h\nu)$ and $\sigma_{p1}°(h\nu)$ in absolute units became available only recently [11]. They are plotted in Fig. 1a. Since these cross-sections are decisive for comparison purposes in subsequent sections, some relevant features should be noted. The $\sigma_{n1}°(h\nu)$ curve rises monotonically from mid-gap, $h\nu = 0.75$ eV, up to the band edge, $h\nu = 1.5$ eV. Some structure is visible which has been attributed to the Γ, L and X minima of the conduction band [10]. In contrast the $\sigma_{p1}°(h\nu)$ curve is peaked at 0.96 eV and possibly reveals another weak peak near 1.27 eV. The energy difference is close to the spin-orbit splitting of the valence band at Γ thus suggesting that this splitting is the cause for the 1.27 eV peak. This is fully confirmed by the photoluminescence excitation data presented in section 3.2. Below $h\nu = 1.08$ eV $\sigma_{p1}°$ dominates $\sigma_{n1}°$, while above this energy the situation is reversed.

The capacitance transients due to a normal deep donor (o/+) level monitored in a photocapacitance experiment are always monotonic. The reason is obvious: The reaction going on during illumination in the space charge region of a junction at low temperatures can be written in the form

$$O \quad \underset{\sigma_{p1}°}{\overset{\sigma_{n1}°}{\rightleftharpoons}} \quad + \tag{1}$$

since carrier recombination is negligible. The rate equations for the above reaction have only monotonic solutions [12] corresponding to transients like that for $h\nu = 1.4$ eV in Fig. 2.

184

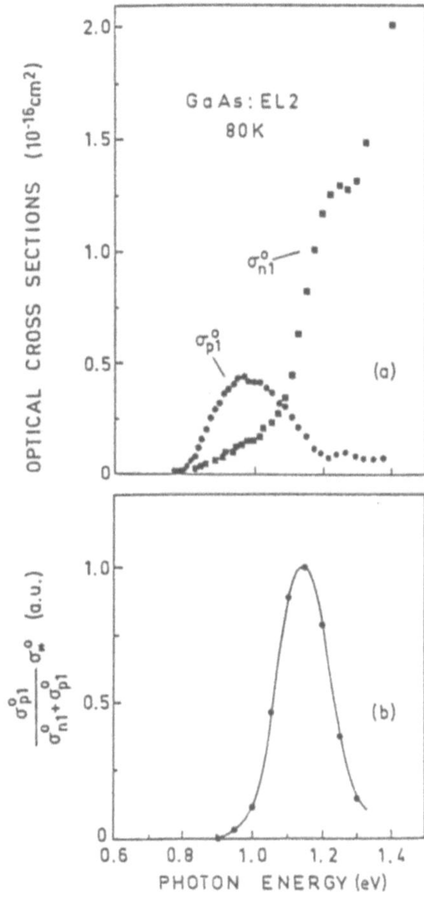

Fig. 1 (a) Spectral dependence of photoionization cross-sections for the EL2 mid-gap level [11]. (b) Spectral dependence of the EL2 persistent photocapacitance quenching effect at 80 K [14].

The unusual optical properties of EL2 were discovered when the EL2 related photocapacitance transients were studied [3, 4]. They manifest only at temperatures below 120 K and only in the photon energy range $1.0 < h\nu < 1.35$ eV. Under such conditions the transients observed are nonmonotonic like that for $h\nu = 1.15$ eV in Fig. 2. The capacitance first increases due to electron emission from EL2°, goes through a maximum but then decreases and reaches a final value close to that before illumination. Switching off the light and subsequently illuminating the diode no longer produces a significant change of its capacitance. Heating of the sample to temperatures above 130 K is required to restore its original photosensitive state. This anomalous phenomenon is called the persistent photocapacitance quenching effect.

The capacitance change ΔC in Fig. 2 is proportional to the concentration of ionized EL2, EL2⁺, since the initial conditions were set such that all centers were neutral before illumination. The fact that ΔC for $h\nu = 1.15$ eV tends to zero for $t \rightarrow \infty$ means that the ensemble of EL2 centers in the space charge region is neutral again after sufficiently long illumination. This in turn suggests that an individual EL2 center is neutral again [13]. Reaction (1) cannot account for this phenomenon unless it contains a "leak". This leak must be to the left of reaction (1) and one can phenomenologically write this in the form

$$O^* \xleftarrow{\sigma_*^{0}} O \underset{\sigma_{p1}^{0}}{\overset{\sigma_{n1}^{0}}{\rightleftharpoons}} + \qquad (2)$$

with an additional optical cross-section σ_*^{0}. The rate equations for reaction (2) have been set up and were found to have nonmonotonic solutions which can account for the nonmonotonic EL2 transients [14]. Physically, the process represented by σ_*^{0} has been interpreted as a structural defect transformation of the normal ground state configuration EL2° into an excited metastable configuration EL2* conserving the neutral charge state. As a consequence after this transformation the (o/+) level of EL2 has disappeared. The new defect configuration appears to be electrically and optically inactive in most experiments [15-20]. Recovery of the normal ground state configuration requires a thermal anneal of the sample. The annealing temperature depends on the free electron concentration in the material [21]. It is around 130 K in the depleted region of a junction or in SI bulk GaAs but decreases to about 50 K for n-conducting samples. Correspondingly the thermal barrier height decreases from 0.38 eV to about 0.06 eV [22, 23].

The spectral shape of the cross-section $\sigma_*°$ has been evaluated from the decreasing part of the nonmonotonic EL2 transients [14] and is shown in Fig. 1b. It corresponds to a relatively narrow band centered at 1.16 eV which is called the EL2 persistent quenching (or bleaching) band. Actually the cross-section thus evaluated does not exactly correspond to $\sigma_*°$. Rather it represents a product of $\sigma_*°$ and $\sigma_{p1}°/(\sigma_{n1}° + \sigma_{p1}°)$. However, inspection of Fig. 1a shows that the latter factor is of order unity and approximately constant at least near the maximum of the bleaching band. Therefore the curve in Fig. 1b is a reasonable approximation for the shape of $\sigma_*°(h\nu)$ and this is confirmed by a direct measurement of $\sigma_*°(h\nu)$ discussed in section 3.1. High resolution photocapacitance studies [24, 25] have revealed that the bleaching band is preceeded by a weak zero phonon line (ZPL) at 1.038 eV. The absolute magnitude of $\sigma_*°(h\nu)$ is about two orders of magnitude lower than $\sigma_{n1}°(h\nu)$ at $h\nu \approx 1.15$ eV as can be directly inferred from the corresponding transient in Fig. 2.

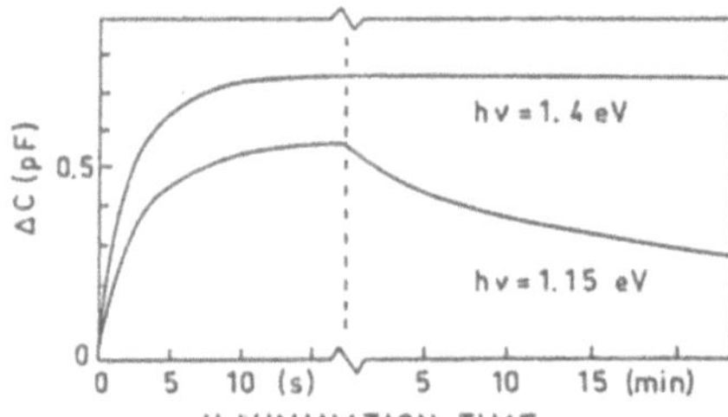

Fig. 2   EL2 photocapacitance transients of n-GaAs Schottky diodes for two photon energies $h\nu$ [4] at 80 K.

The EL2 photocapacitance transients have also been studied under uniaxial stress. A piezo-dichroism has been observed on the decreasing part of the nonmonotonic transients and the results were interpreted as evidence that the EL2 defect has trigonal symmetry [26].

Up to 1985 space charge capacitance experiments on EL2 were interpreted in terms of a single mid-gap donor level (o/+). However, during the last three years convincing evidence that EL2 is actually a double donor has been accumulated. Both thermal [27-29] and optical [27, 29, 30-32] capacitance data locate the second donor level (+/++) at E_c - 1.0 eV and indicate an EL2 level scheme as shown in Fig. 3. An identical level scheme has been previously suggested for the As_{Ga} double donor as will be discussed in section 4.2. Reliable experimental evidence for additional levels of EL2, in particular a shallow donor level, does not exist [1].

186

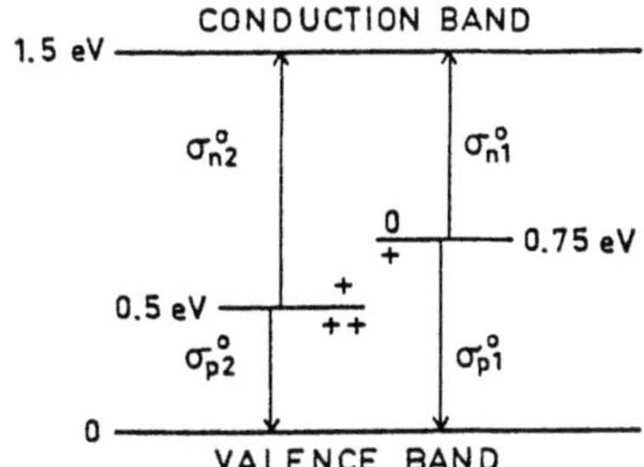

Fig. 3 The two donor levels of EL2 at low temperature as inferred from photocapacitance studies. Observed electron-(hole) transitions are marked by arrows. Note that the same level scheme is valid for the As_{Ga} double donor.

3 Further Optical Manifestations of EL2

3.1 Optical Absorption

Undoped and n-type doped GaAs exhibits a below band gap absorption band, see Fig. 4, first reported in 1962 [33]. Only twenty years later this band was recognized to be EL2 induced [34]. This assignment was based on two observations. First, the shape of the band in n-type GaAs strongly resembles that of the $\sigma_{n1}^{o}(h\nu)$ cross section. Secondly, at low temperatures, the entire band can be completely bleached by illumination with white light, an effect that is reminiscent of the EL2 photocapacitance quenching effect described in the previous section. The spectral shape of the bleaching efficiency has recently been measured using a continuous double beam technique by directly monitoring the decrease in absorption coefficient α at $h\nu = 1.4$ eV under secondary illumination [22]. The bleaching band thus measured at 70 K is identical with that shown in Fig. 1b and probably represents the most accurate and direct measurement of $\sigma_*^{o}(h\nu)$. This is convincing evidence that one and the same defect, EL2, is involved in both types of experiments.

Fig. 4 The EL2 absorption band before and after bleaching. The insert shows the EL2 zero-phonon-line and three phonon replicas [35].

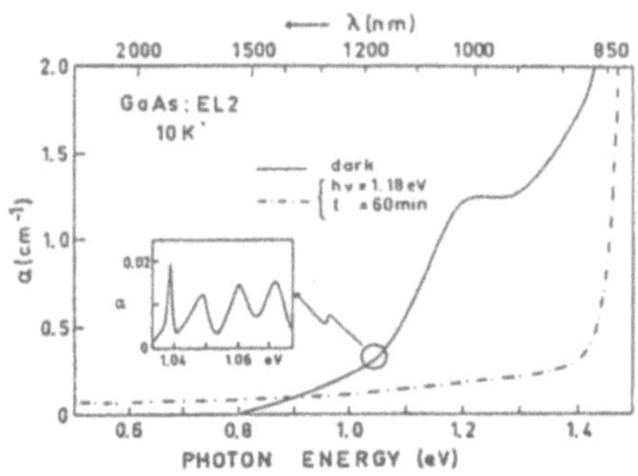

For SI GaAs, changes in the EL2 absorption band shape and intensity not only occur during longtime (~ 10 min) illumination with photon energies $h\nu_s$ within the bleaching band but also for short time (~ 1 s) secondary illumination. These changes are less dramatic than the bleaching effect. However, they are significant and are fully consistent with the optical behavior of the As_{Ga} electron-spin-resonance discussed in section 4.2. They are readily visible in absorption difference spectra where the dark spectrum has been subtracted from the spectrum taken after illumination [35, 36]. Such a difference spectrum for $h\nu_s = 1.4$ eV is shown in Fig. 5. These changes are optically reversible and are not related to the bleaching effect. Rather they reflect a change in occupation of the EL2 mid-gap level and for $h\nu_s = 1.4$ eV in particular a transfer of electrons to traps shallower than mid-gap. The shape of $\Delta\alpha$ in Fig. 5 is expected to be proportional to $\sigma_{p1}° - \sigma_{n1}°$ [36]. This difference, as inferred from the data in Fig. 1a, is plotted as full squares in Fig. 5. The agreement with $\Delta\alpha$ is excellent between 0.8 and 1.0 eV but deviations occur for higher photon energies. They possibly result from contributions of the $\sigma_{n2}°$ transition in Fig. 3.

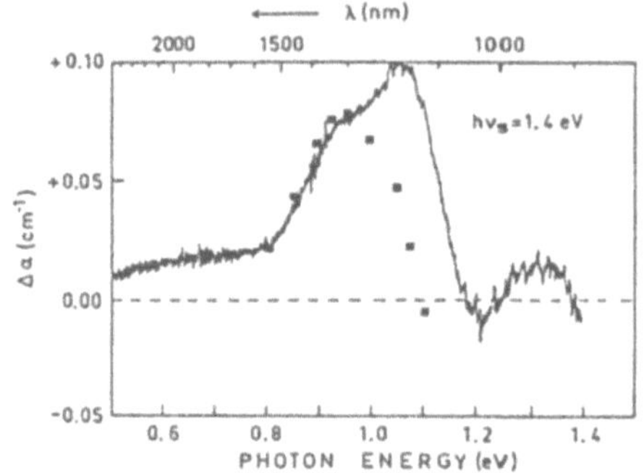

Fig. 5 EL2 absorption difference spectrum at 10 K. The dark spectrum has been subtracted from a spectrum taken after 10 s of secondary illumination with $h\nu = 1.4$ eV [35]. The positive $\Delta\alpha$ below $h\nu \approx 0.8$ eV is not EL2 related and presumably arises from free carrier absorption. The full squares represent $\sigma_{p1}° - \sigma_{n1}°$ according to Fig. 1a adjusted to fit the $\Delta\alpha$ curve at $h\nu = 0.93$ eV.

The EL2 absorption band is of great practical importance since it is currently used for routine determination of the EL2 concentration and for homogeneity assessment of undoped SI GaAs wafers [37, 38].

In 1983 it was recognized that the EL2 absorption band in n-type material, where only EL2° contributes to the absorption, is a superposition of $\sigma_{n1}°(h\nu)$ and an extra band commonly referred to as the EL2 intracenter band [39]. Its position and shape is virtually identical with the bleaching band in Fig. 1b and it is in part responsible for the 1.18 eV peak in the absorption spectrum of Fig. 4. The intracenter contribution to the total absorption at $h\nu \approx 1.17$ eV is significant. It amounts to about 30 % of the total absorption at this energy [39, 40]. This excludes the possibility that the intracenter transition corresponds directly to the normal $\rightarrow$ metastable state transition since $\sigma_*°$ is only a few percent of the $\sigma_{n1}°$ magnitude at $h\nu \approx 1.17$ eV.

The concept of an intracenter band requires the existence of an excited electronic state of the normal neutral EL2 configuration about 1.0 eV above the ground state. This excited state must be resonant with conduction band states and the transition to it is referred to as the intracenter transition. In the following its optical cross section will be denoted as σ_{IC}.
The fact that the intracenter transition manifests in the absorption spectrum [39], the bleaching band [40], and also in the photocurrent spectrum of EL2 [41, 42] suggests a branching in the excited state, i.e. that it can deexcite in three different ways: The normal decay to the ground state of normal EL2° with a probability α. Secondly a spontaneous distortion leading to the metastable configuration EL2* with a probability β. Finally autoionization into free conduction band states with a probability γ. The relative contribution of the intracenter transition to the optical absorption, to the stable $\rightarrow$ metastable state transformation, and to the photocurrent, is then given by $(\alpha + \beta + \gamma)\, \sigma_{IC}°$, $\beta\sigma_{IC}° = \sigma_*°$ and $\gamma\sigma_{IC}°$ respectively [43, 44]. The available data

188

indicate that β is about 1 % of $\alpha + \beta + \gamma$ [40]. In other words, only 1 % of the intra-center transitions lead to the formation of EL2*.

High resolution optical studies have revealed that the intracenter transition gives rise to a weak zero-phonon-line (ZPL) at 1.038 eV preceeding the broad intracenter band, see the insert in Fig. 4. This line was originally discovered in the absorption spectrum [39] but subsequent photocapacitance [24] and absorption measurements showed that it is also present in the bleaching spectrum $\sigma_*^{\,0}(h\nu)$ [40, 45]. In addition it is part of the photocurrent spectrum [41, 42]. The bleaching behavior of the ZPL is the same as that of the broad intracenter band [46] and their intensity ratios are constant in different samples [47]. Taken together these facts indicate that the ZPL and the broad intracenter band belong to the same electronic transition and provide strong support for the branching mechanism discussed in the previous paragraph.

It has been reported [48, 49] that the ZPL exhibits a complicated fine structure. Check measurements by two independent groups have <u>not</u> confirmed this result for a large number of samples [50, 51]. Therefore the splittings reported in Refs. 48 and 49 should neither be taken as evidence for a symmetry of EL2 lower than T_d nor for an EL2 family.

The EL2° ZPL has also been studied under applied uniaxial stress [52]. The observed piezo-splittings have been interpreted in terms of a $^1A_1 \rightarrow {}^1T_2$ transition in tetrahedral (T_d) symmetry. The uniqueness of this assignment has been questioned [53] and it was suggested that the symmetry of the defect involved could also be orthorhombic, C_{2v}. Subsequent stress studies [50] have ruled out this possibility and have confirmed that the ZPL arises from an $^1A_1 \rightarrow {}^1T_2$ transition. This is the most significant result concerning the symmetry of EL2° and indicates that the center in its normal configuration is a simple point defect with tetrahedral symmetry. If the symmetry were trigonal, as suggested [26] on the basis of piezo-photocapacitance data, the trigonal crystal field would split the triply orbitally degenerate $^1T_2$ state into a singlet and a doublet. Transitions to both states would be electric dipole allowed and would have comparable strengths. The magnitude of the splitting would depend on the degree of localization of the $^1T_2$ wavefunctions. For strong localization one would expect a splitting of the order 100 cm^{-1} [54]. For delocalized wavefunctions the splitting may be considerably smaller but it seems unlikely that it would be masked by the width (≈ 7 cm^{-1}) of the ZPL. Therefore the stress splitting of the EL2° ZPL favors a center with tetrahedral symmetry. An argument often cited to support a complex model for EL2 is that "metastable behavior cannot occur for an isolated defect". There is no experimental support for this claim. Rather it is based on the difficulties of visualizing metastable behavior. One possibility for metastability of the isolated As_{Ga} defect has, however, been suggested [55]. It is the Jahn-Teller driven off-center motion of As_{Ga}° in the excited $^1T_2$ state. This model has received considerable support by recent Green's functions calculations [56, 57] which are discussed in detail elsewhere in this volume.

3.2 EL2 Related Luminescence Bands

Four deep photoluminescence bands are commonly observed in undoped SI GaAs [58, 59]. They are broad and almost structureless. Three of them are peaked around 0.65 eV and they strongly overlap. This has greatly complicated their identification and their origin remained controversial for a long time. The issue has been clarified recently by photoluminescence excitation spectroscopy. One band, centered at 0.63 eV, appears to be impurity related and has been assigned to oxygen [59].

In addition to the above oxygen band there exists another band peaked at 0.63 eV and a band peaked at 0.68 eV. They have half widths of about 120 meV. Both bands can be persistently quenched by illuminating the sample with light within the EL2 bleaching band. This fact together with the spectral positions of these luminescence bands suggests that they arise from free carrier recombination with the EL2 mid-gap level. In Fig. 6 the excitation spectra [60, 61] of the 0.63 eV and 0.68 eV band are shown. These spectra reveal striking similarities with the EL2 optical cross-sections $\sigma_{n1}^{\circ}(h\nu)$ and $\sigma_{p1}^{\circ}(h\nu)$, compare Fig. 1a, and in fact can be viewed as independent measurements of the spectral shapes of the cross-sections. These data strongly indicate that the 0.63 eV and the 0.68 eV band result from electron capture at $EL2^{+}$ ($EL2^{+} + e \rightarrow EL2^{\circ} + h\nu$) and from hole capture at $EL2^{\circ}$ ($EL2^{\circ} +$ hole $\rightarrow EL2^{+} + h\nu$) respectively. The fourth luminescence band is peaked at 0.8 eV and has a width of 250 meV. Its origin is less certain. The available data [58] are however consistent with electron recombination at $EL2^{++}$ ($EL2^{++} + e \rightarrow EL2^{+} + h\nu$).

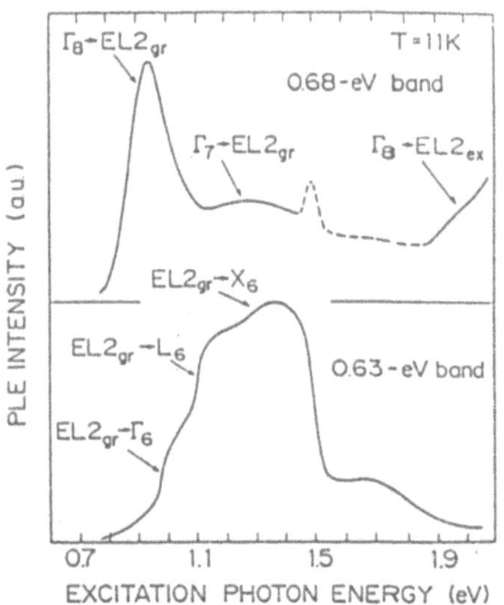

Fig. 6 Excitation spectra [60] of the 0.68 eV and the 0.63 eV EL2 photoluminescence bands. Note the similarity between these excitation spectra and the cross-sections in Fig. 1a.

4 Electron-Spin-Resonance of the As_{Ga} Antisite in Undoped SI GaAs

4.1 Antisite Defects

The data presented in the previous sections provide insight into the symmetry properties of EL2 and favor a defect with tetrahedral symmetry. However, they do not give any information about its chemical nature. A reliable chemical identification of an intrinsic defect can only be based on nuclear labels (isotope effects). Magnetic resonance has the ability to detect such labels via nuclear hyperfine splittings. They are

190

the basis for the identification of a special class of defects in III-V compounds, namely antisites.

The importance of antisites as native intrinsic defects in high-gap III-V compounds has not been appreciated until 1975 [62]. However, it is now well established experimentally that anion antisite defects (a group V atom on a group III site) can, to a large extent, control the properties of device-grade undoped GaP [63], InP [64] and GaAs in particular. Formally these defects are double donors which can compensate residual acceptors thus resulting in so called undoped semi-insulating material. Their singly ionized donor state D^+ has a paramagnetic ground state which is accessible by magnetic resonance techniques.

In 1980 a four-line electron-spin-resonance (ESR) spectrum has been reported in as-grown Bridgman GaAs:Cr using a far infrared (300 GHz) ESR set-up [65]. A spectrum with the same ESR parameters is observed at conventional ESR microwave frequencies (9 - 35 GHz) in as-grown undoped LEC GaAs, see the example shown in Fig. 7. On the basis of the characteristic four-line ^{75}As (I = 3/2) hyperfine splitting it has been identified with As_{Ga}^+ [65]. The spectrum is characterized by a g-factor and a hyperfine coupling constant A , g = 2.04 and A = 2.70 GHz. The value of A is a measure for the As_{Ga}^+ wavefunction localization (spin density) at the central As nucleus [66, 67] and corresponds to about 18 % of the total wavefunction. Very similar localizations at the central nuclei have been evaluated for other native antisites as P_{Ga}^+ in GaP [67, 68], P_{In}^+ in InP [69] and Sb_{Ga}^+ in GaAs:Sb [70]. The major portion of the wavefunction, about 65 %, is localized at the four nearest neighbor ligands for As_{Ga}^+ in GaAs [71] and for P_{Ga}^+ in GaP [67] as well. This strong localization implies that anion antisites are deep level defects.

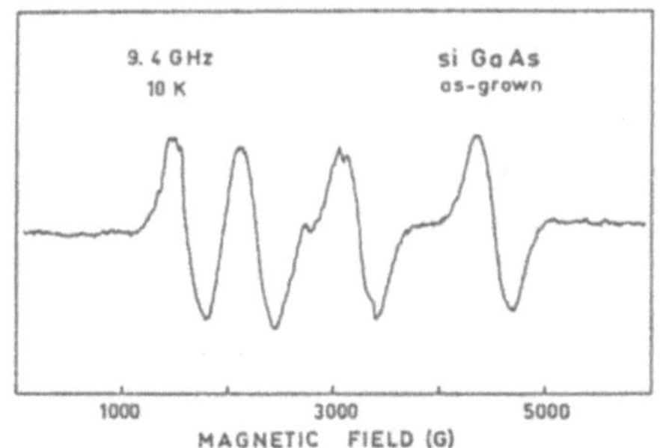

Fig. 7 The ESR spectrum of the As_{Ga}^+ antisite in undoped as-grown GaAs corresponding to an As_{Ga}^+ concentration of about 2×10^{16} cm^{-3}. In situ illumination has been used to enhance the signal by a factor of two. Cavity background signals have been electronically substracted.

For all the antisites mentioned above the ESR spectrum is isotropic which is consistent with an isolated defect. However, the information contained in the spectra is insufficient to definitely exclude association with another defect beyond the first ligand shell. This question has been investigated for P_{In}^+ in InP [72] and As_{Ga}^+ in GaAs [73, 74] by optically detected ENDOR (electron nuclear double resonance). For P_{In}^+ no evidence for an associated defect within the first two ligand shells has been found. On the other hand, the data for As_{Ga}^+ have been interpreted in terms of a trigonal As_{Ga} - As_{int} complex the arsenic interstitial As_{int} sitting beyond the second ligand shell around As_{Ga}. Based on less direct evidence a similar assignment has been made [75]. Nevertheless this assignment is not unanimously accepted since it raises several severe problems: First, no plausible mechanism for the binding of As_{int} to As_{Ga} has been found [76]. Second, the pair is predicted to have a shallow effective mass like level [76] for which experimental evidence does not exist. Third, the pair model is at variance with existing compensation models for undoped SI GaAs since one has to postulate a background acceptor concentration exceeding that of the total mid-gap donor (As_{Ga}/EL2) concentration [77]. Finally, the As_{Ga} defect in as-grown GaAs is stable up to annealing temperatures of 900°C [78, 79], in contrast to what has been stated elsewhere [80]. This high thermal stability is difficult to reconcile with a loosely bound pair.

All these problems do not exist if the spectrum in Fig. 7 is assigned to the isolated As_{Ga} antisite. Therefore the exact microscopic nature of the As_{Ga} defect giving rise to the spectrum in Fig. 7 appears to be an open question. Irrespective of this problem, the photo-ESR data to be presented below, demonstrate convincingly that the As_{Ga} defect represented by the spectrum in Fig. 7, and EL2 as defined by its optical properties are one and the same defect species.

It is noted in passing that the dark equilibrium concentration of As_{Ga}^{+} in standard as-grown SI pBN LEC GaAs, as inferred from ESR, can vary strongly and ranges from $\approx 1 \times 10^{15}$ cm^{-3} to $\approx 1 \times 10^{16}$ cm^{-3}. On the other hand the total As_{Ga} concentration varies much less and is in general slightly larger than 1×10^{16} cm^{-3}.

4.2 Photo-ESR of As_{Ga}

The ESR intensity, which is proportional to the concentration of the paramagnetic charge state of a defect, can often be changed by illuminating the sample <u>in situ</u> with (monochromatic) light. These changes usually result either from photoionization or photoneutralization of the defect in question or from capture of carriers released at other defects. The intensity change can correspond to an enhancement or a quenching of the ESR signal depending on the type of ionization or carrier captured. When the light is switched off the intensity change may decay or be stable. For most ESR centers and for As_{Ga}^{+} in particular the latter situation prevails in SI GaAs. This implies that the material contains electron- and hole traps in concentrations which are a significant fraction of the total As_{Ga} concentration [81-84].

In general, ESR intensity changes involving a charge exchange with other defects are optically reversible. For instance, when the As_{Ga}^{+} ESR is quenched by $h\nu = 0.9$ eV illumination, it can be restored by subsequent $h\nu = 1.4$ eV illumination. For As_{Ga}^{+}, however, a second type of quenching exists which is not reversible optically at sufficiently low temperatures and therefore is referred to as persistent quenching. As for EL2 this quenching is thought to involve a structural change of the As_{Ga}^{o} normal defect configuration.

The first As_{Ga}^{+} photo-ESR results were obtained on plastically deformed GaAs [85]. From the quenching thresholds in SI and the enhancement threshold in p-type material it was concluded that As_{Ga} is a double donor with its first level (o/+) at mid-gap and its second level (+/++) at E_c - 1.0 eV. The fact that As_{Ga} exhibits a mid-gap level like EL2 was the first direct microscopic evidence for the correlation of the two species. This was further supported by the observation that the As_{Ga}^{+} ESR is persistently quenched under $h\nu = 1.17$ eV illumination [86]. It can only be restored by a thermal anneal above 120 K similar to the recovery of EL2° [87].

The photo-ESR data on plastically deformed material [85] reveal excitation - and quenching thresholds but do not contain information about the spectral shapes of the ionization processes involved. Photo-ESR data which could be compared with the EL2 cross-sections have been subsequently reported for <u>as-grown</u> GaAs [81] and Fig. 8 shows the corresponding enhancement and quenching curves for the As_{Ga}^+ ESR. The enhancement is due to the process $As_{Ga}^\circ + h\nu \longrightarrow As_{Ga}^+ + e$ and is stable when the light is switched off. It also does not anneal up to a temperature of 120 K [88]. Thus the electrons released during enhancement must have been captured by <u>deep</u> traps shallower than mid-gap. Such traps have been detected by thermally stimulated current measurements as well [89] but only recently one such trap has been identified [90, 91]. It is the interstitial oxygen - As vacancy pair (or: off center O_{As}) occuring in concentrations in the low 10^{15} cm^{-3} range. The shape of the enhancement curve in Fig. 8a rises monotonically up to the band edge and resembles that of the EL2 $\sigma_{n1}{}^\circ(h\nu)$ curve in Fig. 1a [92].

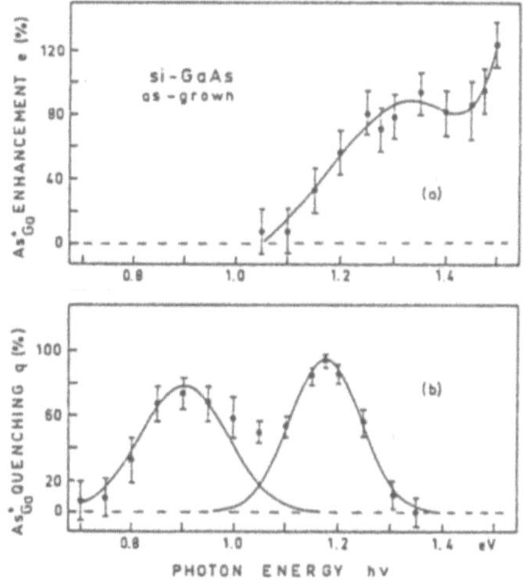

Fig. 8 (a) Spectral dependence of As_{Ga}^+ enhancement efficiency in as-grown GaAs. (b) Spectral dependence of As_{Ga}^+ quenching efficiency [81]

In contrast the As_{Ga}^+ quenching curve in Fig. 8b consists of two bands peaked at $\approx$ 0.9 eV and $\approx$ 1.2 eV respectively. The nature of the quenching within these bands is quite different. For the 0.9 eV band it is only partial, occurs on a time scale of a few seconds and is optically reversible. It corresponds to photoneutralization of As_{Ga}^+ ($As_{Ga}^+ + h\nu \longrightarrow As_{Ga}^\circ + $ hole) the hole released being subsequently trapped at negatively charged acceptors [82]. Note that the shape of the 0.9 eV band up to 1.05 eV is very similar to that of the EL2 cross-section $\sigma_{p1}{}^\circ(h\nu)$ in Fig. 1a. Within the 1.2 eV band the quenching is complete, occurs on a time scale of a few minutes and is persistent. In analogy with EL2 this quenching is interpreted as a two step process involving first photoneutralization of As_{Ga}^+, as before, and then a structural defect rearrangement in the neutral state. The 1.2 eV persistent As_{Ga}^+ quenching band in Fig. 8b and the EL2 bleaching band in Fig. 1b are seen to coincide. The data in Fig. 8 have been confirmed on as-grown samples of varying stoichiometry [93].

The most characteristic property of EL2 is the occurrence of nonmonotonic photo-capacitance transients, compare Fig. 2. A completely equivalent effect has been established for As_{Ga} [81, 88] see Fig. 9. For photon energies between 1.0 eV and 1.4 eV the As_{Ga}^+ ESR transients initially increase, due to photoionization of As_{Ga}°, at a rate that is proportional to $\sigma_{n1}{}^\circ$, but then decrease for prolonged illumination. As already mentioned this decrease can be interpreted as a two step process, occuring at a rate [14] proportional to $\sigma_{p1}{}^\circ \sigma*{}^\circ/(\sigma_{n1}{}^\circ + \sigma_{p1}{}^\circ)$ and involving first neutralization of As_{Ga}^+

193

and then a structural rearrangment of As_{Ga}^{o}. Below $h\nu \approx 1.0$ eV the usual monotonic ESR transients are observed.

As discussed in section 3.1 the persistent EL2 quenching band is preceeded by a weak zero-phonon-line which is related to the neutral state EL2°. The As_{Ga} ESR on the other hand is related to the singly ionized donor As_{Ga}^{+}. If EL2 and As_{Ga} are identical, optically induced changes of the mid-gap level occupation must be accompanied by anticorrelated intensity changes of the zero-phonon-line and the ESR signal. This is exactly what has been observed [89], see Fig.10. This anticorrelation is valid quantitatively [94].

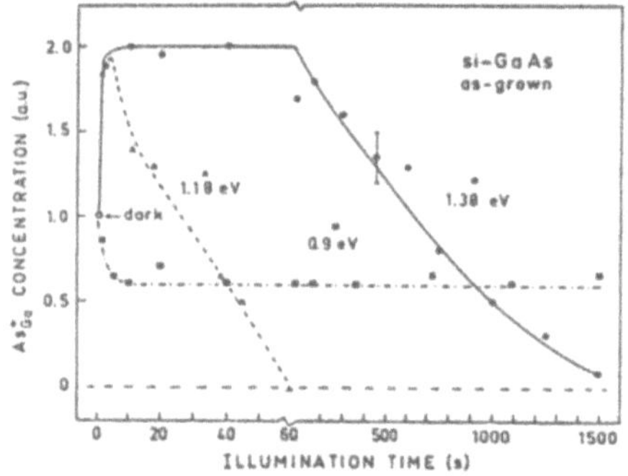

Fig. 9 As_{Ga}^{+} ESR signal changes as a function of illumination time for three different photon energies. The $h\nu = 1.18$ eV and $h\nu = 1.38$ eV transients are nonmonotonic in time [81].

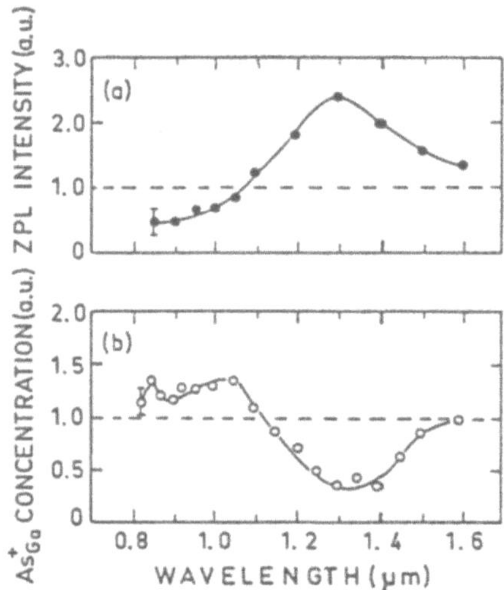

Fig. 10 (a) The EL2° 1.038 eV zero-phonon-line intensity and (b) the As_{Ga}^{+} ESR signal intensity after illumination. Dashed lines represent dark values. Note the anticorrelated behavior of the two signals [89].

In conclusion the photo-ESR data for as-grown GaAs presented above reveal that the optical properties of EL2 and As_{Ga} are virtually the same. At present these results provide the most direct and convincing evidence that EL2 and the As_{Ga} ESR are induced by the same defect. So far conventional photo-ESR evidence for the second As_{Ga} donor level in <u>as-grown</u> GaAs has not been reported, probably because the photoresponse of this level is weaker than that of the mid-gap level. However optically detected ESR has conclusively demonstrated the existence of the E_c - 1.0 eV level in p-GaAs:Zn [95]. Thus the double donor model in Fig. 3 is well established also for the native antisite in as-grown GaAs.

5 The As_{Ga} Magnetic-Circular-Dichroism Absorption

Undoped SI GaAs exhibits a characteristic magnetic-circular-dichroism (MCD) absorption extending from $h\nu \approx 0.8$ eV up to the band edge [96, 97], see Fig. 11a. The MCD is the difference in absorption between right- and left circularly polarized light if the sample is placed in an external magnetic field and if the light propagates along the field axis. The MCD absorption band shape $S(h\nu)$ in Fig. 11a is temperature in-

194

sensitive but its intensity decreases rapidly with increasing temperature. Therefore the MCD absorption is due to a defect which is paramagnetic in its ground state. By tagging the MCD absorption to the As_{Ga}^+ ESR it has been demonstrated that the MCD band in Fig. 11a arises from As_{Ga}^+ [96].

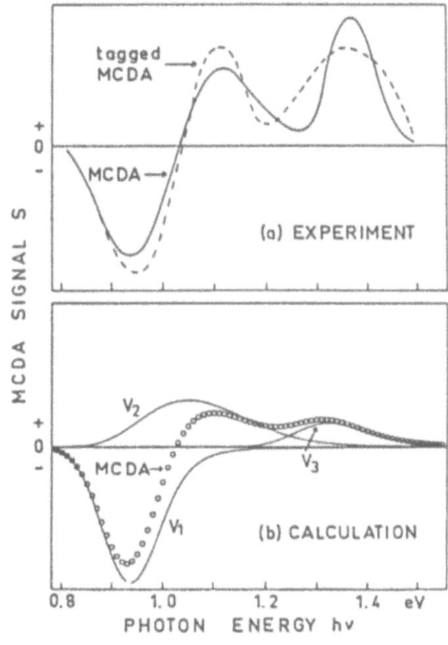

Fig. 11 (a) The As_{Ga}^+ MCD absorption in un-doped as-grown GaAs at a temperature of 4.2 K and an applied magnetic field H = 2 T [96]. (b) Calculated MCD absorption shape (open circles) for the case of the photoneutralization transition $As_{Ga}^+ + h\nu \longrightarrow As_{Ga}^o +$ hole. The individual contributions from the three uppermost valence bands V_1 - V_3, compare Fig. 12, are shown as solid lines [99, 100].

The origin of the optical transition giving rise to the As_{Ga}^+ MCD absorption is controversial. Originally, two internal $A_1 \longrightarrow T_2$ transitions within an isolated As_{Ga}^+ defect have been postulated, the excited T_2 states being resonant with the conduction band [96]. The difficulties associated with this model have been pointed out and an alternative model has been suggested [98-100]. This is shown in Fig. 12a and corresponds simply to the photoneutralization process at the As_{Ga}^+ antisite, i.e. to hole release from As_{Ga}^+ to the three uppermost valence bands V_1 - V_3 near the Brillouin zone center. The photo-ESR data presented in the previous section confirm that this transition is the most prominent one involving As_{Ga}^+. In semiconductors such ionizing transitions give rise to MCD absorption bands as has been demonstrated most clearly for the manganese acceptor in GaAs [102]. On the basis of this model, the shape of the MCD absorption band, $S(h\nu)$, has been calculated [99, 100]. Seven parameters enter the theoretical expression for $S(h\nu)$ six of which are known from <u>independent</u> experiments. This is the (o/+) mid-gap level position, two parameters describing the electron phonon coupling of this level (effective phonon energy and Franck-Condon shift) as well as three parameters describing the valence band structure, namely the masses of the heavy and light holes and the spin-orbit splitting. The only unknown parameter, s, is treated as a fitting parameter. Its physical significance is such that s^{-1} represents the combined effective extension of the bound defect - and Bloch hole wavefunctions. A plot of the calculated MCD shape is shown in Fig. 11b (circles) for s^{-1} = 20 Å. This is the correct order of magnitude for the wave function extension but too much significance should not be attributed to this value since the radial parts of the wavefunctions used to calculate the relevant transition matrix elements are highly simplified [99]. The calculation reproduces the measured MCD shape (and position) surprisingly well, especially if one keeps in mind that only <u>one</u> free fitting parameter is involved.

If the sign of the contribution of the heavy hole band V_1 to the MCD shape is inverted, $S(h\nu)$ is converted into an expression for the $\sigma_{p1}{}^0(h\nu)$ cross-section of As_{Ga}^+. Using exactly the same parameter set as before for the $S(h\nu)$ calculation in Fig. 11b, i.e. no free fitting parameter at all, $\sigma_{p1}{}^0(h\nu)$ of As_{Ga}^+ has been calculated [100]. The result is shown in Fig. 13b and is compared to the measured shape of $\sigma_{p1}{}^0(h\nu)$ of EL2. The agreement is excellent. This is further very strong support for the MCD model of Fig. 12 and closes the circle between the As_{Ga} antisite and EL2 in as-grown GaAs.

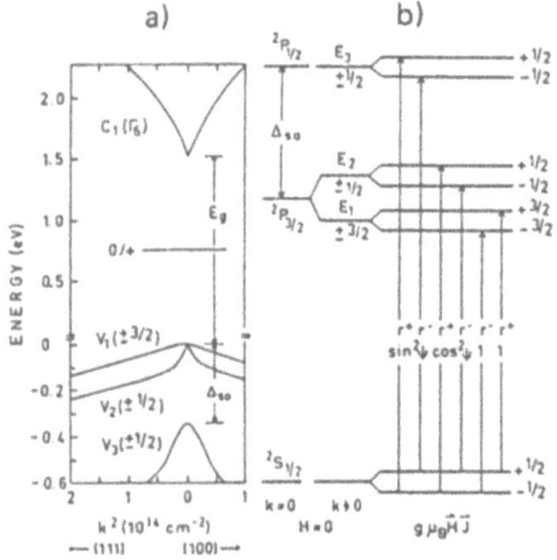

Fig. 12 (a) Band structure level scheme of GaAs near the zone center containing the As_{Ga} mid-gap level. The extension of k values shown is about one tenth of k_{max}. Adopted from Ref. 101. (b) The analogous alkaliatom level scheme for the $As_{Ga}^+ + h\nu \longrightarrow As_{Ga}^0 +$ hole transition in (a). Electric dipole allowed transitions are indicated by arrows. The numbers associated with them are relative transition probabilities [99].

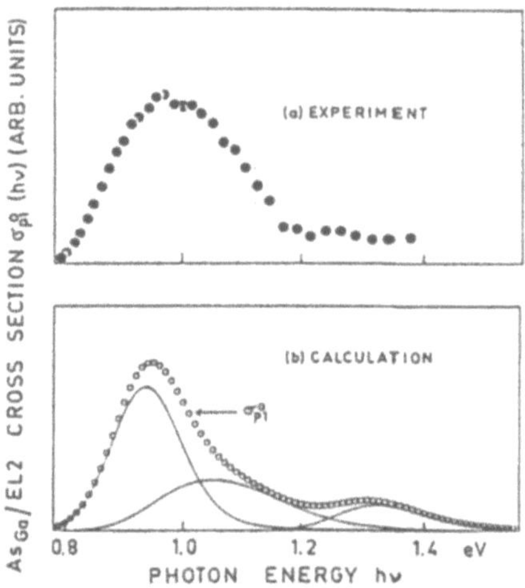

Fig. 13 $(As_{Ga}/EL2)^+$ hole ionization cross-section as a function of photon energy. (a) experimental result according to Ref. 11. (b) calculated $\sigma_{p1}{}^0(h\nu)$ (circles) using the same parameter set as in Fig. 11b. Full lines represent the individual contributions from the valence bands $V_1 - V_3$ [100].

6 Conclusion

The spectroscopic data presented in this article demonstrate that EL2 is the As_{Ga} antisite observed by ESR and related techniques in undoped as-grown GaAs. However, a final consensus whether this is the isolated As_{Ga} defect or an As_{Ga}-As_{int} complex has not been reached. The problem within the first model is the explanation of the ENDOR data. The latter model faces several difficulties that have been discussed, in particular the tetrahedral defect symmetry inferred from the piezo-splitting of the 1.038 eV zero-phonon-line.

196

Acknowledgement

The author thanks F. Eisen, J. Schneider and J. Windscheif for discussions and a critical reading of the manuscript. He is also grateful to A. Meier for her patience during the preparation of the typescript. This work has been supported by Bundesministerium für Forschung und Technologie under contract NT 2766 A.

References

[1]   *G.M. Martin* and *S. Makram-Ebeid* in: "Deep Centers in Semiconductors" Ed. *S.T. Pantelides* (Gordon and Breach, 1986) pp. 389

[2]   *A. M. Huber, N. T. Linh, M. Valladon, J. C. Debrun, G. M. Martin, A. Mitonneau* and *A. Mircea*, J. Appl. Phys. 50, 4022 (1979)

[3]   *D. Bois* and *G. Vincent*, J. Physique 38, L351 (1977)

[4]   *G. Vincent* and *D. Bois*, Solid State Commun. 27, 431 (1978)

[5]   *G.M. Martin, J.P. Farges, G. Jacob, J.P. Hallais* and *G. Poiblaud*, J. Appl. Phys. 51, 2840 (1980)

[6]   *H. Lessoff* private communication 1980

[7]   With one exception As_{Ga}/EL2 data obtained for plastically deformed or particle irradiated material are not considered here.

[8]   When speaking of the As_{Ga} antisite the author refers to the defect that gives rise to the As_{Ga}^{+} electron-spin-resonance spectrum in as-grown undoped GaAs. It is left as an open question whether this is isolated As_{Ga} or As_{Ga} complexed with an As interstitial.

[9]   *D. Bois* and *A. Chantre*, Revue Phys. Appliquée 15, 631 (1980)

[10] *A. Chantre, G. Vincent* and *D. Bois*, Phys. Rev. B23, 5335 (1981)

[11] *P. Silverberg, P. Omling* and *L. Samuelson*, Appl. Phys. Lett. 52, 1689 (1988)

[12] *G.L. Miller, D.V. Lang* and *L.C. Kimerling*, Ann. Review Mater. Sci. 1977 pp. 377

[13] Other conclusions are possible but lead to contradictions.

[14] *G. Vincent, D. Bois* and *A. Chantre*, J. Appl. Phys. 53, 3643 (1982)

[15] A partial or even complete optical recovery has been reported in Refs. 16-20. However it is not clear whether this is a direct optical effect or an indirect one induced by optically generated free carriers, see Ref. 21.

[16] *M. Tajima*, Japanese J. Appl. Phys. 24, L47 (1985)

[17] *H.J. v. Bardeleben, N.T. Bagraev* and *J.C. Bourgoin*, Appl. Phys. Lett. 51, 1451 (1987)

[18] *D.W. Fischer*, Appl. Phys. Lett. 50, 1751 (1987)

[19] *M. Tajima, H. Saito, T. Iino* and *K. Ishida*, Japanese J. Appl. Phys. 27, L101 (1988)

[20] *J.C. Parker* and *R. Bray*, Phys. Rev. B 37, 6368 (1988)

[21] *A. Mitonneau* and *A. Mircea*, Solid State Commun. 30, 157 (1979). The effect described in this paper, also referred to as Auger deexcitation, is not well understood. One could speculate that neutral $EL2^{*}$ captures an electron and that the thermal barrier for the decay of negative $EL2^{*}$ is much smaller than that of neutral $EL2^{*}$. Thus instable $EL2^{-}$ could form which by emitting an electron could form $EL2^{\circ}$. This hypothetical sequence would be an Auger-type process.

[22] *F. Fuchs* and *B. Dischler*, Appl. Phys. Lett. 51, 679 (1987)

[23] *P. Trautmann, M. Kaminska* and *J.M. Baranowski*, Acta Phys. Pol. A71, 269 (1987)

[24] *M. Skowronski, J. Lagowski* and *H.C. Gatos*, Phys. Rev. B 32, 4264 (1985)

[25] *Y. Mochizuko* and *T. Ikoma* in "Semi-Insulating III-V Materials" Eds. *H. Kukimoto, S. Miyazawa* (Ohmsha, 1986) pp. 323

[26] *M. Levinson* and *J.A. Kafalas*, Phys. Rev. B 35, (1987)

[27]   *J. Lagowski, D.G. Lin, T.P. Chen, M. Skowronski* and *H.C. Gatos*, Appl. Phys. Lett. 47, 929 (1985)

[28]   *J. Osaka, H. Okamoto* and *K. Kobayashi*, "Semi-Insulating III-V Materials" Eds. *H. Kukimoto, S. Miyazawa* (Ohmsha, 1986) pp. 421

[29]   *A. Bencherifa, G. Brémond, A. Nouailhat, G. Guillot, A. Guivarch* and *A. Regreny*, Revue Phys. Appliquée 22, 891 (1987)

[30]   *T. Wosinski*, Appl. Phys. A 36, 213 (1985)

[31]   *P. Omling, P. Silverberg* and *L. Samuelson*, Phys. Rev. B 38, 3606 (1988)

[32]   *P. Silverberg, P. Omling* and *L. Samuelson* in "Semi-Insulating III-V Materials" Eds. G. Grossmann, L. Ledebo (Adam Hilger, 1988) pp. 369

[33]   *M.D. Sturge*, Phys. Rev. 127, 768 (1962)

[34]   *G.M. Martin*, Appl. Phys. Lett. 39, 747 (1981)

[35]   *B. Dischler, F. Fuchs* and *U. Kaufmann*, Appl. Phys. Lett. 48, 1282 (1986)

[36]   *F. Fuchs, B. Dischler* and *U. Kaufmann* in "Semi-Insulating III-V Materials" Eds. *H. Kukimoto, S. Miyazawa* (Ohmsha, 1986) pp. 329

[37]   "Defect Recognition and Image Processing in III-V Compounds" Ed. *J.P. Fillard* (Elsevier, 1985)

[38]   "Defect Recognition and Image Processing in III-V Compounds" Ed. *E.R. Weber* (Elsevier, 1987)

[39]   *M. Kaminska, M. Skowronski, J. Lagowski, J.M. Parsey* and *H.C. Gatos*, Appl. Phys. Lett. 43, 302 (1983)

[40]   *F. Fuchs* and *B. Dischler*, Appl. Phys. Lett. 51, 2115 (1987)

[41]   *N. Tsukada, T. Kikuta* and *K. Ishida*, Japanese J. Appl. Phys. 24, L302 (1985)

[42]   *N. Tsukada, T. Kikuta* and *K. Ishida*, Japanese J. Appl. Phys. 25, L196 (1986)

[43]   Here it is assumed that after autoionization of $EL2^\circ$, the electron recombines with $EL2^+$ within a time shorter than a nanosecond, see Ref. 44. Thus $EL2^\circ$ is immediately available for the next absorption cycle.

[44]   *W.W. Rühle, K. Leo* and *N.M. Haegel*, "GaAs and Related Compounds 1987" Eds. *A. Christou* and *H.S. Rupprecht*, Inst. Phys. Conf. Ser. 91, 105 (1988)

[45]   *W. Kuszko* and *M. Kaminska*, Acta Phys. Pol. A69, 427 (1986)

[46]   *F. Fuchs* and *B. Dischler*, private communication

[47]   *M. Skowronski, D.G. Lin, J. Lagowski, M.L. Pawlowicz, K.Y. Ko* and, *H.C. Gatos*, Mat. Res. Soc. Proc. 46, 207 (1985)

[48]   *M.O. Manasreh* and *B.C. Covington*, Phys. Rev. B 35, 2524 (1987)

[49]   *M.O. Manasreh* and *B.C. Covington*, Phys. Rev. B 36, 2730 (1987)

[50]   *K. Bergman, P. Omling, L. Samuelson* and *H.G. Grimmeiss* in "Semi-Insulating III-V Materials" Eds. G. Grossmann, L. Ledebo (Adam Hilger, 1988) p. 397

[51]   *W. Kuszko, M. Jezewski, J.M. Baranowski* and *M. Kaminska*, Appl. Phys. Lett. 53, 2558 (1988)

[52]   *M. Kaminska, M. Skowronski* and *W. Kuszko*, Phys. Rev. Lett. 55, 2204 (1985)

[53]   *T. Figielski* and *T. Wosinski*, Phys. Rev. B 36, 1269 (1987)

[54]   *H. Ennen, U. Kaufmann* and *J. Schneider*, Appl. Phys. Lett. 38, 355 (1981)

[55]   *W. Kuszko, P.J. Walczak, P. Trautman, M. Kaminska* and *J.M. Baranowski*, "Defects in Semiconductors" Ed. *H.J. v. Bardeleben* (Trans Tech, 1986) Materials Science Forum Vols. 10-12, pp. 317

[56]   *J. Dabrowski* and *M. Scheffler*, Phys. Rev. Lett. 60, 2183 (1988)

[57]   *D.J. Chadi* and *K.J. Chang*, Phys. Rev. Lett. 60, 2187 (1988)

[58]   Landolt-Börnstein, New Series, Ed. *M. Schulz* (Springer, 1989) Vol. 22b

[59]   *M. Tajima* in "Semi-Insulating III-V Materials" Eds. *G. Grossmann, L. Ledebo* (Adam Hilger, 1988) pp. 119

[60]   *M. Tajima*, Japanese J. Appl. Phys. 26, L885 (1987)

[61]   *M. Tajima, T. Iino* and *K. Ishida*, Japanese J. Appl. Phys. 26, L1060 (1987)

[62]   *J.A. van Vechten*, J. Electrochem. Soc. 122, 423 (1975)

[63]   *K. Chino, T. Kazuno, K. Satoh* and *M. Kubota* in " Semi-Insulating III-V Materials " Eds. *G. Grossmann, L. Ledebo* (Adam Hilger, 1988) pp. 133

[64]   *T.A. Kennedy, N.D. Wilsey, P.B. Klein* and *R.L. Henry*, Materials Science Forum Vols. 10-12, 271 (1986)

[65]   *R.J. Wagner, J.J. Krebs, G.M. Stauss* and *A.M. White*, Solid State Commun. 36, 15 (1980)

[66]   *J.R. Morton* and *K.F. Preston*, J. Magn. Reson. 30, 577 (1978)

[67]   *U. Kaufmann* and *J. Schneider* in: "Festkörperprobleme XX, Adv. in Solid State Physics", Ed. *J. Treusch* (Vieweg, 1980) pp.87

[68]   *U. Kaufmann, J. Schneider* and *A. Räuber*, Appl. Phys. Lett. 29, 312 (1976)

[69]   *L.H. Robins, P.C. Taylor* and *T.A. Kennedy*, Phys. Rev. B 38, 13227 (1988)

[70]   *M. Baeumler, J. Schneider, U. Kaufmann, W.C. Mitchel* and *P.W. Yu*, Phys. Rev. B March 15 (1989)

[71]   *B.K. Meyer, D.M. Hofmann, F. Lohse* and *J.M. Spaeth* in: "Defects in Semiconductors" Eds. *L.C. Kimerling, J.M. Parsey*, J. Electronic Mat. 14b, 921 (1985)

[72]   *D.Y. Jeon, H.P. Gislason, J.F. Donegan* and *G.D. Watkins*, Phys. Rev. B 36, 1324 (1987)

[73]   *B.K. Meyer, D.M. Hofmann, J.R. Niklas* and *J.M. Spaeth*, Phys. Rev. B 36, 1332 (1987)

[74]   *B.K. Meyer* et al, this volume

[75]   *H.J. v. Bardeleben, D. Stievenard, D. Deresmes, A. Huber* and *J.C. Bourgoin*, Phys. Rev. B 34, 7192 (1986)

[76]   *G.A. Baraff, M. Lannoo* and *M. Schlüter*, Phys. Rev. B 38, 6083 (1988)

[77]   *E.R. Weber* and *M. Kaminska* in: "Semi-Insulating III-V Materials" Eds. *G. Grossmann, L. Ledebo* (Adam Hilger, 1988) pp. 111

[78]   *E.R. Weber* "Semi-Insulating III-V Materials" Eds. *D.C. Look* and *J.S. Blakemoore* (Shiva Ltd. 1984), pp. 296

[79]   *U. Kaufmann, J. Windscheif, M. Baeumler, J. Schneider* and *F. Köhl*, see Ref. 78, pp. 246

[80]   *J.C. Bourgoin, H.J. v. Bardeleben, D. Stievenard*, J. Appl. Phys. 64 , R65 (1988)

[81]   *M. Baeumler, U. Kaufmann* and *J. Windscheif*, Appl. Phys. Lett. 46, 781 (1985)

[82]   *U. Kaufmann*, "GaAs and related Compounds 1987" Eds. *A. Christou, H.S. Rupprecht*, Inst. Phys. Conf. Ser. 91, 41 (1988)

[83]   *M. Baeumler, P.M. Mooney* and *U. Kaufmann*, Materials Science Forum Vols 38-41, 785 (1989)

[84]   This fact casts doubt on the simple three level compensation model for SI GaAs involving only shallow donors and acceptors and the As_{Ga}/EL2 mid-gap level

[85]   *E.R. Weber, H. Ennen, U. Kaufmann, J. Windscheif, J. Schneider* and *T. Wosinski*, J. Appl. Phys. 53, 6140 (1982)

[86]   *E.R. Weber* and *J. Schneider*, Physica 116B, 398 (1983)

[87]   Recovery of EL2° occurs around 130 K but complete recovery of As_{Ga}^+ is not achieved below 250 K. Since different charge states are monitored this does not necessarily mean that different defects are involved.

[88]   *M. Baeumler, U. Kaufmann* and *J. Windscheif*, Mat. Res. Soc. Proc. Vol. 46, 201 (1985)

[89]   *N. Tsukada, T. Kikuta* and *K. Ishida*, Phys. Rev. B 33, 8859 (1986)

[90]   *J. Schneider, B. Dischler, H. Seelewind, P. Mooney, J. Lagowski, M. Matsui, D.R. Beard* and *R. Newman*, Appl. Phys. Lett. April (1989)

[91]   *H.Ch. Alt*, Appl. Phys. Lett. April (1989)

[92]   Contrary to what one expects for a mid-gap donor, no As_{Ga}^+ enhancement is visible below
1.0 eV in Fig. 8a. The reason is that for the sample in question the As_{Ga} (o/+) level is partially
compensated and therefore As_{Ga}^+ quenching competes with As_{Ga}^+ enhancement. If only As_{Ga}^o
is present in thermal equilibrium one also observes an As_{Ga}^+ enhancement between 0.75 eV
and 1.0 eV, see Ref. 93.

[93]   *N. Tsukada, T. Kikuta and K. Ishida*, Japanese J. Appl. Phys. 24, L689 (1985)

[94]   *J. Lagowski, M. Matsui, M. Bugajski, C.H. Kang, M. Skowronski, H.C. Gatos, M. Hoinkis,
E.R. Weber and W. Walukiewicz*, "GaAs and Related Compounds 1987" Eds. *A. Christou,
H.S. Rupprecht*, Inst. Phys. Conf. Ser. 91, 395 (1988)

[95]   *B.K. Meyer, D.M. Hofmann* and *J.M. Spaeth*, J. Phys. C 20, 2445 (1987)

[96]   *B.K. Meyer, J.M. Spaeth* and *M. Scheffler*, Phys. Rev. Lett. 52, 851 (1984)

[97]   *A. Winnacker, Th. Vetter* and *F.X. Zach* " Semi-Insulating III-V Materials " Eds.
G. Grossmann, L. Ledebo (Adam Hilger, 1988) pp. 583

[98]   *U. Kaufmann*, Phys. Rev. Lett. 54, 1332 (1985)

[99]   *U. Kaufmann* and *J. Windscheif*, Phys. Rev. B 38, 10060 (1988)

[100] *U. Kaufmann* and *J. Windscheif*, " Semi-Insulating III-V Materials" Eds. *G. Grossmann,
L. Ledebo* (Adam Hilger, 1988) pp. 343

[101] *J.S. Blakemore*, J. Appl. Phys. 53, R123 (1982)

[102] *M. Baeumler, B.K. Meyer, U. Kaufmann* and *J. Schneider*, "Defects in Semiconductors" Ed.
G. Ferenczi (Trans Tech, 1989) Materials Science Forum Vols. 38-41, pp. 797

On the Charge State of the EL2 Mid Gap Level in Semi-Insulating GaAs from a Quantitative Analysis of the Compensation

Bruno Meyer, Klaus Krambrock, Detlev Hofmann and Johann-Martin Spaeth

Universität Paderborn, Experimentalphysik, Warburger Str. 100 A, 4790 Paderborn, Federal Republic of Germany

Summary: We present quantitative results on the concentrations of intrinsic and extrinsic donors and acceptors in semi-insulating GaAs and their role in the compensation mechanism. The existence of intrinsic cation antisite defects in As-rich GaAs in significant concentrations indicates that the EL2 mid gap level is indeed positively charged and hence would confirm the structure model of EL2 derived previously by ODENDOR experiments as being an Arsenic antisite-Arsenic interstitial pair defect.

1 Introduction

The microscopic structure of the dominant defect in undoped semi-insulating (s.i.) GaAs EL2 is still under controversial discussion. Numerous papers have been presented during the last decade presenting experimental and theoretical studies on the behaviour and structure of this deep level defect [1]. Apart from the technological importance the mysterious physical properties such as metastability have stimulated efforts to identify the EL2 defect. The present discussion focusses on the possible models discussed most at present: in one EL2 is identified with an isolated Arsenic on a Gallium site, As_{Ga}, which is a double donor in GaAs. Consequently its two energy levels in the gap are connected to the charge states $+/++$ at $E_{vb}+0.54$ eV and $0/+$ at $E_{vb}+0.76$ eV (mid gap). This model was recently favoured by theoretical calculations [2,3], in which the light induced metastability of the EL2 defect can be explained with an As antisite. Optically detected electron nuclear double resonance (ODENDOR) investigations [4] clearly as well as combined deep level transient spectroscopy (DLTS) and electron spin resonance (ESR) measurements [5] and other experiments [6,7], however, have lead to the identification of EL2 with an As antisite–As interstitial pair defect. The interpretation of the ligand quadrupole and hyperfine interactions on the paramagnetic EL2 defect at $E_{vb}+0.54$ eV, in which the hyperfine interaction with several neighbour shells could be resolved, points to a positively charged As interstitial separated by approximately two bond lengths from the As_{Ga} [4]. Consequently, the energy levels should correspond to the charge states $+++/++$ ($E_{vb}+0.54$ eV, paramagnetic) and $++/+$ ($E_{vb}+0.76$ eV, mid gap, diamagnetic). There are also theoretical calculations in support of this model [8]. However, there remain questions as a consequence of this identification, which are not understood at present. One is to understand in which way the interstitial is bound and another one how to understand the compensation behaviour of EL2, which is a triple donor according to this model.

201

2 Compensation and the Role of EL2

In undoped s.i. GaAs grown by the LEC (liquid encapsulated Czochralsky technique from BN crucibles the deep donor EL2 compensates the residual acceptors. These are the extrinsic impurities, mainly C and Zn, and grown-in defects involving intrinsic acceptors. In the "state of the art" material the concentration of the shallow acceptors $N_{s.a.}$ (intrinsic and extrinsic) exceeds the concentration of the shallow donors $N_{s.d.}$, which are mostly S and Si. $(N_{s.a.}\text{-}N_{s.d.})$ is approximately in the range $2 - 5 \cdot 10^{15} \text{cm}^{-3}$ [9]. If the excess acceptors are compensated by EL2, the mid gap level is no longer completely occupied. Also ionized EL2 defects are present in the concentration range close to the number of residual acceptors. DLTS experiments on n-type GaAs show that EL2 is the dominant trap in as-grown GaAs, donor type defects with their energy levels in the upper half of the band gap (EL3, EL5 and EL6) appear in concentrations lower by a factor 10 to 100 [10].

3 Experimental

The occupation of both charge states of EL2 in s.i. material can be determined quantitatively. The occupation of the mid gap level can be determined from the infrared absorption intensity of the zero phonon line (ZPL) (see Fig.1 a, b) [11]. The paramagnetic charge state can be determined from ESR [12] or from the magnetic

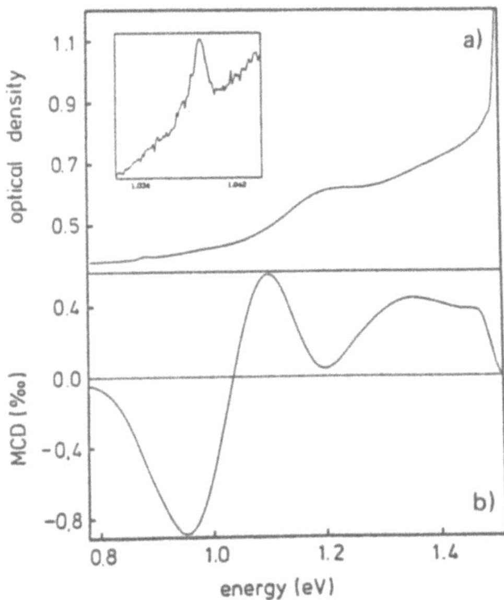

Fig. 1 (a) Infrared absorption of the mid gap EL2 defects (inset: zero phonon line of the intracenter transition), (b) the magnetic circular dichroism of the absorption of the paramagnetic EL2 defects.

202

circular dichroism of the absorption (MCD) measurements after calibration (see Fig.1c). The concentrations of the residual intrinsic and extrinsic acceptors can be determined by local vibrational mode (LVM) or Raman spectroscopy [13]. We used s.i. "state of the art" material from different vendors as well as p-type Gallium-rich material grown by the LEC technique.

4 Experimental Results

4.1 Quantitative Analysis of EL2 in Both Charge States

Fig.2 shows a comparison of the respective concentrations of EL2 midgap (determined from the intensity of the ZPL line at 1.039 eV) and the paramagnetic EL2 (from the MCD intensity at 0.98 eV calibrated by ESR) measured for 10 samples. There is a one-to-one relation between the concentrations of the mid gap and the paramagnetic EL2 defects. The total EL2 concentration in most of the samples ranges between $2 - 4 \cdot 10^{16} \mathrm{cm}^{-3}$. Typical concentrations of the paramagnetic EL2 defects are around $0.3 - 1.5 \cdot 10^{16} \mathrm{cm}^{-3}$. These values can be compared with the total amount of C and Zn present (see Table 1) [14].

All samples were grown from stoichiometric melts except W1 (slightly Gallium-rich). ITC2 was subjected to ITC (Inverted Thermal Conversion) treatment ($1200°C$ annealing 12 hours under equilibrium As-pressure followed by a rapid quench) and subsequent annealing at $800°C$ (1h) to restore EL2 (ITC1 is the as-grown reference sample) [10].

The concentration of the paramagnetic EL2 was always found to be above the concentration of the extrinsic acceptors C and Zn. Therefore, to account for this difference also intrinsic acceptors must be present in the material in significant concentrations (see last column in Table 1). This quantitative result contradicts the three level compensation model established so far taking into account only shallow extrinsic donors and acceptors [15].

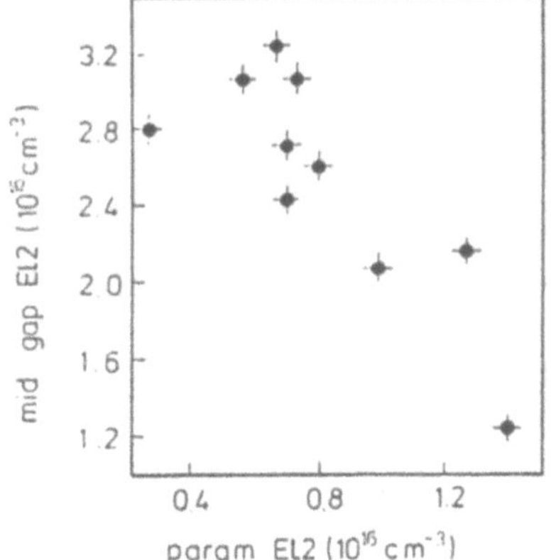

Fig. 2

Concentration of the mid gap EL2 (determined from the intensity of the ZPL at 1.039 eV) versus concentration of the paramagnetic EL2 (determined of the MCD at 0.98 eV after calibration by ESR)

sample	mid gap EL2 10^{15}	param. EL2 10^{15}	[C + Zn] 10^{15}	(param. EL2-[C+Zn]) 10^{15} (cm^{-3})
R1	30.6	7.3	3.7	3.6
R2	28.0	15.8	3.0	12.8
I2	20.7	9.9	4.6	5.3
I3	32.4	6.6	2.4	4.2
W1	5.2	4.9	0.5	4.4
W2	21.6	12.7	5.4	7.3
ITC1	28.0	2.7	n.d.	2.7
ITC2	12.4	14.0	n.d.	14.0

4.2 Identification of Intrinsic Acceptors in s.i. GaAs

When the EL2 in s.i. GaAs is transformed into its metastable state free holes are released. The free holes can be trapped at intrinsic and extrinsic acceptors thereby changing their charge states. Neutral C and Zn acceptors are observed in the electronic Raman scattering experiments upon this process [13]. Also the grown-in intrinsic acceptors with energy levels in the lower half of the band gap, now become visible. Because of the hole capture they are brought into the ESR/MCD active charge states. Examples are the defects called FR1 to FR3 and BE1 detected by conventional ESR. Their structure is not completely understood, a relation with cation antisite defects Ga_{As} is sometimes assumed [16]. Fig.3 (curve a) shows the MCD signal of a Ga_{As} related intrinsic acceptor with trigonal symmetry which appears when EL2 is persistently bleached [17]. From photo-MCD investigations its energy levels have been determined to be at $E_{vb}+0.1$ eV and $E_{vb}+0.5$ eV. The total MCD shown in Fig.3 (curve a) is only due to this defect, which was established by measuring the excitation spectrum of its optically detected electron spin resonance (ODESR) signal ("tagged MCD") [17]. In the MCD spectrum the ionisation transition from the lower level to the conduction band dominates, the intensity ratio of the band located at 1.46 eV to the band at 1.1 eV is 4.7 to 1. It is observed with about the same signal intensity in many different s.i. samples (see Table 2 last column). This suggests that this defect is not a result of a particular thermal treatment but a major intrinsic defect as is EL2. Its influence on the occupations of the EL2 levels is confirmed by spatially resolved measurements.

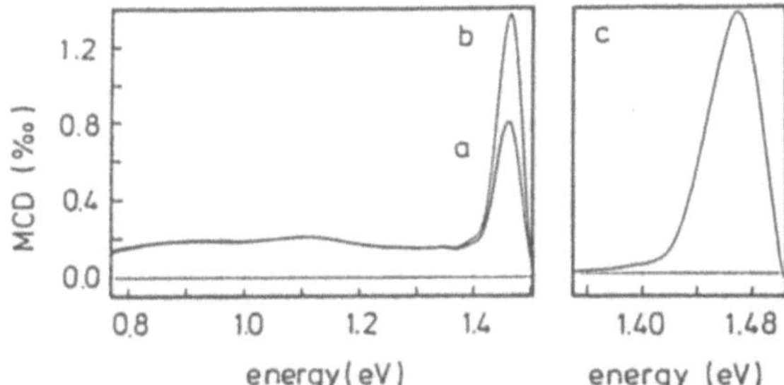

Fig. 3 (a) MCD spectrum of the trigonal Ga$_{As}$ acceptor pair after EL2 bleaching at T<10 K with a tungsten lamp at E=1.16 eV,   (b) MCD after annealing at T=70 K and subsequent irradiation at T<10 K with a tungsten lamp for 3–4 minutes at E=1.16 eV, (c) is the difference MCD spectrum of (a) and (b).

4.3 Spatially Resolved NIR Absorption and MCD Measurements

The near infrared (NIR) absorption of the EL2 mid gap level defect is used to image the two dimensional non-uniform EL2 defect distribution [18]. In as-grown s.i. wafers, which were not subjected to any thermal treatment, the NIR absorption is W-shaped across the wafer. For such material the MCD, which monitors the concentration of the paramagnetic defects, is anticorrelated (M-shaped), the total EL2 concentration being rather homogenous (see Fig.4) [19]. The fluctuations in the NIR absorption and the MCD should therefore be caused by an inhomogenous distribution of the compensated acceptor having a M-shaped profile. Indeed, the trigonal acceptor follows the distribution of the paramagnetic EL2 (Fig.5). It was

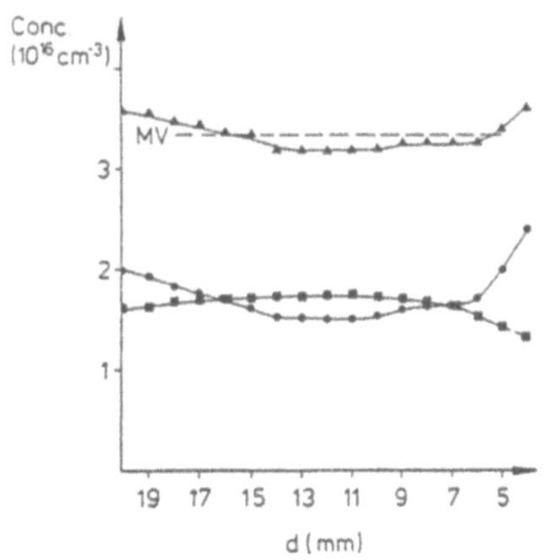

Fig. 4

Spatially resolved MCD (paramagnetic EL2, squares) and IR absorption (mid gap EL2, circles) measurements across half a s.i. as-grown GaAs wafer (triangles: total EL2 concentration, MV: mean value)

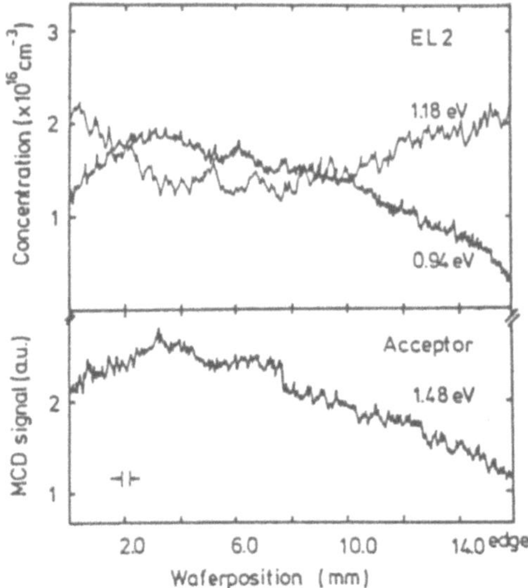

Fig. 5 Spatially resolved IR absorption at 1.18 eV (mid gap EL2) and MCD at 0.94 eV (paramagnetic EL2) and MCD at 1.48 eV (trigonal Ga_{As} acceptor pair)

measured with a one dimensional linescan over half a wafer diameter, the spatial resolution being $300\mu m$. For this particular wafer the total amount of C and Zn was again of the order of $4-5\cdot10^{15}cm^{-3}$. This implies that the fluctuations across the wafer of the order of $1\cdot10^{16}cm^{-3}$ are determined by the spatial variation of the trigonal acceptor and not by the shallow ones.

4.4 The EL2 Transformation–Influence on Intrinsic Acceptors

In s.i. GaAs the transfer of the EL2 into its metastable state and the release of free holes with their subsequent trapping at the acceptors occurs in a short time regime of the order of 30 to 60 seconds at T< 10K. The MCD of the trigonal acceptor pair appears with its MCD spectrum characterised by the typical intensity ratio of the 1.46 eV to that at 1.1 eV mentioned above. This intensity ratio changes when further annealing the samples at T=70 K for 10 minutes and after subsequent irradiation (3-4 minutes) with light of E=1.16 eV energy at T< 10K, which is needed for an efficient transformation of the mid gap EL2 into its metastable state. Surprisingly, only the high energy MCD band increases in intensity, the low energy band remains constant. An additional MCD band with its transition energy around 1.47 eV appears (Fig.3, curve b).

By subtracting the MCD spectrum of the trigonal acceptor pair one obtains a single paramagnetic MCD band (Fig.3, curve c). Its shape is very similar to the electron

206

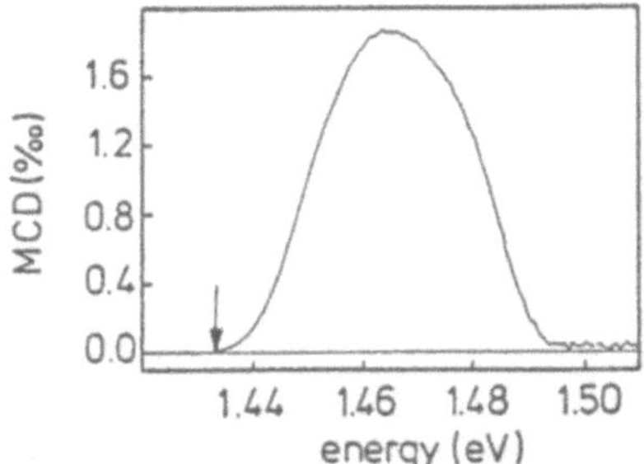

Fig. 6 MCD spectrum of the photoionisation transition of the single ionized Ga_{As} acceptor after subtraction of other MCD spectra ("tagged MCD")

ionisation transition of the Mn acceptor ($Mn^- \Longrightarrow Mn^0 - e_{cb}$) in GaAs with its energy level at $E_{vb} + 113$ meV [20]. From its onset at 1.44 eV a binding energy of 78 ± 4 meV can be inferred, when attributing the MCD band to the ionsation transition $A^- \Longrightarrow A^0 + e_{cb}$ for a double acceptor with A^- being the paramagnetic charge state. The formation of a shallow defect upon EL2 bleaching has been reported and will be reviewed briefly. After prolonged irradiation (of the order of 5-10min.) a trap appeared at 85K in the thermally stimulated current spectrum, its amplitude increasing with illumination time [21]. Its binding energy was calculated to be 50 ± 10 meV and it was concluded not to be due to a residual extrinsic acceptor. Photo-Hall measurements clearly revealed that p-type conductivity was connected with this trap. Kuszko [21] concluded that this trap resulted from the photodissociation of the metastable EL2 state into a neutral Ga vacancy and the neutral As interstitial i.e. $As_{Ga}^{0\cdot} \Longrightarrow V_{Ga}^0 + As_i^0$. The p-type conductivity should be connected with the properties of the Ga vacancy, which is an acceptor in GaAs. In optically induced photoconductivity measurements [22] a near band gap photo-current spectrum was obtained with a spectral shape very similar to the MCD spectrum found in Gallium-rich GaAs (see below). It shows an onset at 1.43 eV with a maximum at 1.48 eV (full circles in Fig.7). This photo-generated photo-current peak is again associated with p-type conductivity. One can speculate that the observation of this defect in s.i. material is connected with the structural changes of EL2 when transformed into

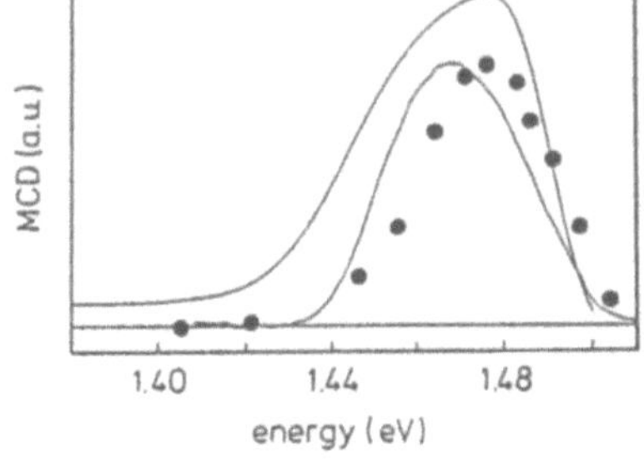

Fig. 7

MCD spectra of the trigonal Ga_{As} acceptor pair and the 78/203 meV Ga_{As} acceptor in comparison with the photo-current spectrum after [22] (dots)

the metastable state. The association of this trap with the metastable EL2 state is unlikely, since we observe it also in Gallium-rich material (see below). The crystal used in the experiment with a melt composition As/As+Ga of 0.45 contains at most 10^{14}cm^{-3} EL2.

4.5 Identification of Intrinsic Acceptors in p-Type GaAs

The isolated cation antisite defect Ga_{As} is a double acceptor in GaAs with energy levels at 78 (0/-) and 203 meV (-/- -) above the valence band as determined from Raman and FIR (Fourier Infrared) measurements in Gallium–rich material [23,24]. However, a quantitative analysis of the defect concentration was missing so far. In samples where the background concentration of the extrinsic shallow acceptors C and Zn exceeds that of the deep level defects, DLTS experiments could be performed. The two levels of Ga_{As} were observed by DLTS in Gallium-rich GaAs (As/As +Ga=0.45) quite recently. They were found at 55 and 130 meV above the valence band (see Fig.8) and appear with the same concentration. After correction for the electrical field dependence (Poole Frenkel effect) the level positions 78 and 203 meV are calculated [25].

Based on the DLTS experiments, the Raman measurements can now be calibrated and thus made quantitative. They were performed on the same sample for which DLTS was measured [25] and monitor the transitions originating from the neutral and singly ionized defect ground states to the excited states. For MCD and ESR investigations this material is not suited since the Ga_{As} defect is in its neutral, diamagnetic charge state. The Fermi level is pinned at the shallow extrinsic acceptors with energy levels around 30 meV above the valence band.

To observe the paramagnetic singly negative charge state of Ga_{As}, the Fermi level must be above the 0/- level (E_{vb}+0.078 eV) but below the -/- - level at E_{vb}+0.203 eV. In order to shift the Fermi level accordingly, the material had to be compensated, which was done by electron irradiation with low doses. The shift of the Fermi level is again monitored by Raman measurements. As soon as the Ga_{As} defect was in its singly negative charge state, the MCD spectrum shown in Fig.6 after subtraction of underlying MCD spectra ("tagged MCD") due to other defects is observed. From

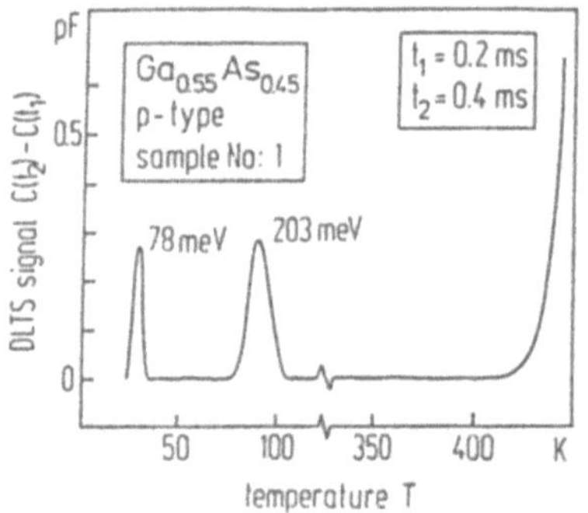

Fig. 8

DLTS spectrum of Gallium-rich GaAs: Ga$_{0.55}$ As$_{0.45}$ versus temperature

208

the temperature and magnetic field dependence of the MCD the spin value of the defect can be obtained if its g-factor can be determined independently e.g. by ESR or ODESR [26]. However, it was not possible to observe an ODESR spectrum in the magnetic field range up to values corresponding to $g \simeq 0.5$ and using microwave powers up to 1W. This may be expected, if the ground and excited state of the corresponding defect can be described within the effective mass theory (EMT). For this type of defects the ESR can only be observed by the application of uniaxial stress [27]. Assuming a spin-value of S=J=3/2 as is the case for the EMT acceptors C and Zn the Boltzmann polarisation of the ground state as measured in the MCD can be fitted with a g-factor of 1.3 ± 0.2, which is in the range of 0.6 (C, Zn) and 2 (expected for the deeper acceptors dominated by a central cell potential). This behaviour along with the binding energy of 78 ± 4 meV indicates the presence of the isolated Ga_{As} antisite defect in p-type and in s.i. material (see Fig.3b and Fig.6). The observation of Ga_{As} in s.i. material is an unexpected result. We note, however, that the first experiments (photoluminescence and FIR absorption) which showed the presence of the two levels 78 and 203 meV were undertaken in crystals grown with 0.5 As atom fraction compensated with Si [24].

4.6 A Quantitative Estimate of the Acceptor Concentration

We now adress the question of the concentration of the intrinsic acceptor defects i.e. GaAs and the trigonal Ga_{As}-acceptor pair. Based on the assumption that the single MCD band is caused by the isolated 78/203 meV acceptor, the quantitative analysis uses the correlation of Raman and DLTS experiments described in Sect.4.5 [25], in which the Raman spectrum of the neutral Ga_{As} defects was calibrated by means of the DLTS experiment. Prior to electron irradiation the concentration of the 78/203 meV acceptor was $(5 \pm 1) \cdot 10^{16} cm^{-3}$. After electron irradiation the Raman transitions of the neutral Ga_{As} defects vanished completely. The Fermi level was now located between $E_{vb}+78$ meV and $E_{vb}+203$ meV. Therefore the MCD signal at 1.47 eV in this irradiated material is due to the same $(5 \pm 1) \cdot 10^{16} cm^{-3}$ Ga_{As} defects in the singly negative charge state. With this calibration the concentration of defects in s.i. material was determined (Table 2).

Table 2

sample	midgap EL2 10^{15}	param. EL2 10^{15}	[C + Zn] 10^{15}	Ga^-_{As} 10^{15}	trig. Ga^-_{As} $10^{15}(cm^{-3})$
R1	30.6	7.3	3.7	1.5	3.4
R2	28.0	15.8	3.0	7.4	12.0
I2	20.7	9.9	4.6	5.8	7.6
W1	5.2	4.9	0.5	6.2	13.0
W2	21.6	12.7	5.4	7.1	5.5

The trigonal acceptor could not get be observed by conventional ESR. Conventionial ESR would have allowed a determination of its concentration. The observed spatial variations of the mid gap EL2 indicate concentrations in the order $5 - 8 \cdot 10^{15} \mathrm{cm}^{-3}$. The MCD intensity of the high energy band in different s.i. samples reflects its concentration. In a first crude estimate we assume that the optical cross section (oscillator strength) of the trigonal Ga_{As}-acceptor pair and the 78 meV Ga_{As}-acceptor are identical for the respective transition to the conduction band ($A^- \Longrightarrow A^0 + e_{cb}$). This would underestimate the concentration of the trigonal Ga_{As}-acceptor, since one can expect that for the deeper center i.e. the trigonal one, the oscillator strength would be lower. Moreover, the MCD reflects only the paramagnetic charge state of the trigonal Ga_{As}-acceptor. Therefore, the estimate of the concentration of the trigonal acceptor is a lower limit. With these assumptions the concentration given in Table 2 were obtained.

Except for the Gallium-rich sample W1 there is a very good agreement between the concentration of the trigonal Ga_{As} acceptor and the concentration of the paramagnetic EL2 not compensated by the shallow acceptors C and Zn (compare columns 3 and 5 in Table 2). If the trigonal acceptor were the major compensating acceptor, this could be expected, since the holes released by the metastable transformation of EL2 are captured at the deep intrinsic acceptors. Our experimental findings on this capture process differ from those reported in the literature on the photoneutralisation of the shallow extrinsic acceptors using Raman spectroscopy [28]. The appearance of neutral C and Zn acceptors in the range $5 \cdot 10^{15} \mathrm{cm}^{-3}$ after bleaching EL2 with a halogen lamp or with a Nd YAG laser as often done would easily be seen in the MCD spectrum at longer wavelength (0.6-0.8 eV). From MCD investigations on a GaAs crystal intentionally doped with Zn their respective MCD transition e.g. $Zn^0 \Longrightarrow Zn^- + h_{vb}$ has been observed [29]. The MCD corresponding to $2 \cdot 10^{16} \mathrm{cm}^{-3}$ Zn^0 is by a factor 10-20 stronger at 0.8 eV compared to the MCD of the trigonal acceptor at the same energy. This would allow to detect easily $1 - 2 \cdot 10^{15} \mathrm{cm}^{-3}$ neutral shallow extrinsic acceptors (Fig.9).

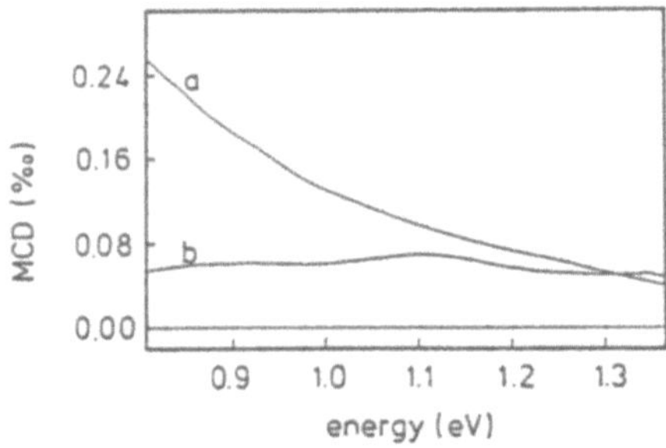

Fig. 9 Comparison of the MCD spectrum of a GaAs crystal doped with Zn ($1.4 \cdot 10^{16} \mathrm{cm}^{-3}$) (a) to the MCD spectrum of the trigonal acceptor (b). The MCD of the Zn doped crystal is divided by a factor of 7, i.e. it represents the MCD of neutral Zn acceptors of the concentration $2 \cdot 10^{15} \mathrm{cm}^{-3}$.

After the bleaching procedure the Nd YAG laser light is removed and the samples are studied under thermal equilibrium conditions, whereas in the Raman experiments the laser light is persistently maintained on the sample necessary for the Raman detection mechanism. It is conceivable that under these experimental conditions a different light induced equilibrium (quasi Fremi level) is present. The study of the material using different techniques (Raman, FIR, ESR and MCD), especially exchanging samples, will help to understand the experimental conditions for the observation and non-observation of the intrinsic and extrinsic acceptors.

5 Discussion and Conclusion–Charge State of the EL2 Mid Gap Level

The connection between the compensation mechanism and the charge state of the mid gap EL2 level is outlined in the following: For the isolated antisite model the concentration of the singly ionized EL2 should be lower or equal to the number of residual acceptors. For the triple donor model, in which the mid gap level is positively charged, As^0_{Ga}-As^+_i, the concentration of the acceptors must be comparable to the total concentration of EL2, at least larger than that of the paramagnetic EL2. Our experiments support the triple donor model (see Table 2). One has to note that the MCD investigations measure only the number of the paramagnetic acceptors A^-. The concentrations in Table 2 are therefore only the lower limits. The total number depends on the fractional occupation f of the acceptor levels, i.e. $N_{tot}=fA^-+(1-f)A^{--}$.

Additional intrinsic donors deeper than the shallow ones such as S and Si are not present in s.i. GaAs in significant concentrations. From DLTS experiments on conducting samples it is concluded that only the EL6 level appears in comparable concentrations (i.e. $1-2\cdot10^{15}$cm^{-3}) [30] (it is unclear at present whether EL6 is a donor or an acceptor) [31]. However, there are still other acceptors present in the material close or above this value. Data are available for the deep acceptor iron $(1-5\cdot10^{15}$cm$^{-3})$ [32] and, last not least, the FR1 to FR3 defects mentioned above, of which no concentration could be determined.

Our quantitative analysis makes the isolated antisite model on the basis of the compensation behaviour of the EL2 defect very unlikely. It is not based on a specific defect identification technique, as ODENDOR, DLTS/ESR or DLTS under stress [4,5,6,7] which all showed the presence of a complex defect, i.e. As_{Ga}-As_i. The results presented here call for a positively charged mid gap level. A close spatial correlation between the EL2 donor and the intrinsic acceptors seems natural. Indeed, the mobility data of the "state of the art" s.i. GaAs has been interpreted in terms of a dipolar scattering mechanism (neutral donor acceptor pair) [33].

Acknowledgement

We would like to thank J. Wagner (IAF Freiburg) for the Raman measurements on our samples and M. Fockele for preparing this manuscript.

References

[1] Recent developements in the study of the EL2 defect in GaAs edited by *H.J. Bardeleben*, and *B. Pajot*, Rev.Phys.Appl.**23**(France), p.727-869 (1988)

[2] *D.J. Chadi*, and *K.J. Chang*, Phys.Rev.Lett.**60**, p.2187 (1988)

[3] *J. Dabrowski*, and *M. Scheffler*, Phys.Rev.Lett.**60**, p.2183 (1988)

[4] *B.K. Meyer, D.M. Hofmann, J.R. Niklas*, and *J.-M. Spaeth* Phys.Rev.**B36**, p.1332 (1987)

[5] *H.J. Bardeleben, D. Stievenard* and *J.C. Bourgoin*, Appl.Phys.Lett.**47**, p.970 (1985)

[6] *M. Levinson*, and *J.A. Kafalas*, Phys.Rev.**B35**, p.9383 (1987)

[7] *J.C.Culbertson, U. Strom*, and *S.A. Wolf*, Phys.Rev.**B36**, p.2962 (1987)

[8] *M.J. Caldas*, and *A. Fazzio*, Proceedings of the 15th International Conference on Defects in Semiconductors, p.119, Budapest (1988)

[9] *D.E. Holmes, R.T. Chen*, and *C.G. Kirkpatrick*, Appl.Phys.Lett.**40**, p.46 (1982)

[10] *C.H. Kang, J. Lagowski*, and *H.C. Gatos*, J.Appl.Phys.**62**, p.3482 (1987)

[11] *J. Lagowski, M. Bugajski, M. Matsui*, and *H.C. Gatos*, Appl.Phys.Lett.**51**, p.511 (1987)

[12] *J. Lagowski, M. Matsui, M. Bugajski, C.H. Kang, M. Skowronski, H.C. Gatos, M. Hoinkins, E.R. Weber*, and *W. Walukiewicz*, Proceedings of the 14th International Symposium on GaAs and related compounds, p.395, Creta (1987)

[13] *J. Wagner, M. Ramsteiner*, and *H. Seelewind*, J.Appl.Phys.**64**, p.802 (1988)

[14] *J. Wagner*, and *H. Seelewind*, priv. comm.

[15] *G.M. Martin, J.P. Farges, G. Jacob, J.P. Hallais*, and *G. Poiblaud*, J.Appl.Phys.**51**, p.2840 (1980)

[16] *U. Kaufmann, M. Baeumler, J. Windscheif*, and *W. Wilkening*, Appl.Phys.Lett.**49**, p.1254 (1986)

[17] *K. Krambrock, B.K. Meyer*, and *J.-M. Spaeth*, Phys.Rev.**B39**, p.1973 (1989)

[18] *P. Dobrilla*, and *J.S. Blakemore*, J.Appl.Phys.**58**, p.208 (1985)

[19] *B.K. Meyer, D.M. Hofmann, M. Heinemann*, and *J.-M. Spaeth*, Proceedings of the 14th International Symposium on GaAs and related compounds, p.391, Creta (1987)

[20] *M. Baeumler, B.K. Meyer, U. Kaufmann*, and *J. Schneider*, Proceedings of the 15th International Conference on Defects in Semiconductors, p.797, Budapest (1988)

[21] *W. Kuszko, P.J. Walczak, P. Trautmann, M. Kaminska*, and *J.M. Baranowski*, Proceedings of the 14th International Conference on Defects in Semiconductors, p. 317, Paris (1986)

[22] *J. Jiminez, P. Hernandez, J.A. Saja*, and *J. Bonnafe*, Phys.Rev.**B35**, p.3832 (1987)

[23] *J. Wagner, M. Ramsteiner*, and *R.C. Newman*, Sol.Stat.Comm.**64**, p.459 (1987)

[24] *K.R. Elliot*, Appl.Phys.Lett.**42**, p.274 (1983)

[25] *G. Roos, A. Schöner, G. Pensl, K. Krambrock, B.K. Meyer, J.-M. Spaeth*, and *J. Wagner*, Proceedings of the 15th International Conference on Defects in Semiconductors, p.951, Budapest (1988)

[26] *A. Görger, B.K. Meyer*, and *J.-M. Spaeth*, Proceedings of the 5th Conference on Semi-insulating III-V Materials, p.331, Malmö, Sweden (1988)

[27] *F. Mehran, T.N. Morgan, R.S. Title*, and *S.E. Blum*, J.Magn.Res.**6**, p.620 (1972)

[28] *K. Wan*, and *R. Bray*, Phys.Rev.**B32**, p.5265 (1985)

212

[29]   *K. Krambrock, B.K. Meyer,* and *J.-M. Spaeth,* Proceedings of the 5th Conference
       on Semi-insulating III-V Materials, p.159, Malm ö, Sweden (1988)
[30]   *G.M. Martin, A. Mitonneau,* and *A. Mircea,* Electron.Lett.**13**, p.191, (1977)
[31]   *S. Makram-Ebeid,* and *P. Boher,* Rev.Phys.Appl.**23**, p.847, (1988)
[32]   *E.R. Weber,* and *M. Kaminska,* Proceedings of the 5th Conference on Semi-insulating
       III-V Materials, p.111, Malmö, Sweden (1988)
[33]   *H.J. Bardeleben, D. Stievenard, H.R. Zelsmann,* and *J.C. Bourgoin,* Proceedings
       of the 5th Conference on Semi-insulating III-V Materials, p.127, Malmö, Sweden
       (1988)

Deep Donor Levels (DX Centers) in III–V Semiconductors: Recent Experimental Results

Patricia M. Mooney

IBM Research Division, T.J. Watson Research Center, P O Box 218, Yorktown Heights, NY 10598 USA

Summary: The DX center, the lowest energy state of the donor in $Al_xGa_{1-x}As$ with $x \geq 0.22$, is responsible for the reduced conductivity as well as the persistent photoconductivity observed in this material at low temperature. Extensive studies of the properties of this deep level in Si-doped AlGaAs are reviewed here. It has been shown that the characteristics of the DX center remain essentially unchanged when it is resonant with the conduction band ($x \leq 0.22$) with the result that, independent of other compensation mechanisms, the DX center limits the free carrier concentration in Si-doped GaAs to a maximum of about $2 \cdot 10^{19}$ cm^{-3}. Recent measurements suggesting that the lattice relaxation involves the motion of the Si atom from the substitutional site toward an interstitial site are also presented. Evidence for the negative U model, that the DX level is the two electron state of the substitutional donor, is discussed.

1 Introduction

Independent of the donor dopant species the conductivity of n-type AlGaAs, when the AlAs mole fraction is $x \geq 0.22$, is controlled by a deep donor whose energy position *roughly* follows the L-valley of the conduction band [1–7]. Persistent photoconductivity (PPC) which is observed in AlGaAs at low temperature [1,3] has also been attributed to this deep donor state [8,9]. In the model originally proposed by Lang and co-workers [8,9] the deep level is a highly localized state with a strong coupling to the crystal lattice. Capture and emission of an electron to this level occur by a multiphonon process and thus are thermally activated. Capture of an electron results in a relaxation of the crystal lattice, a rearrangement of the atoms near the donor. At low temperature the capture rate is extremely slow resulting in PPC; once electrons are photoexcited to the conduction band from the DX level they remain in the conduction band even after the exciting light is turned off. A measure of the lattice relaxation is seen in the large energy difference between the photoionization energy and the binding energy of the level (Frank-Condon shift). In order to account for such a large lattice relaxation (LLR) energy, it was proposed that the origin of this electronic state is a defect complex consisting of the donor and another unknown defect, hence the name DX center. More recently others have proposed that the DX center is the isolated donor atom; both LLR [10–12] and small lattice relaxation (SLR) models have been suggested [13–17]. It has also been proposed that this level has negative U [18,19]. In this review it will be argued that the experimental evidence strongly supports a LLR model. Evidence for negative U and the microscopic configuration of the DX level will also be discussed.

Because AlGaAs has properties which make it attractive for many types of heterojunction device structures, e.g. lasers and several types of transistors, there has been an explosion of research on epitaxial growth of various device structures using AlGaAs. There has been much new work on the physics of this deep level not only because of its interesting and unusual properties but also because a fundamental understanding of this deep level is needed in order to evaluate the usefulness of AlGaAs for various device structures. Deep donor levels with similar properties are also found in many other wide band gap semiconductors, including other III–V alloys such as GaAsP [20], GaInAs [21], and InAlAs [22], which are also interesting for heterojunction device structures. Other examples include some II–VI materials such as Cl-doped CdZnTe [23,24] as well as ionic crystals such as In-doped CdF_2 [23]. The discussion in this review will be focused on experimental results for GaAs and AlGaAs; however, deep donors with LLR are a general phenomenon occurring in a variety of materials.

2 Properties of the DX Level in AlGaAs with x ≥ 0.22

A schematic diagram of the DX level binding energy from temperature dependent Hall measurements performed in many laboratories [1–7] plotted with the energy of the three conduction band minima in AlGaAs is shown in Fig. 1 a. The DX level *roughly* follows the L-valley, moving below the bottom of the conduction band when x = 0.22 and going deeper into the gap as x increases. Beyond the crossover point from direct to indirect gap the DX level lies closer to the bottom of the conduction band. Similar behavior was seen when the band structure was modified by the application of hydrostatic pressure [25]. Deep level transient spectroscopy (DLTS) measurements of the change in occupation of the DX level in $Al_{0.74}Ga_{0.26}As$ with applied hydrostatic pressure suggest that the DX level roughly follows the L-valley even for high AlAs mole fraction [26]. A level which roughly follows the L-valley was also observed by photoluminescence excitation spectroscopy [27]. Based on these data it was suggested that the DX level is made up of states from the L conduction band minimum as is predicted by effective mass theory for shallow donors [2,3,27,28].

Recent measurements of the pressure coefficient of the DX level demonstrate that the pressure coefficient of the DX level is not the same as that of the L-valley [29,30] indicating that the DX level cannot be an effective-mass-like level made up only of L-valley states. It has been proposed that the DX level, like other deep localized states, is made up of states from the whole conduction band and that it's energy level follows the average density of

216

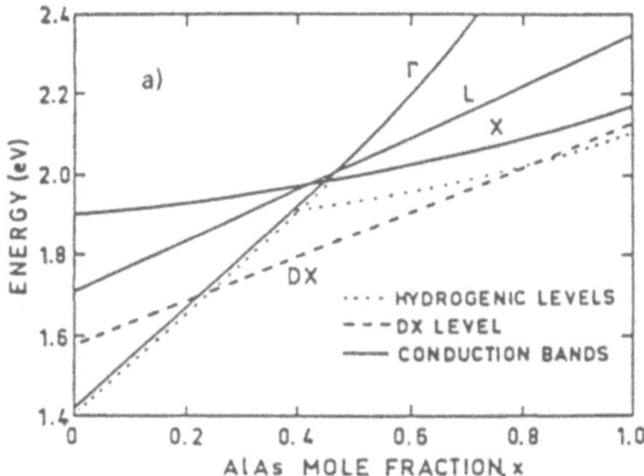

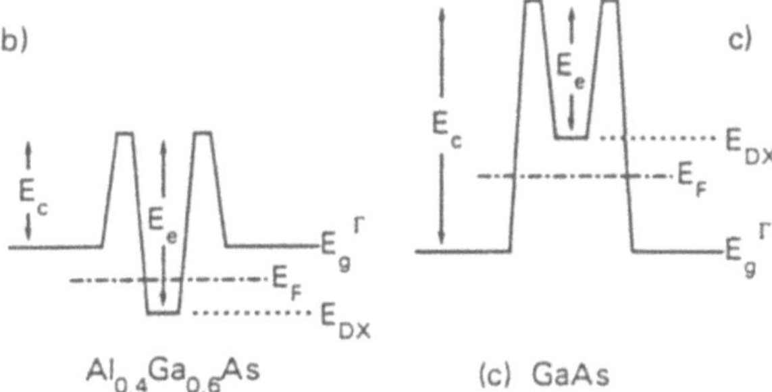

Fig. 1 (a) Schematic diagram showing the alloy composition dependence of the three conduction band valleys (solid lines), the DX level (dashed line) and the hydrogenic states associated with the Γ- and X-valleys in Si-doped AlGaAs. Diagrams indicating the capture and emission barriers when (b) the DX level lies below the bottom of the conduction band and (c) it is resonant with the conduction band states.

states in the conduction band which lies close to the bottom of the L-valley [18]. This explains the alloy composition dependence of the DX level shown in Fig. 1 a.

Hall data in direct gap samples have been interpreted in terms of two levels, a deep donor and a shallow one [4,5] and DLTS measurements have also indicated the presence of both deep and shallow levels [31]. The Hall measurements [4,5] were modeled by assuming different chemical species for the two donors, but far infrared absorption measurements show that the same chemical species gives rise to both the DX level and a shallow state lying close to the Γ conduction band minimum [32]. The same conclusion is reached through a proper thermodynamic analysis of Hall data [33]. For x $\geq$ 0.22, the DX level is the lowest state of the donor and is therefore the occupied state. Hydrogenic levels associated with the X-valley have also been observed by Hall [34] and infrared absorption [35] measurements. These hydrogenic states are indicated schematically on Fig. 1a by dotted lines.

The activation energy for thermal emission, measured by constant capacitance DLTS, is independent of the alloy composition and doping concentration [36–38]. As shown in Table 1 the emission energy depends on the donor species [39]; in the case of silicon doping it is 0.43 eV. Consistent with the existence of a thermally activated capture cross section, the emission energy is much larger than the binding energy (Fig. 1b). The DLTS peak is broadened and sometimes shoulders have been observed but only a single broad peak was resolved in our samples when x $\geq$ 0.14.

In AlGaAs the local environment of the donor atoms varies due to the random distribution of the group III atoms. This could result in a distribution of emission rates rather than a single emission rate for a deep level whose electronic wave function is spatially localized. Calleja, et al. have investigated the alloy composition dependence of the non-exponential character of the emission transients in samples where only a single DLTS peak is observed for the DX level [37,40]. The deviation from a single exponential is a maximum near x = 0.5 and decreases rapidly as x is increased or decreased as expected for an alloy effect [41]. The transient is unchanged when hydrostatic pressure is applied to the sample confirming that the non-exponential nature of the transient is unrelated to the alloy band structure. Similar conclusions have been reached from other experiments [42,43]. The sensitivity of the DX center emission kinetics to the local environment of the Si atom demonstrates the highly localized nature of this level. As will be discussed in detail in section 4, discrete DLTS peaks corresponding to different local environments around the donor atoms have been observed in the dilute alloy.

The kinetics of electron capture at the DX level have been studied in samples with different doping concentrations and over a wide range of alloy compositions as well as with applied hydrostatic pressure [37,44].

The temperature dependence of the capture cross section is conventionally written $\sigma = \sigma_\infty \exp(-E_c/kT)$, where E_c is the capture activation energy measured with respect to the bottom of the conduction band. It was shown that a much better fit to the observed capture transients results from assuming $\sigma = \sigma_\infty \exp -(E_c - E_F)/kT$ where E_F is the quasi Fermi level [38]. Such an expression results if capture occurs through an intermediate state lying high in the conduction band. The movement of the quasi-Fermi level during the capture explains the observed broadening of the capture transient in heavily doped samples. Better agreement between the model and the experimental data were achieved when an inhomogeneous broadening was included, consistent with the alloy effects discussed in the previous section. The alloy composition dependence of E_c is plotted in Fig. 2. The large capture bar-

218

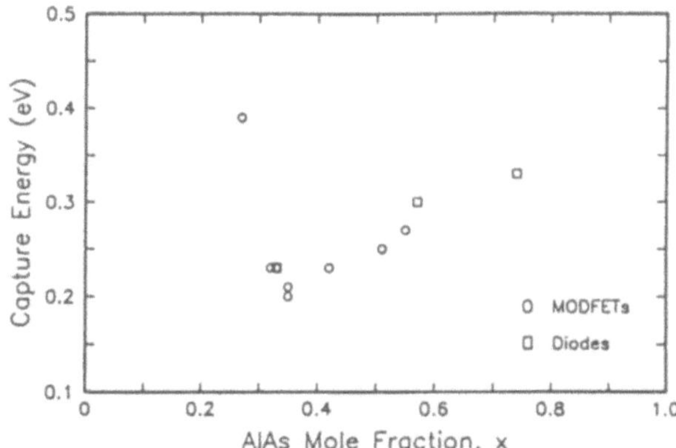

Fig. 2 Alloy composition dependence of the thermal capture energy of the DX level in Si-doped AlGaAs.

Table 1 Chemical Trends for Substitutional Donors in AlGaAs

Donor	$E_c^{a)}$ (eV)	E_b (eV)	E_0 (eV)
S	0.28 [b,c]	—	—
Se	0.28 [b,c]	0.14 [c]	0.85 [c]
Te	0.28 [b,c]	0.14 [c]	0.85 [c]
Si	0.43 [b,c,d]	0.21 [d]	1.45 [f]
Ge	0.33 [b]	—	—
Sn	0.21 [b]	0.02 [e]	1.11 [c]

a) $E_b = E_c - (E_L - E_\Gamma)$
b) Ref. 39
c) Ref. 45
d) Ref. 38
e) Ref. 46
f) Ref. 48

rier is indicative of an electron capture process involving a large number of phonons (multiphonon capture). As shown in Table 1 the magnitude of the capture barrier depends on the donor species [45,46].

Measurements of the time for the trap to become half full at 115 K can be used to determine the magnitude of the average value of the effective capture cross section assuming $\sigma = 1/nv\tau$, where n is the free carrier concentration, v is the average thermal velocity of the electrons, and τ is the time for the trap to become half occupied. Fig. 3 shows the temperature dependence of σ for the largest (x = 0.27) and the smallest (x = 0.35) measured values of E_c. The point on each curve indicates the value at 115 K, the solid line

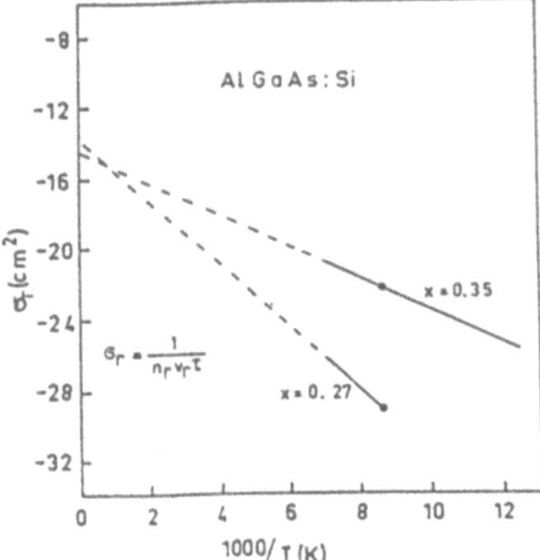

Fig. 3 Temperature dependence of the thermal capture cross section of the DX level in Si-doped AlGaAs.

indicates the temperature range over which the activation energy has been measured, and the dashed line is an extrapolation of the data. The values of the effective capture cross sections fall between these two curves. They extrapolate to a value of σ_∞ equal to 10^{-14}–10^{-15} cm^2 which is comparable to what is found for other deep levels with multiphonon capture processes [47].

Two important pieces of information can be learned from such a plot. First, the extremely small magnitude of the thermal capture cross section at low temperature explains the PPC observed in AlGaAs. Once the electrons are optically excited from the DX level at low temperature, they don't have enough thermal energy to surmount the capture barrier and thus they remain in the conduction band for very long times. The suppression of other capture mechanisms, e.g. a radiative capture process, even when the magnitude of the thermal capture cross section is 10^{-30} cm^2, is strong evidence for a large lattice relaxation. Second, we note that the data are exponential over the entire measurement temperature range down to 80 K. The temperature at which the radiative capture process becomes dominant over the thermal capture process indicates the energy of the phonon modes involved in the lattice relaxation [47]. In this case the thermal capture process dominates even at 80 K, which implies a phonon energy of less than 10 meV.

An important property of the DX level is the magnitude of the lattice relaxation energy when an electron is captured. A key experiment to determine the lattice relaxation energy is the measurement of the photon energy dependence of $\sigma_n^0(h\nu)$, the photoionization cross section. The difference between the optical ionization energy and the donor binding energy is the lattice relaxation energy [47]. Like the thermal capture and emission energies, the photoionization energy depends on the dopant species (see Table

220

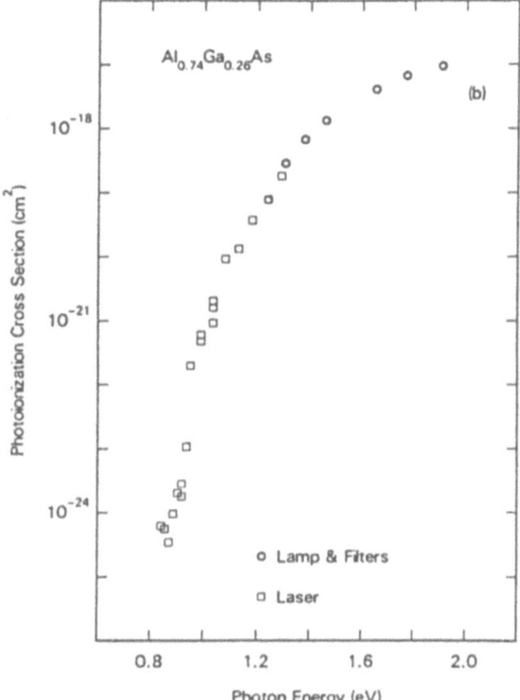

Fig. 4 Photoionization cross section of the DX level in Si-doped AlGaAs.

1) [45]. Extensive studies of σ_n^0 in Te-doped AlGaAs [9] show that there is little dependence on the alloy composition. The large difference between the photoionization threshold and the donor binding energy as well as the temperature dependence of σ_n^0 are in agreement with a multiphonon capture model. Since the thermal emission energy for the DX level in Si-doped AlGaAs is much larger than that in Te-doped AlGaAs, σ_n^0 can be measured over a much larger temperature range. Thus measurements of the temperature dependence of σ_n^0 in Si-doped AlGaAs provide a more rigorous test of the LLR model. Extensive measurements of σ_n^0 in Si-doped AlGaAs show little dependence of the cross section on alloy composition, and a large threshold energy and temperature dependence in excellent agreement with the LLR model [48]. The optical ionization energy of the DX level in Si-doped AlGaAs was found to be 1.25 eV [45,48]. In these experiments σ_n^0 was determined from the time constant of the voltage transient when the sample was exposed to monochromatic light at constant capacitance. The magnitude of the cross section was measured over about 3 orders of magnitude. Recently the sensitivity of the photocapacitance technique was improved and σ_n^0 was measured over 8 orders of magnitude (Fig. 4) [49]. No photocapacitance transient which could be attributed to the DX level was observed for photon energies below 0.8 eV.

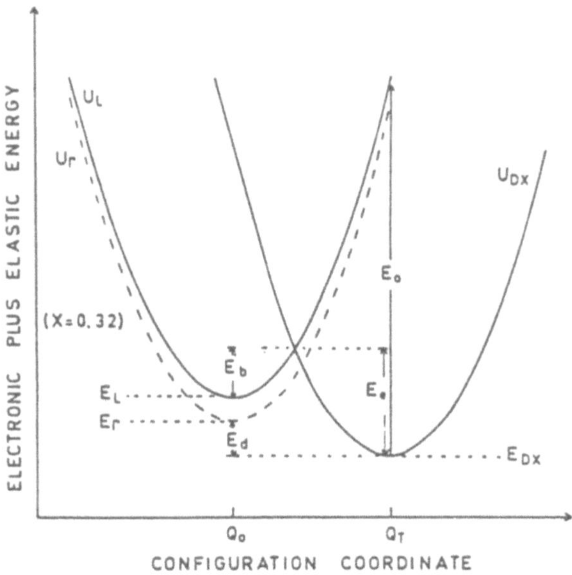

Fig. 5 Configuration coordinate diagram of the DX level in Si-doped AlGaAs. (See text for explanation.)

A detailed analysis of the photon energy dependence of σ_n^0 of the DX level in Ref. 49, using a model which is independent of the microscopic structure of the DX center, gives a value for the optical ionization energy of 1.3–1.6 eV. The large uncertainty is due to the scatter in the data. This value is slightly higher than that found from an analysis of the earlier measurements, no doubt because of the increased range of data for σ_n^0. Analysis of the temperature dependence of σ_n^0 indicates that the dominant phonon mode involved in the lattice relaxation process is 5.6 ± 1.0 meV. This is excellent agreement with the thermal capture data from which it is inferred that the energy of the dominant phonon mode must be less than 10 meV. Thus two independent experiments, measurements of the thermal capture kinetics and the optical emission kinetics, demonstrate that the DX level has a large lattice relaxation energy and that the lattice relaxation process invloves a large number of very low energy phonons.

The four energies discussed above, which characterize the DX level, can be summarized in a configuration coordinate diagram shown in Fig. 5. The vertical axis represents the total energy, electronic plus elastic, and the lattice distortion is represented on the horizontal axis by a single coordinate Q. The parabolas centered at Q_0 represent the undistorted lattice with the ionized donor at the substitutional site and the electron in the conduction band. The occupied DX level is represented by the parabola shifted to the

222

right. In order to facilitate comparisons among samples doped with different donor species, we eliminate the alloy composition dependence of the capture energy by expressing it as $E_c = E_b + (E_L - E_\Gamma)$ in both Fig. 5 and Table 1. Thus the L-valley is indicated in Fig. 5 as a reference energy, not to suggest that capture necessarily occurs through a state of the L-valley. Optical transitions occur in a vertical direction on a configuration coordinate diagram. Note that due to the large shift in Q for the occupied DX state (large lattice relaxation) there is no radiative capture path, consistent with the observation of thermally activated (non-radiative) capture down to the lowest temperature at which the capture can still be measured.

3 The DX Level in GaAs and AlGaAs with x ≤ 0.22

As shown in Fig. 1 a, extrapolating the energy position of the DX level to smaller values of x suggests that the DX level lies about 170 meV above the bottom of the conduction band in GaAs [50]. At the usual doping concentrations its occupation in thermal equilibrium is negligible and therefore the level is not observed. However, applying hydrostatic pressure to GaAs modifies the conduction band structure in much the same way as increasing the AlAs mole fraction does. Thus above a critical pressure the DX level, which moves *roughly* with the L-valley of the conduction band, is pushed below the bottom of the conduction band into the gap; both PPC and a DLTS peak are then observed [51–53]. The DLTS peak is narrower and the thermal emission energy is lower than what is observed in AlGaAs; however, the characteristic temperature dependent capture of the DX level is observed [52]. The concentration of electrons trapped at the DX level was found to be equal to the Si doping concentration [51,52]. In another experiment the photon energy dependence of the photoionization cross section for Te-doped GaAs under pressure was found to be identical to that in Te-doped AlGaAs [54].

Heavily doped GaAs samples with the Fermi level in the range 140–240 meV above the bottom of the conduction band were prepared in order to study the characteristics of the DX center in GaAs without applied hydrostatic pressure [55]. A DLTS spectrum is shown in Fig. 6. As was found for GaAs under pressure, the DLTS peak is narrower and shifted to lower temperature compared to that in the alloy. The thermal emission energy, 0.33 eV, agrees well with what was measured in lightly doped GaAs under pressure indicating that the pressure coefficient of the emission energy is negligible. (This result is confirmed by measurements of the pressure coefficient of the emission energy in Te-doped AlGaAs [29].) The emission transient is exponential, consistent with the absence of alloy effects in GaAs [40]. This also demonstrates the strong spatial localization of the level which is not measurably perturbed by neighboring Si atoms, either donors or acceptors, at an average distance of $\simeq$ 35 Å. As expected, a thermally activated capture rate was observed. The DLTS measurements indicate a significant occupa-

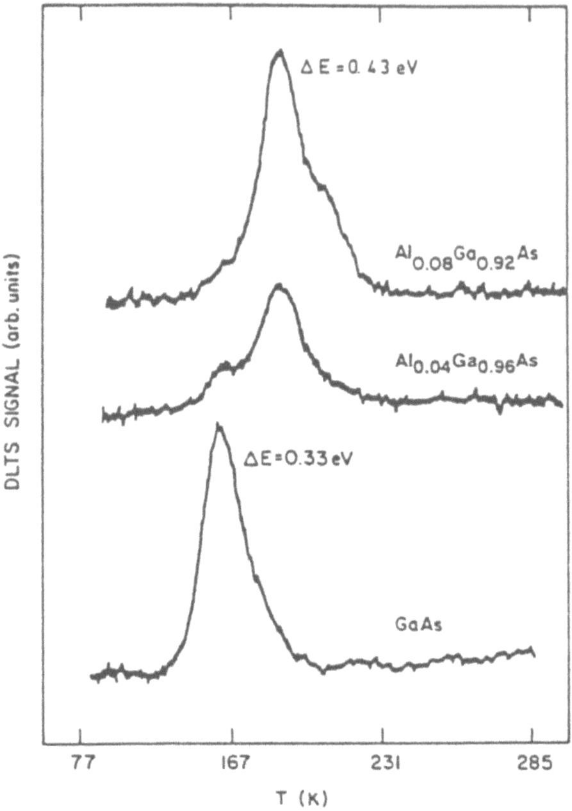

Fig. 6   DLTS spectra for Si-doped GaAs and dilute AlGaAs showing the discrete shift of the emission rate when Al is present.

tion of the DX level reaching $n_{DX}/n_0 \simeq 0.06$ at $n_0 = 1.1 \cdot 10^{19}$ cm^{-3}, where n_{DX} is the concentration of electrons trapped at the DX level. As n_0 increases, an increasing fraction of the additional electrons are lost to the DX level. A straightforward extrapolation shows that the Fermi level reaches the DX level at $n_0 = 1.5 - 2 \cdot 10^{19}$ cm^{-3}. If $N_{DX} > (N_D - N_A)$, this will be the maximum possible value of n_0 in Si-doped GaAs. A similar limit is inferred from hydrostatic pressure experiments [56,57].

The DX level is a highly localized state even when it is resonant with the conduction band. This occurs because the DX level is isolated from the extended states of the conduction band by capture and emission barriers as indicated in Fig. 1 c, and thus the transition rates between the DX level and the band states are extremely small, consistent with a large energetic relaxation of the lattice.

224

4 Microscopic Structure of the DX Center

The experiments reviewed above strongly support a large lattice relaxation model for the DX level and demonstrate that the DX level is a localized state with strong coupling to the crystal lattice. The observation of the DX level in GaAs rules out alloy fluctuation models [16,17] for the capture barrier. A number of experiments strongly suggest that the DX level is not a defect complex but is simply the isolated donor atom. Near the direct-indirect crossover composition, where the binding energy of the DX level is largest, all electrons are trapped at the DX level at low temperature [31]. In lightly doped GaAs under about 25–30 kbar of hydrostatic pressure all electrons are trapped at the DX level; in GaAs without applied pressure all the donors have the properties of shallow effective mass-like levels derived from substitutional impurities [51–53]. The most direct evidence that all the donors are isolated substitutional atoms comes from measurements of the local vibrational modes of *ionized* silicon atoms in heavily doped GaAs grown at very low temperature by molecular beam epitaxy (MBE) [56,57]. In a sample with a free carrier concentration of $1.1 \cdot 10^{19}$ cm$^{-3} \simeq 90\%$ of the Si is Si_{Ga}; Si-Si pairs and Si complexes (Si-X and Si-Y) occur in very low concentrations. With about 9 kbar of hydrostatic pressure applied, more than 25% of the electrons were persistently trapped at the DX level as indicated by measurements of the PPC in the same sample [56]. We conclude from this that the lattice relaxation which occurs when electrons are captured at the DX level is a distortion of the atomic configuration around the isolated Si atom.

Recent DLTS measurements in dilute AlGaAs alloys suggest that the silicon atom moves from the group III site towards an interstitial site as has been proposed independently in Refs. 18 and 58. DLTS spectra for Si-doped GaAs and AlGaAs with x = 0.04 and 0.08 are shown in Fig. 6 [59,60]. Notice that the same DLTS peak dominates the spectra of both the x = 0.04 and 0.08 samples; it is shifted to higher temperature and has a larger emission energy compared to the GaAs peak. Note also the presence of the GaAs peak in the spectrum for the x = 0.04 sample and the appearance of a third peak at higher temperature in the x = 0.08 sample. This third peak dominates the DLTS spectra in samples with x $\geq$ 0.14. The emission energy of this peak is also 0.43 eV (see Fig. 2) but the prefactor differs from that of the dominant peak in the x = 0.04 and 0.08 samples by an order of magnitude [60]. The discrete shifts in the emission rate as the alloy composition is varied cannot be due to a band structure effect since the band structure changes continuously with the alloy composition. It must therefore be due to an alloy effect. Simulated DLTS spectra consisting of the sum of three exponential transients compare well with the experimental spectra for samples of GaAs and AlGaAs with AlAs mole fraction up to x = 0.27 [59,60]. The prefactors and emission energies used in the simulations were determined from the measured DLTS data for samples with different dominant peaks, so the only free parameter in the simulations was the relative initial amplitude of the three transients.

The large shift in the temperature of the dominant emission peak and the small number of peaks observed suggest a relaxed configuration in which the group III atoms (Ga or Al) are first nearest neighbors to the Si donor atom. When the Si is substitutional on the group III site, other group III sites are second nearest neighbors. However, when the silicon atom moves along the <111> direction away from one As atom and squeezes between three other As atoms toward an interstitial site, it will have a vacancy and 3 group III atoms as nearest neighbors. This is exactly the configuration corresponding to the localized DX state proposed in Refs. 18 and 58. The observed barriers for capture and emission (Figs. 1 b and 1 c) are related to the motion of the Si atom between these two positions when electrons are captured or emitted. In the case of a group VI donor which is substitutional on the As site, it has been suggested that the lattice relaxes by the motion of a group III atom. This proposal is consistent with the data in Table 1 which shows that the lattice relaxation energy varies with the donor species in the case of the group VI donors, but is the same for all the group VI donors. These DLTS results are the strongest evidence available for such a model. Experiments which give direct evidence of atomic positions and which can be correlated with electrical measurements on the same sample are needed to confirm these models.

Recently it has been suggested that the DX level has negative U, i.e. that the DX level is the two electron state of the donor [18,19]. For a donor having positive U the two electron level lies above the one electron level due to the Coulomb repulsion between the two electrons. In order that the two electron level lie below the one electron level (negative U), the total energy must be lowered due to a lattice relaxation. The ground state of an uncompensated system of such negative U donors would be half the donors positively charged (ionized) and half negatively charged. Capacitance measurements detect only a net change in charge and cannot distinguish between the emission of a single electron from all the donor atoms or two electrons from half of them. Moreover, all of the experimental results discussed above are essentially unchanged when interpreted in this model, provided the rate limiting process in the capture and emission kinetics is the transition between the neutral and negative charge states. It is therefore difficult to distinguish between positive and negative U donors.

An experiment which should determine the charge state of the DX level is electron paramagnetic resonance (EPR) [19]. In the case of the Si donor, the neutral state (one electron) is paramagnetic. If the negative charge state (two electrons) is a singlet with spin 1/2, it is diamagnetic and one would not observe an EPR spectrum for the DX level. It has been pointed out however that the lowest energy level of the negative charge state is expected to be a triplet state with spin 1 [61]. Because of the axial symmetry due to the displacement of the Si atom in the <111> direction, a large zero field splitting between the $m = 0$ level and the $m = \pm 1$ levels is expected. Provided this splitting is larger than the microwave energy, no EPR signal would be observed, since a transition with $\Delta m = 2$ is not allowed. Therefore one would not expect to observe an EPR signal for the DX level if it has negative U.

226

A Si-donor related resonance has been observed by EPR in indirect gap material [62,63] and in AlGaAs with x ≥ 0.35 by optically detected magnetic resonance (ODMR) [64,65]. Detailed ODMR studies of this spectrum led to the conclusion that it arises from the hydrogenic level associated with the X-valley of the conduction band, not from the DX level [66]. This assignment was confirmed by photo-EPR measurements showing that the intensity of the EPR line increases when electrons are photoionized from the DX level to the conduction band (Fig. 7) [67]. Since there is no capture barrier between the shallow hydrogenic states and the conduction band states, electrons fall into the hydrogenic levels at low temperature. Apparently similar EPR results are observed in Sn-doped AlGaAs [68]. In all of the spin resonance experiments no spectrum which could be attributed to the DX level was observed. This is consistent with the negative U model but does not prove it. Since EPR line widths of localized deep levels in GaAs are very broad, typically between 350 and 1200 G, it cannot be ruled out that the line width of the DX level is so broad that its amplitude is too small to be observed in the samples measured [67].

Another experimental test for negative U could come from Hall measurements. Measurements of the temperature dependence of the free carrier concentration can be fit equally well by both positive and negative U models [33]. However, it is expected that electron mobility measurements could distinguish between positive and negative U models, since the scattering in the case of all donors in the neutral charge state at low temperature is ex-

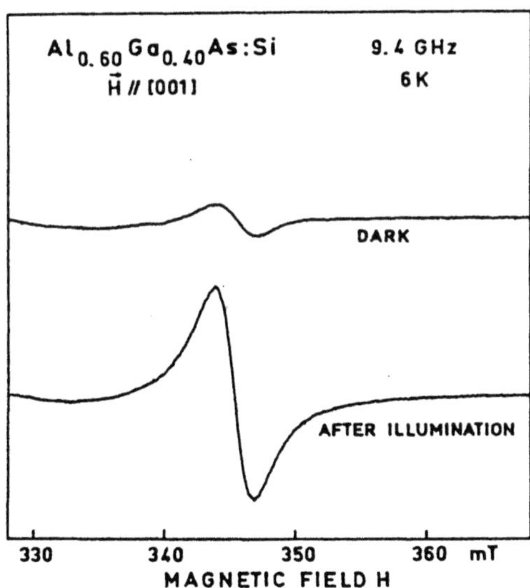

Fig. 7 EPR spectrum of the Si-related donor state in Al$_{0.6}$Ga$_{0.40}$As. The signal amplitude increases when electrons are photoexcited from the DX level indicating that this signal comes from the X-valley hydrogenic level.

pected to be different from the case with half the donors negatively charged and half ionized [19]. The data and analysis to date are inconclusive but work on these experiments continues.

Recently the time dependence of the near-band-edge luminescence in Al-GaAs was measured in order to test the negative U model for the DX level [69]. Assuming that half the Si atoms are negatively charged and half are positively charged at low temperature one would expect to see a strong luminescence at t = 0 when the sample is exposed to laser light. On the other hand if all the Si atoms are in the relaxed state, i.e. each has a single electron, one would expect no recombination via shallow donor states at t = 0. In fact what was observed is a slow increase of the photoluminescence signal from zero to a steady state value, suggesting that the DX level is a single electron state. Only upon warming the sample is it returned to the initial condition of no luminescence. This is the only experimental result to date which seems to contradict the negative U model.

5 Conclusions

Deep donors with large lattice relaxation and PPC at low temperature have been observed in many different materials. Perhaps the most studied such material is AlGaAs, where all substitutional donors, both group IV and group VI, have deep states known as DX levels. Evidence has been presented here demonstrating that the DX level is a deep localized state which undergoes a large lattice relaxation when electrons are captured. Evidence that the lattice relaxation occurs by the motion of an atom from the group III lattice site toward an interstitial site has been presented. Experiments designed to test proposals that the DX level has negative U have also been discussed. Direct evidence for the atomic configuration of the relaxed DX level and conclusive evidence for or against the negative U model are still needed.

References

[1] R.J. Nelson, Appl. Phys. Lett. **31**, 351 (1977)

[2] A. Saxena, Phys. Stat. Sol. (b) **105**, 777 (1981)

[3] N. Chand, T. Henderson, J. Clem, W.T. Masselink, R. Fisher, Y.C. Chang, and H. Morkoç, Phys. Rev. B **30**, 4481 (1984)

[4] E.F. Schubert and K. Ploog, Phys. Rev. B **30**, 7021 (1984)

[5] M.O. Watanabe and H. Maeda, Japn. J. Appl. Phys. 21, L675 (1982)

[6] T. Ishikawa, J. Saito, S. Sasa, and S. Hiyamizu, Japn. J. Appl. Phys. **21**, L675 (1982)

[7] T. Ishibashi, S. Tarucha, and H. Okamoto, Japn. J. Appl. Phys. 21, L476 (1982)

[8] D.V. Lang and R.A. Logan, Phys. Rev. Lett. 39, 635 (1977)

[9] D.V. Lang, R.A. Logan, and M. Jaros, Phys. Rev. B **19**, 1015 (1979)

[10] K.L. Kobayashi, U. Uchida, and H. Nakashima, Japn. J. Appl. Phys. **24**, L928 (1985)

[11] T.N. Morgan, Phys. Rev. B **43**, 2664 (1986)

228

[12]   A. Oshiyama and S. Ohnishi, Phys. Rev. **B 33**, 4320 (1986)

[13]   H.P. Hjalmarson and T.J. Drummond, Appl. Phys. Lett. **48**, 656 (1986)

[14]   E. Yamaguchi, Japn. J. Appl. Phys. **25**, L643 (1986)

[15]   J.C.M. Henning and J.P.M. Ansems, Semicond. Sci. Technol. **2**, 1 (1987)

[16]   T. Baba, T. Mizutani, and M. Ogawa, Japn. J. Appl. Phys. **22**, L627 (1984)

[17]   L.-Å. Ledebo, Semi-insulating III–V Materials, ed. by H. Kukimoto (OHMSHA, Tokyo 1986), p. 543

[18]   D.J. Chadi and K.J. Chang, Phys. Rev. Lett. **61**, 873 (1988)

[19]   K. Khachaturyan, E.R. Weber, and M. Kaminska, Materials Science Forum, Vols. 38–41, (Trans. Tech. Publications, Switzerland 1989) p. 1067

[20]   R.A. Craven and D. Finn, J. Appl. Phys. **50**, 6334 (1979)

[21]   K. Kitahara, M. Hoshino, K. Kodama, and M. Ozeki, Japn. J. Appl. Phys. **25**, L534 (1986)

[22]   K. Nakashima, S. Nijima, Y. Kawamura, and H. Asahi, Phys. Stat. Sol. (a) **103**, 511 (1987)

[23]   B.C. Burkey, R.P. Khosla, J.R. Fisher, and D.L. Losee, J. Appl. Phys. **47**, 1095 (1976); K. Kachaturyan, M. Kaminska, and E.R. Weber, unpublished

[24]   J.M. Langer, J. Phys. Soc. Japan **49**, (1980) Suppl. A, p. 207

[25]   A. Saxena, J. Phys. C: Solid State Phys. **13**, 4323 (1980)

[26]   E. Calleja, A. Gomez, and E. Muñoz, Appl. Phys. Lett. **52**, 383 (1988)

[27]   J.C.M. Henning, J.P.M. Ansems, and A.G.M. Nijs, J. Phys. C: Solid State Phys. **17**, L915 (1984)

[28]   J. Bourgoin and A. Moger, Appl. Phys. Lett. **53**, 749 (1988)

[29]   M.F. Li, W. Shan, P.Y. Yu, W.L. Hansen, E.R. Weber, and E. Bauser, Appl. Phys. Lett. **53**, 1195 (1988)

[30]   S. Azema, V. Mosser, J. Camassel, R. Piotrzkowski, J.L. Robert, P. Gibart, J.P. Contour, J. Massiá, and A. Marty, Materials Science Forum, Vols. 38–41, (Trans. Tech. Publications, Switzerland 1989) p. 857

[31]   M.O. Watanabe, K. Morizuka, M. Mashita, Y. Askizawa, and Y. Zota, Japn. J. Appl. Phys. **23**, L103 (1984)

[32]   T.N. Theis, T.F. Kuech, L.F. Palmateer, and P.M. Mooney, Inst. Phys. Conf. Ser. **74**, 241 (1984)

[33]   T.N. Theis, Materials Science Forum, Vols. 38–41, (Trans. Tech. Publications, Switzerland 1989) p. 1073

[34]   M. Mizuta and K. Mori, Phys. Rev. **B 37**, 1043 (1988)

[35]   J.E. Dmowski, J.M. Langer, J. Raczyńska, and W. Jantsch, Phys. Rev. **B 38**, 3276 (1988)

[36]   M. Tachikawa, M. Mizuta, and H. Kukimoto, Japn. J. Appl. Phys. **23**, 1594 (1984)

[37]   P.M. Mooney, E. Calleja, S.L. Wright, and M. Heiblum, in: Defects in Semiconductors, ed. by H.J. von Bardeleben, Materials Science Forum, Vols. 10–12 (Trans. Tech. Publications, Switzerland 1986) p. 417

[38]   P.M. Mooney, N.S. Caswell, and S.L. Wright, J. Appl. Phys. **62**, 4786 (1987)

[39]   O. Kumagi, H. Kawai, Y. Mori, and K. Kaneko, Appl. Phys. Lett. **45**, 1322 (1984)

[40]   E. Calleja, P.M. Mooney, S.L. Wright, and M. Heiblum, Appl. Phys. Lett. **49**, 657 (1986)

[41]   L. Samuelson, 13th Int. Conf. on Defecta in Semiconductors, ed. by L.C. Kimerling and J.M. Parsey, Jr. (The Metalurgical Society of the AIME, Warrendale, PA 1985) p. 101

[42]   M. Takikawa and M. Ozeki, Japn. J. Appl. Phys. **24**, 303 (1985)

[43]   L. Dobaczewski and J.M. Langer, in: Defects in Semiconductors, ed. by H.J. von Bardeleben, Materials Science Forum, Vols. 10–12 (Trans. Tech. Publications, Switzerland 1986) p. 399

[44]   N.S. Caswell, P.M. Mooney, S.L. Wright, and P.M. Solomon, Appl. Phys. Lett. **48**, 1093 (1986)

[45]   D.V. Lang and R.A. Logan, Inst. Phys. Conf. Ser. **43**, 433 (1979)

[46]   P.K. Battacharya, A. Majerfeld, and A.K. Saxena, Inst. Phys. Conf. Ser. No. **45**, 199 (1979)

[47]   D.V. Lang, in: Deep Centers in Semiconductors, ed. by S.T. Pantelides (Gordon and Breach Science Publishers, New York 1986), p. 498

[48]   R. Legros, P.M. Mooney, and S.L. Wright, Phys. Rev. **B 35**, 7505 (1987)

[49]   P.M. Mooney, G.A. Northrop, T.N. Morgan, and H.G. Grimmeiss, Phys. Rev. **B 37**, 8298 (1988)

[50]   T.N. Theis, in: Defects in Semiconductors, ed. by H.J. von Bardeleben, Materials Science Forum 10–12 (Trans. Tech. Publications, Switzerland 1986) p. 393

[51]   M. Tachikawa, T. Fujisawa, H. Kukimoto, G. Oomi, and S Minomura, Japn. J. Appl. Phys. **24**, L893 (1985)

[52]   M. Mizuta, M. Tachikawa, H. Kukimoto, and S. Minomura, Japn. J. Appl. Phys. **24**, L143 (1985)

[53]   M. Tachikawa, M. Mizuta, H. Kukimoto, and S. Minomura, Japn. J. Appl. Phys. **24**, L821 (1985)

[54]   M.F. Li, P.T. Yu, E.R. Weber, and W. Hansen, Phys. Rev. **B 36**, 4531 (1887)

[55]   T.N. Theis, P.M. Mooney, and S.L. Wright, Phys. Rev. Lett. **60**, 361 (1988)

[56]   D.K. Maude, J.C. Portal, L. Dmowski, T. Foster, L. Eaves, M. Nathan, M. Heiblum, J.J. Harris, and R.B. Beall, Phys. Rev. Lett. **59**, 815 (1987)

[57]   L. Eaves, T.J. Foster, D.K. Maude, G.A. Toombs, R. Murray, R.C. Newman, J.C. Portal, L. Dmowski, R.B. Beall, J.J. Harris, M.I. Nathan, and M. Heiblum, Inst. Phys. Conf. Ser. No. **91**, 355 (1988)

[58]   T.N. Morgan, Materials Science Forum, Vols. 38–41, (Trans. Tech. Publications, Switzerland 1989) p. 1079

[59]   P.M. Mooney, T.N. Theis, and S.L. Wright, Appl. Phys. Lett. **53**, 2546 (1988)

[60]   P.M. Mooney, T.N. Theis, and S.L. Wright, Materials Science Forum, Vols. 38–41, (Trans. Tech. Publications, Switzerland 1989) p. 1109

[61]   T.N. Morgan, private communication

[62]   R. Böttcher, S. Wartewig, R. Bindemann, G. Kühn, and P. Fisher, Phys. Stat. Sol. (b) **58**, K23 (1973)

[63]   S. Wartewig, R. Böttcher, and G. Kühn, Phys. Stat. Sol. (b) **70**, K23 (1975)

[64]   T.A. Kennedy, R. Magno, E. Glaser, and M.G. Spencer, Mat. Res. Soc. Symp. Proc. Vol. 104, 555 (1988)

[65]   J.C.M. Henning and J.P.M. Ansems, ICDS15, Budapest (1988)

[66]   E. Glaser, T.A. Kennedy, and B. Molnar, Conf. on Shallow Impurities, Linkoping, Sweden (1988)

[67]   P.M. Mooney, W. Wilkening, U. Kaufmann, and T.F. Kuech, Phys. Rev. B (in press)

[68]   H.J. von Bardeleben and J.C. Bourgoin, unpublished

[69]   W. Jantsch, G. Brunthaler, K. Ploog, J.E. Dmochowski, and J.M. Langer, Materials Science Forum, Vols. 38–41, (Trans. Tech. Publications, Switzerland 1989) p. 1131

Chemical Binding, Stability and Metastability of Defects in Semiconductors

Matthias Scheffler

Fritz-Haber-Institut der Max-Planck-Gesellschaft, Faradayweg 4-6,
D-1000 Berlin 33, Federal Republic of Germany

Summary: The paper describes recent theoretical studies of the electronic and structural properties of defects in semiconductors. In particular we discuss iron-acceptor pairs and chalcogen pairs in silicon, the As_{Ga}-As_i pair in GaAs and the As_{Ga} antisite defect. The results explain mechanisms which give rise to pair formation and they show possibilities of structural metastabilities.

1 Introduction

In two earlier review papers we discussed the electronic structure [1], the chemical bond between a defect and the crystal and lattice distortions [2] of isolated point defects in semiconductors. The present paper deals with defect-defect interactions with special emphasis on metastabilities. Four different examples of defect systems, for which parameter-free density-functional theory (DFT) calculations have been performed in the last years, are described. The methods and techniques of these studies have been different but are not discussed here, because we focus on the results. The interested reader is referred to the original papers. The following Section 2 sketches the basic theory and defines what we call the configurational coordinate diagram. Section 3 describes the iron-aluminum pair in silicon, which at this time is the best understood metastable defect, Then in Section 4, chalcogen pairing in silicon is considered, to discuss the effects of covalent defect-defect interactions [3]. Section 5 summarizes results for a defect pair in GaAs, namely the distant As-antisite As-interstitial pair [4]. This center has recently attracted much attention, because it was "experimentally identified" [5, 6] as the famous EL2 center. Our theoretical results rise some doubts [4] to this identification and its suggested metastability [7]. In Section 6 we discuss calculations of the isolated As_{Ga} antisite in GaAs. In contrast to what was hitherto expected about this defect, the calculations predict that it can undergo an optically inducible structural transition, ending as a $V_{Ga}As_i$ defect pair [8, 9]. The latter being a metastable configuration. The analysis of the theoretical results suggests that this new type of metastability might be important also for other centers in semiconductors. Section 7 concludes this paper.

2 Basic Aspects

Within the adiabatic approximation for the electrons of the quantum-mechanical, interacting many-atom problem, the positions and dynamics of the nuclei are described by the Hamiltonian [11]

$$H = \sum_i \frac{-\hbar^2}{2M_i} \nabla^2_{R_i} + U^{static}(\{R_i\}) , \tag{1}$$

where $U^{static}(\{R_i\})$ is the electron ground-state total energy, calculated for the interacting many-electron system with a static, external field

$$v_{ext.}(r;\{R_i\}) = \frac{1}{4\pi\varepsilon_0} \sum_i \frac{e^2 Z_i}{|r - R_i|} . \tag{2}$$

Here Z_i and R_i are the atomic numbers and positions, respectively. If the electron ground-state particle density is $n(r)$ and the magnetization density is $m(r)$, U^{static} is given by

$$U^{static}(\{R_i\}) = \frac{1}{8\pi\varepsilon_0} \sum_{\substack{i,j \\ i \neq j}} \frac{e^2 Z_i Z_j}{|R_i - R_j|} + \int v_{ext.}(r;\{R_i\}) \, n(r) \, d^3r$$

$$+ \frac{1}{8\pi\varepsilon_0} \int \int \frac{e^2 n(r) n(r')}{|r - r'|} \, d^3r d^3r' + F[n,m] , \tag{3}$$

where the functional $F[n,m]$ contains the kinetic energy as well as the exchange-correlation interaction of the electrons. $U^{static}(\{R_i\})$ is the static contribution to the thermodynamic internal energy U. In a complete thermodynamic description the zero-point and temperature-induced vibrations, which follow from eq. (1), must be considered as well [12, 13]. The ground-state densities $n(r)$ and $m(r)$ and $F[n,m]$ can be evaluated using density-functional theory [14-16]. Usually a self-consistent approach is applied to evaluate $n(r)$ and $m(r)$. Approximate densities also give accurate results, provided that the functional $F[n,m]$ is evaluated without destroying the variational principle [17-19]. On the other hand, approximations of U^{static} in terms of a superposition of two- and three-body potentials, of valence-force models, or of an estimate by empirical tight-binding theory work only in a limited region of configuration space $\{R_i\}$, and it is not clear when such approximations break down.

In our calculations we use the Kohn-Sham Ansatz [15, 16]

$$F[n,m] = T_s[n] + E_{xc}[n,m] , \tag{4}$$

232

with T_s the kinetic-energy functional of non-interacting Fermions and E_{xc} the exchange-correlation functional. Although the functional T_s is not known explicitly, its correct evaluation is important and can be indeed achieved by solving the Kohn-Sham equation [15, 16]. The main approximation in our calculations of $n(r)$, $m(r)$ and $U^{static}(\{R_i\})$ is the local-spin-density approximation (LSDA) to E_{xc}, or if $m(r)$ is small and spin splitting negligible, the local-density approximation (LDA). This treatment of exchange and correlation is exact for interacting, many-electron systems of constant densities. Experience over the last years has shown that the jellia-derived exchange-correlation functional is a good approximation also for the description of the solid state. The approximation gets less accurate if densities are very localized, as for example in f-electron systems. In general, a cautious interpretation of theoretical results is necessary if two ground states of different symmetry (or different magnetization) are close in energy. This result will then still be correct, but it may be uncertain which of them is the true ground state.

Eq. (3) should be evaluated under the constraints that the number of electrons

$$N = \int n(r)\ d^3r \tag{5}$$

as well is the total spin,

$$S = \int m(r)\ d^3r \tag{6}$$

is fixed. A different number of electrons N, i.e. a different charge state, or a different spin S, will give rise to a different total-energy surface $U^{static}(\{R_i\})$. Each of these total-energy surfaces is an electronic ground state for the constraint given by N and S. If the $\{R_i\}$ have a point or translation symmetry, then for each irreducible representation of the symmetry group there is a different many-electron wave function. The total-energy surfaces of these states will be different, and in general the absolute minima of these different total-energy surfaces will be at different points in configuration space $\{R_i\}$. If these points are well separated this gives rise to bi- or multistabilities.

The evaluation of eqs. (3)-(6) requires complicated methods and techniques. This is in particular so for low-symmetry systems, as for example defects, where Bloch's theorem is not valid. Such methods and techniques have been developed only during the last years (see Ref. 1, 20-32 and references therein). Up to date the applicability of these methods and techniques is still limited to special systems and new ideas and improvements of the theory are still important. Several controllable approximations are necessary in an actual first-principles calculation. If carefully applied they will not significantly affect the results for $n(r)$, $m(r)$ and $U^{static}(\{R_i\})$, but they induce certain numerical inaccuracies. The most accurate method developed so far is the self-consistent pseudopotential Green-function method [1, 20, 21, 24-26] which, if used together with first-principle, norm-conserving pseudopotentials [33-35], is essentially exact. In the LMTO Green-function method [22] the approximation of spherical potentials is introduced. Because of the variational principle in DFT, this is usually not a severe approximation, but it prevents to evaluate defect-induced lattice distortions. Cluster methods suffer from more severe (sometimes uncontrollable) problems: They impose artificial boundary conditions to the wave-functions and they localize the wave functions and charge densities to the size of the cluster, which can cause a wrong description of covalent binding. The super-cell approach also suffers from these problems of cluster approximations, but the boundary conditions are better, and with modern techniques it is now possible to take a cell size of more than fifty atoms. This allows for a systematic test of cell-size induced inaccuracies. Because of the high complexity of first-principles methods there is always a risk that some aspects may be overlooked in the calculations. It is therefore good to know if two independent groups, hopefully using different techniques, arrived at the same theoretical results.

From the knowledge of $U^{static}(\{R_i\})$, the forces and force constants with respect to nuclear displacements can be evaluated and in this way it is now also possible to evaluate the vibrational contributions to the internal energy, to determine entropies and to evaluate any desired thermodynamic potential [12, 13, 36]. Two configurations which have similar energy may differ in their configurational, vibrational and electron-hole entropies. Then at $T>0$ K one of them will be favored over the other. Because of space limitations we consider in this paper only low temperature situations. Thus, entropy driven or hindered effects will be neglected.

$U^{static}(\{R_i\})$ defines an energy surface over the multidimensional space of all atomic coordinates. A configurational coordinate diagram is a section of this total-energy surface along a line of a generalized coordinate which serves to describe a collective nuclear rearrangement. For example, for perfect crystals with one atom per unit cell, a configurational coordinate may be the lattice parameter of a chosen Bravais lattice. The corresponding total-energy curve (configurational coordinate diagram) will usually have only one minimum. For a defect system one interesting configurational coordinate may be the position of the impurity. If the positions of all other nuclei are relaxed according to the impurity position, we move along a valley of the total-energy surface $U^{static}(\{R_i\})$. In general, the total energy as a function of the impurity position will have several minima. The lowest is the $T=0$ K stable geometry and the others are metastable. If the energy differences between stable and metastable structures are small, it will be possible to excite the defect into one or more metastable minima. In fact, it is also possible that defects are frozen in at metastable configurations, so that the stable one is practically not present. The main purpose of a configurational coordinate diagram is to explain how a system changes from one minimum to another. Usually this is shown in a plot with only one configurational coordinate. One should not forget, however, that the complete picture can be much more complicated, because many ways are possible to go from one point in configuration space to another.

The relative stabilities (or lifetimes) of metastable structures depend on the barriers between them and the stable configuration and on the possible channels to overcome these barriers. Metastable systems can have a considerable lifetime. We remind on the diamond crystal, which is a metastable crystalline structure of carbon, the stable one being graphite. Barriers can be overcome by thermal energy but also by other forms of energy, as for example light, because an excited electronic state and also an ionized system can have a different total-energy

surface, driving the system to a different structure from where the true, stable
ground state can be possibly reached more easily (see Section 6 below).

3 Iron-Aluminum Pairs in Silicon

The pairs of substitutional aluminum with interstitial iron are the best
understood metastable centers and there is no significant dispute on the
mechanisms of pairing and metastability. Therefore these systems serve as the
prototype example of iron-acceptor pairs [37-44] and for some general aspects of
metastable defects. A too simple generalization of the properties of donor-acceptor
pairs to other metastable systems is, however, not possible.

Because of its high solubility and diffusion coefficient [45] iron is often present in
Si samples and it is easily introduced during heat treatment. Because of its low
migration energy (0.69 eV [45]) it is readily involved in defect reactions [41, 42].
Using EPR on Al doped Si samples van Kooten et al. [43, 44] could identify two
different paramagnetic FeAl pairs. One was oriented in the $<111>$ direction the
other in the $<100>$ direction. The pairing reaction takes place even at room
temperature and heating the samples to 90° C destroys the pairs [43, 44]. In
agreement with a DLTS analysis of Chantre and Bois [39], the two pairs are
interpreted as interstitial Fe which occupies the nearest and the second-nearest
neighbor T_d interstitial site away from a substitutional Al. The $<111>$ oriented
pair then has a separation of about 2.35 Å and the $<100>$ oriented pair has a
separation of about 2.72 Å.

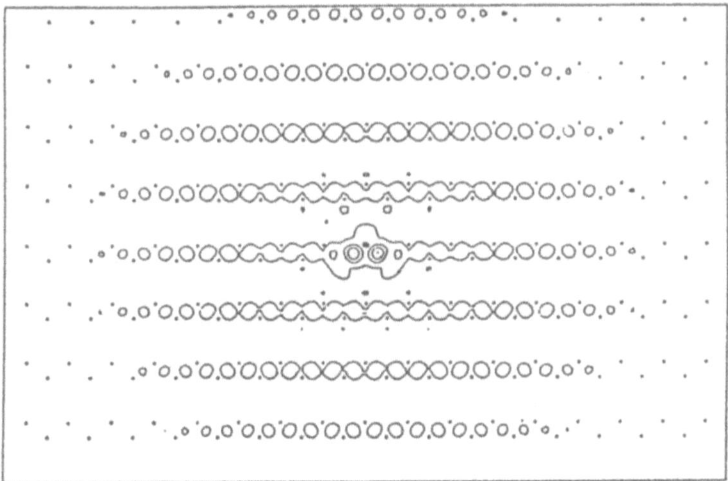

Fig. 1   Wave function (squared) of substitutional aluminum in silicon in the (110)
plane. It is calculated using a Green-function approach [46]. Small dots denote the
nuclei. Note that the wave function follows essentially the zig-zag bonding chain of
the crystal.

We start with a short description of the isolated partners of the pair and then discuss the defect-defect interaction. Isolated substitutional aluminum in silicon is a shallow acceptor, with its energy level at 0.07 eV above the valence-band edge [40]. The wave function is sketched in Fig. 1. It originates from the valence-band states at Γ and extends over more than hundred atoms, essentially following the zig-zag bonding chains of the crystal. The defect-induced potential can be described in the language of a Green-function approach: In a perfect Si crystal one proton of a Si atom is removed, or compensated by a negative charge. This transmutes the Si into an Al nucleus and changes the *external* potential (eq. (2)) by $e^2/4\pi\varepsilon_0 r$. The electrons of the Si valence band will screen this negative charge so that (in an average) the *effective* potential attains a $e^2/4\pi\varepsilon_0\varepsilon r$ behavior. Because we are dealing with a shallow defect (extended wave function) the number of electrons in the shallow level, i.e. the defect charge state, plays only a minor role and affects the potential only at distances larger than about 20-50 Å. At closer distances self-consistent Green-function calculations show that screening in Si is very efficient and that the simple description in terms of a static dielectric constant, ε, is approximately valid already for distances larger than one interatomic spacing [2].

Iron in silicon occupies the tetrahedral interstitial site [37]. The electronic structure of interstitial iron can be understood in terms of the free atom transition-metal 3d orbitals, which are split by the crystal field and some covalent interaction with the crystal valence and conduction band states [24, 25]. Because of the strong localization of the 3d-like states it is important to take the spin-spin interaction into account. Then the iron 3d level gives rise to t_2-up, e-up, t_2-down and e-down states with increasing energy [47, 24] (see Fig. 2).

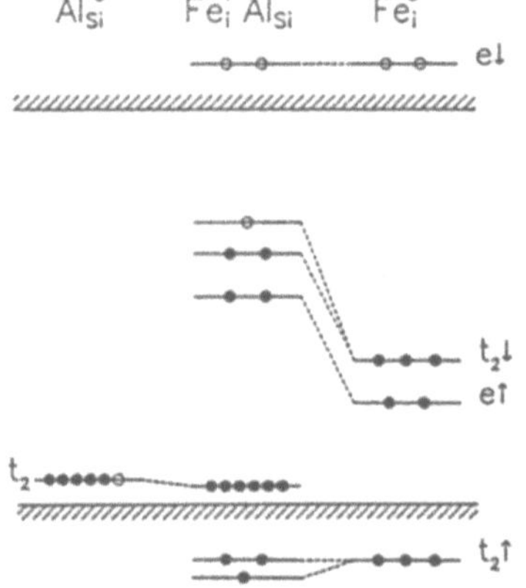

Fig. 2
Single-particle energies and the interaction of substitutional aluminum (left) and interstitial iron (right) to form a nearest-neighbor pair (middle).

The iron-induced deep level is due to the t_2-down state, which in the neutral impurity is filled with three electrons. This electronic structure gives rise to only one transition-state level in the gap, namely the 0/+ transition where the occupancy of t_2-down changes from three to two. The transition level is calculated at 0.25 eV above the valence band [24], which is close to the experimental result of 0.39 eV [40]. When one electron is removed from the t_2-down state, the single-particle energies move down in energy. Therefore, the +/2+ transition is not possible.

Whereas the effect of spin polarization is important for iron, it is irrelevant for the extended aluminum-induced wave functions. In aluminum doped samples, one of the iron 3d electrons will be transferred to the shallow aluminum state (Fig. 1) and the positively charged iron interstitial will be attracted by the aluminum $e^2/4\pi\varepsilon_0\varepsilon r$ potential. From the above description of the wave functions of the two partners, it is clear that there is practically no overlap between the very extended aluminum state, which avoids the interstitial region (Fig. 1), and the very localized interstitial iron 3d-like state. Because of these special conditions and because of the efficient screening of defects in silicon (see above and Ref. 2) the binding mechanism of the pair is largely ionic in character with only little covalent contributions. Recent cluster calculations on a similar pair, namely FeB [48], arrived at a different picture. We believe, that the cluster calculations overestimate the covalent effects, because the acceptor wave function is artificially localized to the cluster size of twenty five Si atoms.

In the pair the iron-like levels are shifted to higher energies because of the repulsive interaction with the screened Al potential (see Fig. 2). Therefore, in difference to isolated Fe impurities, the pair has two iron-like transition levels in the gap: The pair +/0 level, which corresponds to the Fe 2+/+ transition, and the pair 0/− level, which corresponds to the Fe +/0 transition. For the <111>

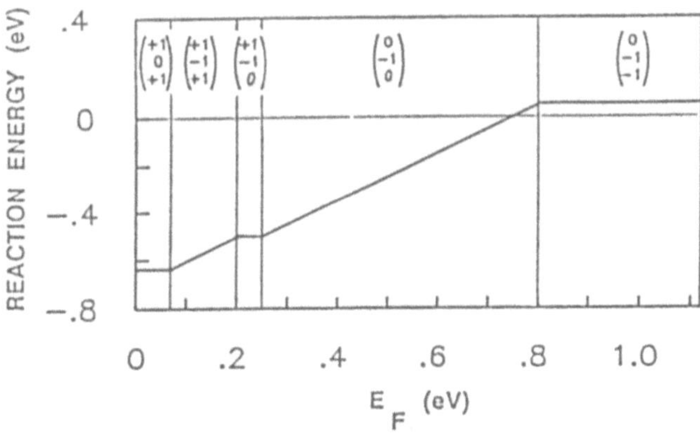

Fig. 3   Reaction energy for the formation of a nearest neighbor Fe_iAl_{Si} impurity pair in silicon (see eq. (7)). The charges of the different species (l, m, n) in eq. (7) are given at the top of the figure. Zero of E_F is the top of the valence band. A negative reaction energy corresponds to an attractive interaction (exothermal process).

pair the two levels are calculated at 0.2 and 0.8 eV above the valence band. These values are obtained without a self-consistent calculation for the pair, but taking only the self-consistent Al defect potential into account as it acts at the nearest neighbor T_d site.

The reaction energy of pair formation is displayed in Fig. 3. The considered reaction reads

$$Fe_i^{(l)} + Al_{Si}^{(m)} + (l+m)\, e^-(E_F) \rightleftharpoons (Fe_i Al_{Si})^{(n)} + n\, e^-(E_F). \tag{7}$$

It is important to note that the Fermi level acts as a reservoir of electrons and electrons can be exchanged in the reaction. $e^-(E_F)$ denotes an electron at the Fermi level. Therefore the binding energy (the energy of the right side of the reaction eq. (7) minus the left side) depends on the Fermi energy. A negative reaction energy means that pairs are energetically more favorable than the dissociated centers. Still, because of the higher configurational entropy of dissociated defects, some unpaired point defects will be present as well. Figure 3 shows that two important aspects can favor or disfavor the formation and stability of such pairs. If the pair is neutral and the two dissociated components are oppositely charged, the pair will be formed because interstitial Fe^+ is attracted by the aluminum $e^2/4\pi\varepsilon_0\varepsilon r$ potential over a long distance, and the Coulomb interaction also holds the pair together. In the ionic model the binding energy of the Fe^+Al^- pair is $e^2/4\pi\varepsilon_0\varepsilon S_1$. For the <111> pair and with $\varepsilon = 11.8$ and $S_1 = 2.35$ Å this gives 0.50 eV. If pair formation is accompanied by an exchange of electrons with the Fermi level, there is an additional and possibly more important (indirect) interaction [3]: If for a given Fermi energy the pair has more electrons than the two dissociated components, pair formation is favored with increasing Fermi energy. If the pair holds less electrons than the dissociated components, pair formation gets less favorable with increasing Fermi energy (see the energy range 0.07-0.2 eV and 0.25-0.8 eV in Fig. 3). In fact, at $E_F > 0.7$ eV the Fe^+Al^- pair becomes unstable. In particular neutral iron is not expected to pair with aluminum.

Using the law of mass action [49, 50], the energies of Fig. 3 can be used to estimate the pair concentration relative to that of the dissociated, isolated point defects, in thermodynamic equilibrium. And from the energy difference of the <111> and <100> pairs the relative concentration of the two pairs can be obtained. For the Fe^+Al^- pair the ionic model gives a difference of the Gibbs free energy between the two pair configurations of $\Delta G = 0.07$ eV - 0.4 $k_B T$. Then at room temperature and in thermodynamic equilibrium the concentration ratio of <111> to <100> pairs should be 14:1. Thus, 7% of the pairs should be <100> oriented. For the $Fe^{2+}Al^-$ pair the difference in Gibbs free energy for a Fe^{2+} ion in the field of the aluminum is changed to 0.14 eV-0.4 $k_B T$ for the two pairs. Then the concentration ratio gets very large, namely 295:1. Thus, <100> oriented pairs are practically absent. If, in an actual experiment, thermodynamic equilibrium is reached or not depends largely on the diffusion constant of the mobile partner and on the capture radius [50-54]. For iron-aluminum pairs the measurements of Chantre and Bois [39] and van Kooten et al. [43, 44] showed that these conditions are such that pairs are formed at room temperature after some hours.

The metastability of the pair was systematically studied by Chantre and Bois [39]. Their analysis is fully in agreement with our microscopic picture, which only adds some details to their explanation. From Fig. 3 and from the above discussion it follows that the Fermi level can be used to control the relative concentration of the stable and metastable structures. Chantre and Bois indeed observed that the charge state of the defect during sample cool-down to lower temperature controls such reversible transmutation behavior between the two pairs. In thermodynamic equilibrium the concentration of $<111>$ FeAl centers should be always higher than that of $<100>$ pairs. Van Kooten et al. [43, 44] noted that this was not obvious from the EPR intensities. The authors argued, however, that a reliable determination of defect concentrations is not possible from EPR.

4 Sulfur Pairs in Silicon

Covalent defect-defect interactions, pair formation and dissociation were studied by Weinert and Scheffler [3] for the chalcogens in silicon. Three different pair structures were considered: Both constituents occupying substitutional sites, both occupying interstitial sites, and the mixed substitutional-interstitial geometry. In all three cases the principal axis of the complex was the threefold-symmetric $<111>$ axis, and the separation was the nearest-neighbor distance (i.e. 2.35 Å).

It had been shown earlier [23] from total-energy calculations that isolated chalcogen point defects in silicon would kick out a Si atom and occupy the substitutional site. In the calculations for the pair the following reaction was therefore considered:

$$S_{Si}^{(m)} + S_{Si}^{(m)} + 2m\,e^-(E_F) \rightleftharpoons S_2^{(n)} + n\,e^-(E_F). \tag{8}$$

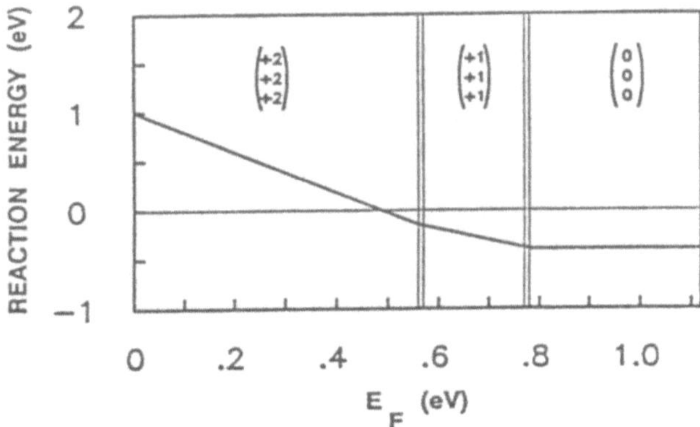

Fig. 4 Reaction energy for the formation of a nearest-neighbor substitutional sulfur impurity pair in silicon (see eq. (8)). The charges of the different species (m, m, n) in eq. (8) are given at the top of the figure. Zero of E_F is the top of the valence band. A negative reaction energy corresponds to an attractive interaction (exothermal process).

The reaction energy for the pair with *both partners at substitutional sites* is shown in Fig. 4 as a function of the Fermi level. For n-type material both the isolated impurities and the pair are neutral. Because the highest occupied defect-induced level is completely filled, the interaction between distant isolated sulfur impurities is practically absent. The attractive interaction starts only at smaller separations. The binding of the nearest-neighbor sulfur pair can be understood in terms of a free S_2 molecule which is placed into a di-vacancy [3]. It is then found that the interaction with the six di-vacancy dangling bonds weakens the S–S bond considerably compared to the gas-phase situation. The calculations show that these pairs are stable only if the Fermi level is in the upper half of the gap. For other Fermi-level positions such pairs are metastable. These results seem to be consistent with EPR and ENDOR measurements which identified sulfur and selenium pairs [55, 56]. The measurements could, however, only determine the symmetry and thus they could not distinguish between the pairs where both partners occupy substitutional sites and those where both occupy interstitial sites. This missing identification step is provided by the total-energy calculations [3] which show that the substitutional pairs should be dominant. Recently it became also possible to evaluate the hyper-fine and ligand hyper-fine fields of isolated point defects and defect pairs [57]. These calculations for the chalcogens show the importance of many-electron screening, and if compared to ENDOR data they prove that these measurements are due to substitutional-substitutional pairs.

A pair where *both partners occupy interstitial sites* is also bound by covalent interactions. The highest occupied wave functions of isolated sulfur interstitials are somehow similar (although more extended) to atomic sulfur 3p orbitals. Therefore the binding in the interstitial-interstitial pair is similar to that in the gas phase S_2 molecule. According to the reaction eq. (8) the relevant process requires that at first two substitutional sulfur impurity atoms go to interstitial positions and then form the pair. The cost to bring the two chalcogens from substitutional to interstitial sites is higher than the pair binding. As a consequence, interstitial-interstitial pairs should not exist in thermodynamic equilibrium in significant concentration.

The electronic structure of the *mixed geometry*, where one sulfur atom occupies a substitutional site and the other an interstitial site is summarized in Fig. 5. It results from the interaction between the a_1 level of the substitutional and the t_2 level of the interstitial [23]. Because the interstitial t_2 level is below the substitutional a_1 level, two electrons are transferred from the more substitutional-type to the more interstitial-type wave function. Thus, a chalcogen interstitial, which is usually believed to act as a donor, can also act as an acceptor and take two additional electrons. As a consequence, the binding of this system has some ionic character. Substitutional sulfur is a deep defect. Therefore a simple treatment in terms of purely Coulomb interactions, as that described in Section 3, is not appropriate and overestimates the binding energy by about a factor of 2. The energy gain due to pairing is smaller than the cost to bring a chalcogen atom from its normal substitutional site to the interstitial position. Only in n-type material are these two energies of similar size and we cannot exclude from the calculations that substitutional-interstitial pairs may exist.

If we compare the mechanisms which yield complex formation in semiconductors
to those giving rise to molecule formation in normal chemistry we find that
differences are mainly due to the following reasons.

(i) Chemical reactions in solids do not require charge conservation, because the
Fermi level can take or give electrons, if needed. Therefore it can depend on the
position of the Fermi energy if complex formation is exothermal or endothermal
(see eqs. (7) and (8) and Figs. 3 and 4).

(ii) The wave function of the highest occupied level of an isolated defect is usually
qualitatively different (spatial distribution and degree of degeneracy) from that
of the free atom. To give an example: The electronic structure of a neutral
substitutional sulfur impurity in Si has a closed shell and therefore the
interaction between distant centers may be better compared to argon than to
sulfur atoms.

(iii) Impurities in semiconductors can generally exist at different sites
(substitutional and interstitial). This can allow two atoms of the same kind to
form a partly ionic bond.

(iv) The impurity-impurity equilibrium distance is largely influenced by the host-
crystal structure.

(v) Because of dielectric screening and the larger extent of wave functions the
strengths of impurity-impurity ionic and covalent interactions are reduced
compared to gas-phase ion-ion or atom-atom interactions. Furthermore, if a deep
level shifts upon impurity-impurity interaction and approaches the conduction
band (or the valence band) it simultaneously gets more delocalized because of
increasing hybridization with the band states.

5 The Distant As_{Ga} -- As_i Pair in GaAs

The distant antisite-interstitial pair was recently proposed to be the defect behind
the famous EL2 level [5]. In particular an ODENDOR analysis [6] seems to
confirm and to specify this interpretation: Meyer et al. [6] concluded to see a 4.88

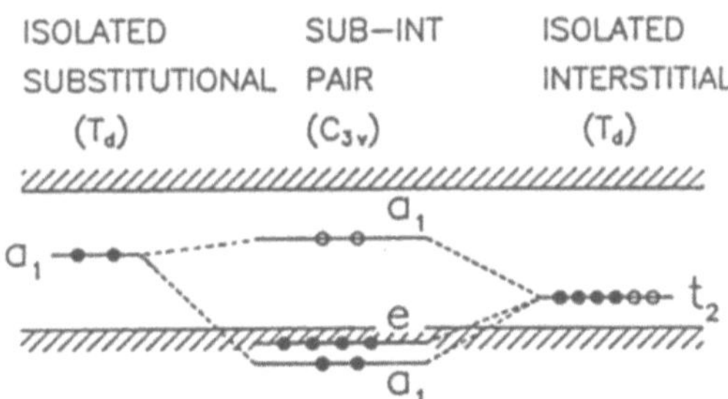

Fig. 5   Single-particle energies and interaction of a substitutional chalcogen (left)
and an interstitial chalcogen (right) to form a nearest-neighbor pair (middle).

Å separated <111> oriented As_{Ga}^{+} -- As_i^{+} pair. Thus, the interstitial member is not paramagnetic and placed at the distant tetrahedral site. The authors identified this pair with the paramagnetic state of EL2 (see also Ref. 69).

The antisite-interstitial pair has been investigated theoretically by semi-empirical model calculations [58, 59, 7] and by parameter-free self-consistent studies [4]. We investigated in particular the positive charge state of the pair allowing the interstitial member to move along the [111] axis. In accordance with the results of Baraff et al. [58, 59] we find that the pair has a total-energy minimum when the interstitial is near the hexagonal site, 4 Å away from the antisite (see Fig. 6). The double positive pair is paramagnetic and its total energy minimum is close to the hexagonal interstitial site too. The interaction between the two partners at this quite long distance is small so that the energy levels of the pair are very close to those of the isolated components. We expect a binding at closer separation, similar to that discussed above in Section 4 for the substitutional - interstitial sulfur pair in Si (see Fig. 5). At the distances shown in Fig. 6 and in particular at the ENDOR derived 4.88 Å separation we find that the pair is practically unbound.

With the so far established understanding of pair formation and stabilities (see the above Sections and Ref. 3 and 4) we concluded that [8, 4] the 4.88 Å separated As_{Ga} -- As_i pair cannot exist in significant concentration. We thus question the

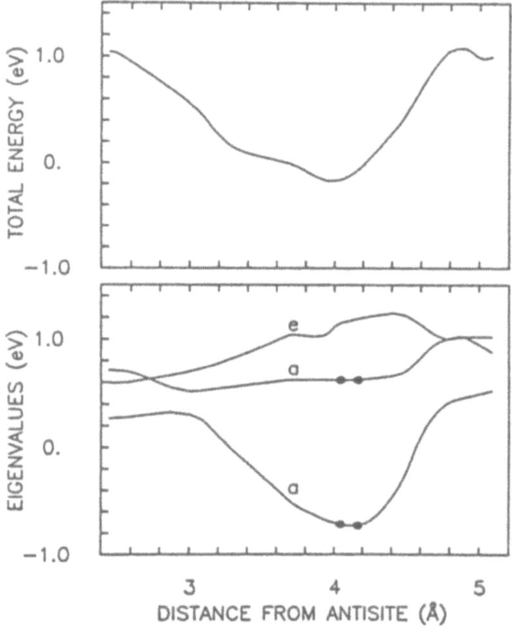

Fig. 6   Single-particle energies (with respect to the top of the valence band) and total energy as functions of the distance between an As-interstitial and an As-antisite. Lattice distortions are neglected. The pair's charge state is singly positive.

242

statement that this pair is identical with EL2 [5-7] as this would imply that the pair should be the dominant intrinsic deep donor. No mechanism has been suggested so far which brings the two constituents together and no mechanism is known which can be responsible for a strong binding.

It should be noted that a correlation of the paramagnetic EPR and ENDOR with the EL2 absorption does not imply the exact structural identity of these centers. In fact, similarly to what we discussed in Section 3, it is likely that a positively charged As_{Ga} antisite, which undoubtedly is at the core of the EPR [60, 61] and ENDOR [6] centers, may complex differently than the neutral charge state. Independent of the EL2 identification we are left with the question what is seen in paramagnetic resonance [61, 6]. We had speculated earlier [8, 4] that the ENDOR studies [6] are due to an As_{Ga} - X complex where X stands for an acceptor, not just an As_i.

6 The As_{Ga} Defect in GaAs

First studies of the structural properties of the As_{Ga} antisite in GaAs were performed by Bachelet and Scheffler [62, 63] five years ago. Figure 7 shows the electronic structure of the As_{Ga} antisite and how it can be understood in terms of the interaction of a Ga-vacancy and a free As-atom (see also Ref. 64, 8 and 9). In the neutral ground state the electronic configuration of the As_{Ga} antisite is $a_1^2t_2^0$. It was found [62, 63] that the defect behaves rather normal, when in this electronic configuration. For the excited electronic configuration $a_1^1t_2^1$ a structural instability was predicted, and it was pointed out [62, 63] that this instability resembles the properties of the metastable transition of EL2. It was also noted (see the discussion of Fig. 4 in Ref. 62) that the electronic structure of Fig. 7 and the instability of the electronic $a_1^1t_2^1$ configuration should be a general property of deep substitutional donors, not just of the As_{Ga} antisite. A careful total-energy calculation of low-symmetry defects was not possible at that time, but was recently achieved by Dabrowski and Scheffler [8, 9] and Chadi and Chang [10] and will be described below.

Figure 8 summarizes results [8, 9] for the As_{Ga}-antisite, when the central As-nucleus is displaced in the <111> direction. Displaced geometries thus correspond to a $V_{Ga}As_i$ defect pair with varied separation. The defect symmetry is now C_{3v} and the T_d t_2 state of Fig. 7 splits into two states now labeled as 2a and 1e. Figure 8 shows the single-particle energies of the 1a, 2a and 1e states (top) and three total-energy curves obtained for the three electronic configurations $1a^22a^0$ (labeled as F, which stands for fundamental), $1a^12a^1$ (labeled as E, which stands for excited) and $1a^02a^2$ (labeled as M, which stands for metastable). For the E total-energy curve, the Jahn-Teller effect *starts* the distortion, but already after a small displacement we find that the different wave functions (namely the 1a and the 2a single-particle states) will mix. It is most likely that the excited system falls back down to the F-curve, the ground-state total energy. Then the system ends again as a tetrahedral As_{Ga} antisite. However, Fig. 8 shows that for displaced geometries also another electronic configuration, namely $1a^02a^2$ becomes possible. Thus, once excited to the E-curve, the system has a certain probability to change to the M-curve. Then the arsenic defect atom will end

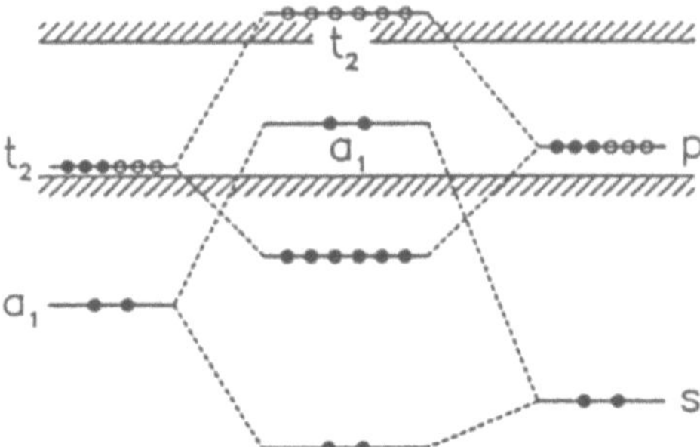

Fig. 7 Single-particle energies and interaction of a neutral Ga vacancy in GaAs (left) and a free As atom (right) which results in a As_{Ga}-antisite defect (middle).

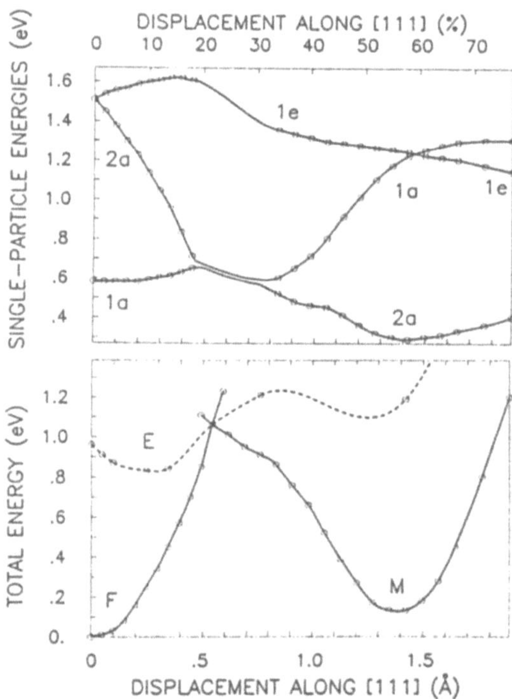

Fig. 8 Single-particle energies with respect to the valence-band edge (top), and total energies of the S=0 ground states (curves F and M) as functions of the position of the arsenic defect atom (bottom). Zero displacement refers to the tetrahedral As-antisite configuration. The total-energy curve labelled E is an electronic excited state with electronic configuration $1a^1 2a^1$. These calculations were performed with a basis set of E_{cut} = 8 Ryd, keeping all neighbors of the displaced arsenic atom at their perfect crystal positions (see Ref. 8).

244

considerably far away (about 1.4 Å) from the initial, central gallium site. We refer to this metastable atomic configuration as the gallium vacancy - arsenic interstitial pair. The As-interstitial is about 1 Å away from the T_d interstitial site. It is therefore chemically bound to only three arsenic atoms.

As the arsenic defect atom leaves the gallium site, its bond with one arsenic neighbor which is left behind is stretched and it almost breaks when the defect enters the barrier region. This is shown in Fig. 9. The barrier of the structural transition is reached when the arsenic atom passes through the (111) plane of three As neighbors. In the metastable configuration the arsenic defect atom (now

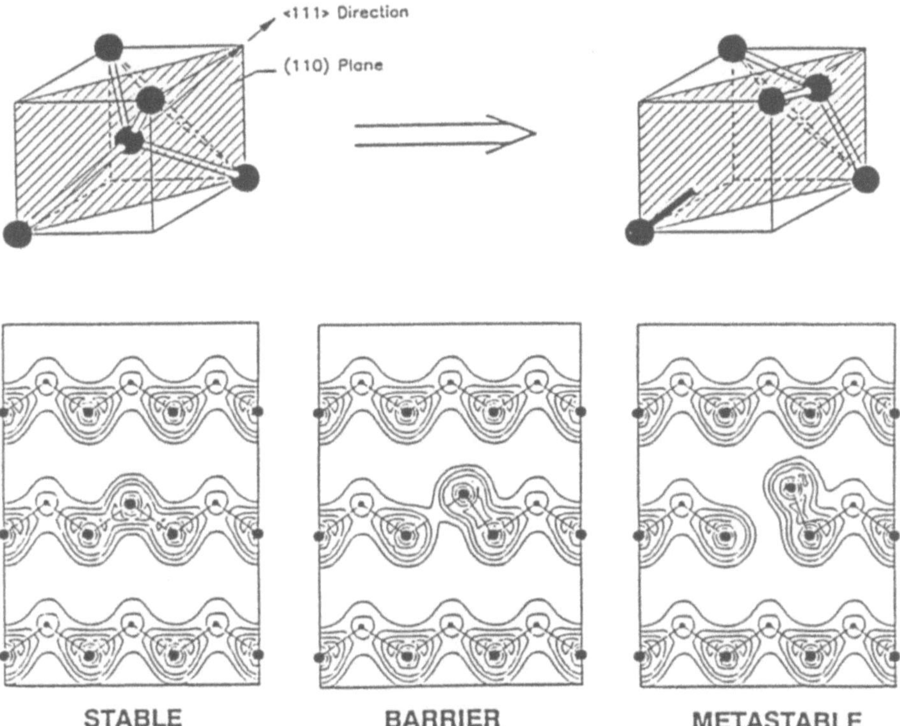

Fig. 9 The metastability of the As-antisite defect. Top: atomic structure, and bottom: The electron density in the (110) plane. Thick dots represent As atoms. Small dots represent Ga atoms. The left side shows the fundamental configuration, where the arsenic defect atom is bound to four nearest neighbors (only two are in the displayed (110) plane). The middle panel shows the barrier region. The right pictures correspond to the metastable situation (the $V_{Ga}As_i$ defect pair). Here the arsenic defect is bound to three arsenic neighbors (only one of them is in the (110) plane). The black "dangling bond" in the top right picture indicates the vacancy state (labeled 2a in Fig. 8), which is responsible for the barrier between the metastable and the fundamental configuration.

an interstitial) binds to these three atoms, similar to the bonding in crystalline grey arsenic. In the vacancy region there is one broken bond, which is filled with two electrons. The calculations of Fig. 8 give a barrier for the neutral ground state of 0.92 eV between the minimum of the metastable configuration (the $V_{Ga}As_i$ pair) and the fundamental configuration (the As_{Ga} antisite). We note that these calculations are performed with a smaller basis set and without allowing the nearest neighbor atoms to relax [8, 9]. If these two constraints are removed we obtain a barrier of about 0.4 eV.

According to the calculations [8-10] the four-fold coordinated As_{Ga} antisite and the metastable configuration with the threefold coordinated As_i have similar total energy. This result is indeed plausible for a group V element. The origin of the barrier between the two configurations is, however, not obvious. It may be understood by the fact that the covalent radius of an As atom is 1.2 Å. Therefore the As_i is too "thick" to pass easily through the (111) plane of the three As atoms. This argument is not complete and it cannot explain why for other charge states the barrier will in fact disappear (see Ref. 8, 9). The main reason for the barrier is the filled vacancy-like dangling bond schematically shown in Fig. 9. This state is antibonding with respect to the arsenic interstitial and its energy (the 2a level in Fig. 8) increases if the As-interstitial is moved from the metastable configuration towards the vacancy. The occupied vacancy dangling orbital therefore contributes to the repulsion between the constituents of the metastable pair. It is therefore obvious that the barrier will change if one electron is removed from this level. For a negatively charged $V_{Ga}As_i$ pair the barrier was found to be significantly reduced [8, 9] and an electron-induced regeneration $V_{Ga}As_i \rightarrow As_{Ga}$ should be likely. For a positively charged As_{Ga} the calculations [8, 9] predict that it should not exhibit metastable behavior.

The optically inducible transition to the metastable state competes with two other processes: At small displacements it is possible that the excited electron of the As_{Ga} goes to the conduction band $(1a^12a^1 \rightarrow 1a^12a^0 + e^-)$ or that it goes back to the 1a state $(1a^12a^1 \rightarrow 1a^22a^0)$. These two processes bring the arsenic defect atom back to the fundamental configuration, i.e. the tetrahedral As_{Ga} antisite. Because of these competitors the probability of the metastable transition will be small and it should be sensitive to local stress and other perturbations. It also depends sensitively on the conduction-band structure.

There is no clear, unambiguous experimental verification of the above described As_{Ga} metastability. However, the comparison of the theoretical results for the isolated arsenic antisite and the $As_{Ga} \rightleftharpoons V_{Ga}As_i$ metastability with a carefully compiled list of 17 experimentally established properties of the EL2 defect (see Ref. 8 and 9) revealed clear similarities. We therefore identified the mechanism of the EL2 metastability with that of the As_{Ga} antisite.

Although the described type of metastability should be present in principle for all substitutional deep donors, we cannot predict without a full calculation when it really will be observable: The height of the barrier and the energy difference of the metastable and stable configurations will be different for different systems. A very similar system as the As_{Ga}-antisite is a negatively charged Si_{Ga} defect in GaAs. The number of electrons is the same as in the neutral As_{Ga}. Indeed, according to calculations of Chadi and Chang [65], this defect shows a $Si_{Ga} \rightleftharpoons$

$V_{Ga}Si_i$ metastable behavior. The properties of the Si_{Ga} or $V_{Ga}Si_i$ defect seem to explain the properties of the famous DX centers in GaAlAs and in GaAs under pressure [65, 66]. The $As_{Ga} \rightleftharpoons V_{Ga}As_i$ type of process may be also important for defect diffusion. With this respect we note a similar phenomenon, found in group IV A elements (e.g. Ti and Zr), which are the transition-metal counterparts of the group IV B elements (e.g. C and Si). Group IV A elements show a transition from the bcc to the hexagonal structure, the so called ω-phase, in which the atoms are trigonally bonded. In the bcc phase a so called ω-embryo can be formed, which explains certain self-diffusion properties in these materials [67, 68].

7 Summary

In this paper we described defect-defect interactions and mechanisms of defect pair formation. Only point defects and defect pairs have been considered, because larger complexes have not been studied so far by first-principles calculations. Three classes of stable $\rightleftharpoons$ metastable configurations were discussed:
1) The bound pair versus the dissociated, isolated point defects,
2) bound pairs with different separations, and,
3) as a special case of No. 2, the $As_{Ga} \rightleftharpoons V_{Ga}As_i$ process.
The described calculations show that defect metastabilities can be much more important than this was often expected. It is also found that the pairing or dissociation is largely determined by the defect charge state and the crystal Fermi level. The concentrations of different atomic configurations are given by the law of mass action, assuming that the system is in thermodynamic equilibrium. However, often thermodynamic equilibrium will be not reached, because metastable configurations can have significant lifetimes. Then the concentrations can be affected by the history of the sample, and a higher temperature thermodynamic-equilibrium state can be frozen in. Furthermore, for compound semiconductors, as for example GaAs, the crystal environment can be crucial for the existing defect structures, because the concentration of intrinsic defects and their complexes depend on the Ga or As chemical potentials. These are controlled by the crystal environment, i.e. if Ga metal is present or if the sample is in an As_2 or As_4 gas atmosphere. Both, the electron chemical potential (i.e. the Fermi energy) and the atomic chemical potential, can affect defect-formation energies by more than one eV [12].

The $As_{Ga} \rightleftharpoons V_{Ga}As_i$ process, discussed in Section 6 was only recently found to be a likely reaction. In fact, this new type of metastability should be common to many substitutional defects. It resembles properties of the diamond $\rightleftharpoons$ graphite metastability.

We feel that with the theoretical results discussed in this paper we are just at the beginning of a microscopic understanding of defect complex formation in semiconductors and of defect metastabilities.

Acknowledgements

I gratefully acknowledge the collaboration with F. Beeler, C.M. Weinert and J. Dabrowski on some of the above discussed subjects. I thank H. Overhof, U. Scherz and R. Gillert for reading the manuscript.

References and Footnotes

[1]   M. Scheffler, Festkörperprobleme XXII ed. by P. Grosse, (Vieweg, Braunschweig, 1982), p. 115.

[2]   M. Scheffler, Physica 146 B, 176 (1987).

[3]   C. M. Weinert, and M. Scheffler, Phys. Rev. Letters 58, 1456 (1987).

[4]   J. Dabrowski, and M. Scheffler, Materials Science Forum 38-41, 51 (1989).

[5]   H. J. von Bardeleben, D. Stievenard, D. Deresmes, A. Huber, and J. C. Bourgoin, Phys. Rev. B 34, 7192 (1986).

[6]   B. K. Meyer, D. M. Hofmann, J. R. Niklas, and J.-M. Spaeth, Phys. Rev. B 36, 1332 (1987).

[7]   C. Delerue, M. Lannoo, and D. Stievenard, Phys. Rev. Letters 59, 2875 (1987).

[8]   J. Dabrowski, and M. Scheffler, Phys. Rev. Letters 60, 2183 (1988).

[9]   J. Dabrowski, and M. Scheffler, submitted to Phys. Rev. B (1989).

[10]   J.D. Chadi, and K.J. Chang, Phys. Rev. Letters 60, 2187 (1988)

[11]   In the adiabatic approximation the wave functions are written as $\psi(\{r_i\},\{R_j\})$ = $\phi(\{r_i\},\{R_j\})$ $X(\{R_j\})$, with $\phi(\{r_i\},\{R_j\})$ beeing the ground state of the electronic Hamiltonian of a fixed nuclear structure. If electron-phonon interactions are neglected, the Hamiltonian of eq. (1) has to be solved only for $X(\{R_i\})$.

[12]   M. Scheffler, and J. Dabrowski, Phil. Mag. A 58, 107 (1988).

[13]   S. Biernacki, and M. Scheffler, submitted to Phys. Rev. (1989).

[14]   P. Hohenberg, and W. Kohn, Phys. Rev. 136, B 864 (1964).

[15]   W. Kohn, and L.J. Sham, Phys. Rev. 140, A 1133 (1965).

[16]   The Inhomogeneous Electron Gas, ed. by N.H. March and S. Lundqvist, (Plenum, New York, 1984).

[17]   J.R. Chelikowsky and S.G. Louie, Phys. Rev. B 29, 3470 (1984).

[18]   J. Harris, Phys. Rev. B 31, 1770 (1985).

[19]   H.M. Polatoglou, and M. Methfessel, Phys. Rev. B 37, 10403 (1989).

[20]   M. Scheffler, J.P. Vigneron, and G. B. Bachelet, Phys. Rev. Letters 49, 1765 (1982); Phys. Rev. B 31, 6541 (1985).

[21]   G.A. Baraff, and M. Schlüter, Phys. Rev. B 30, 1853 (1985).

[22]   O. Gunnarsson, O. Jepsen, and O.K. Andersen, Phys. Rev. B 27, 7144 (1983).

[23]   F. Beeler, M. Scheffler, O. Jepsen, and O.K. Andersen, Phys. Rev. Letters 54, 2525 (1985).

[24]   F. Beeler, O.K. Andersen, and M. Scheffler, Phys. Rev. Letters 55, 1498 (1985); Phys. Rev. B, in print.

[25]   J. Bernholc, N.O. Lipari, and S.T. Pantelides, Phys. Rev. Letters 41, 895 (1978); Phys. Rev. B 21, 3545 (1980).

[26] R. Car, P.J. Kelly, A. Oshiyama, and S.T. Pantelides, Phys. Rev. Letters **52**, 1814 (1984); **54**, 360 (1985).

[27] Y. Bar-Yam, and J.D. Joannopoulos, Phys. Rev. Letters **52**, 1129 (1984); Phys. Rev. B **30**, 1844 (1984).

[28] R. Car, and M. Parrinello, Phys Rev. Letters **55**, 2471 (1985).

[29] G.L. Chiarotti, F. Buda, R. Car, and M. Parrinello, to be published.

[30] Ch.G. Van de Walle, Y. Bar-Yam, and S.T. Pantelides, Phys. Rev. Letters **60**, 2761 (1988).

[31] K.C. Pandey, Phys. Rev. Letters **57**, 2287 (1986).

[32] S. Froyen, and A. Zunger, Phys. Rev. B **34**, 7451 (1986).

[33] G.P. Kerker, J. Phys. C **13**, L189 (1980).

[34] D.R. Hamann, M. Schlüter, and C. Chiang, Phys. Rev. Letters **43**, 1494 (1979).

[35] G.B. Bachelet, D.R. Hamann, and M. Schlüter, Phys. Rev. B **26**, 4199 (1982).

[36] U. Scherz, D. Weider, and M. Scheffler, to be published.

[37] G.W. Ludwig and H.H. Woodbury, in Solid State Physics 13, ed. by F. Seitz and D. Turnbull (Academic, New York, 1962), p. 223.

[38] P.J. Dean, Prog. in Sol. St. Chem. **8**, 1 (1973).

[39] A. Chantre, and D. Bois, Phys. Rev. B **31**, 7979 (1985).

[40] H. Feichtinger, J. Oswald, R. Czaputa, P. Vogl, and K. Wünstel, Proceedings of the 13th International Conference on Defects in Semiconductors, ed. by L.C. Kimerling and J.M. Parsey, Jr. (The metallurgical Society of AIME, 1984), p. 855.

[41] K. Wünstel, and P. Wagner, Appl. Phys. A **27**, 207 (1982).

[42] K. Graff and H. Pieper, J. Electrochem. Soc. **128**, 669 (1981).

[43] J.J. van Kooten, G.A. Weller, and C.A.J. Ammerlaan, Phys. Rev. B **30**, 4564 (1984).

[44] J.J. van Kooten, Ph.D. thesis, Amsterdam (1987).

[45] E.R. Weber, Appl. Phys. A **30**, 1 (1980).

[46] J.P. Vigneron et al., unpublished results.

[47] H. Katayama-Yoshida, and A. Zunger, Phys. Rev. Letters **53**, 1256 (1984); Phys. Rev. B **31**, 7877 (1985).

[48] L.V.C. Assali, and J.R. Leite, Phys. Rev. B **36**, 1296 (1987).

[49] O. Madelung, Festkörpertheorie Vol. III (Springer, Berlin 1972).

[50] F.A. Kröger, The Chemistry of Imperfect Crystals (North-Holland, Amsterdam 1974).

[51] H. Reiss, C.S. Fuller, and F.J. Morin, Bell Syst. Tech. J. **35**, 535 (1956).

[52] W.H. Shepherd, and J.A. Turner, J. Phys. Chem. Solids **23**, 1697 (1962).

[53] L.C. Kimerling, J.L. Benton, and J.J. Rubin, in Defects and Radiation Effects in Semiconductors (Inst. Phys. Conf. Ser. 59, 1981), p. 217.

[54] L.C. Kimerling, and J.L. Benton, Physica **116B**, 297 (1983).

[55] R. Wörner, and O.F. Schirmer, Solid State Commun. **51**, 665 (1984).

[56] S. Greulich-Weber, J.R. Niklas, J.-M. Spaeth, J. Phys. Cond. Matter **1**, 35 (1989).

[57] H. Overhof, M. Scheffler, and C.M. Weinert, Materials Science Forum **38-41**, 293 (1989); and to be published (1989).

[58] G.A. Baraff, and M. Lannoo, Revue Phys. Appl. **23**, 817 (1988).

[59] G.A. Baraff, M. Lannoo, and M. Schlüter, Phys. Rev. B **38**, 6003 (1988).

[60] E.R. Weber, H. Ennen, U. Kaufmann, J. Windscheif, J. Schneider, and T. Wosinski, J. Appl. Phys. **53**, 6140 (1982).

[61] M. Wattenbach, J. Kröger, C. Kisielowski-Kemmerich, and H. Alexander, Materials Science Forum **38-41**, 73 (1989).

[62] M. Scheffler, F. Beeler, O. Jepsen, O. Gunnarsson, O.K. Anderesen, and G.B. Bachelet, Proceedings of the 13th International Conference on Defects in Semiconductors, ed. by L.C. Kimerling and J.M. Parsey, Jr. (The metallurgical Society of AIME, 1984), p. 45.

[63] G.B. Bachelet, and M. Scheffler, Proceedings of the 17th Int. Conf. on the Physics of Semiconductors, ed. by J.D. Chadi and W.A. Harrison (Springer, New York, 1985), p. 755.

[64] G.B. Bachelet, M. Schlüter, and G.A. Baraff, Phys. Rev. B **27**, 2545 (1983).

[65] J.D. Chadi, and K.J. Chang, Phys. Rev. Letters **61**, 873 (1988).

[66] P. Mooney, this volume.

[67] J.M. Sanchez, and D. de Fontaine, Phys. Rev. Letters **35**, 227 (1975).

[68] U. Köhler, and Ch. Herzig, Phil. Mag. A **58**, 769 (1988).

[69] At this coference B. Meyer told that the 4.88 Å separation reported in Ref. 6 may be not very accurate. He emphasized, however, that the C_{3v} symmetry of the complex and the existance of an As-interstitial close to the As_{Ga} can be safely identified from the ENDOR data.

A New Look at the Reliability of Thin Film Metallizations for Microelectronic Devices

Rolf E. Hummel

Department of Materials Science and Engineering, University of Florida, Gainesville, FL 32611, USA

Summary: A brief review on the failure modes in thin film metallizations is given emphasizing electromigration, thermomigration, and thermal grooving of the grain boundaries. The most important parameters which influence the failures, such as grain boundaries, grain boundary gradients, temperature gradients, activation energy for ion migration, etc., are discussed. Alternate methods which improve the electromigration resistance in metallizations such as films consisting of gold or copper, or of ionized cluster beam deposited aluminum are described. Finally, a model is presented which enables one to evaluate electromigration under pulsed conditions.

1 Introduction

About a quarter of a century has passed since the first reports on electrical failures in connecting stripes, which link the individual parts of integrated circuits, became public. The connections in question consist of thin metallic films, commonly made of aluminum, which are only about 1 μm thick and thus have to sustain current densities which are in normal operation several orders of magnitude larger compared to those in domestically used wires. The failures have been observed to be related to holes, mainly near the cathode side of the thin metal stripe (Fig. 1) and, to a lesser extent, to hillocks near the anode (Fig. 2). Because of the specific location of holes and hillocks it was quickly suspected that a momentum exchange between the accelerated electrons and the metal ions plays a dominant role in these failures which cause a migration of some metal ions from the cathode towards the anode [1–8]. This *electron wind effect* had been observed before in bulk metals and was known by the name of electrotransport or electromigration [9]. Much has been learned in the past 25 years about electrotransport and other related failure mechanisms in thin film metallizations and as a consequence certain improvements of the electromigration resistance have eventually been accomplished. Nevertheless it is almost embarrassing to admit that a real solution of the problem, that is, a complete elimination of all failures has not yet been found. This is aggravated by the fact that the number of individual devices which industry accommodates on a single chip (that is, the device density) is steadily increasing. As a consequence, the active elements on a chip are becoming smaller and the connecting stripes

251

Fig. 1 Scanning electron micrograph of void formation (black areas) in an aluminum film (approximately 1000 Å thick) which was passed for 10 hours by a direct current whose current density was 2×10^5 A/cm^2. The shown area is close to the cathode side of the stripe, see Fig. 3.

Fig. 2 Scanning electron micrograph of hillock formation in tin. The parameters are similar as listed in the figure caption of Fig. 1 (anode side).

252

need to be narrower which in turn causes a further increase in current density and thus potential for failure. Industry has learned to design its way around this problem but eventually has recognized that the lifetime of a computer chip has its limitations unless completely new designs or concepts are found and applied. The following pages are a short summary of our present understanding of various failure modes in thin film metallizations. In addition, some new avenues will be outlined which could lead to an improved electromigration resistance and thus to a higher reliability of microelectronic devices.

2 Critical Parameters

2.1 Grain Boundaries and Grain Boundary Gradients

Electrotransport (that is, the migration of metal ions under the influence of an applied electric field) as well as thermotransport (that is, the migration of matter under the influence of a temperature gradient) occur predominantly via grain boundaries, at the temperatures of actual device operation (100–200°C). Thus, grain boundaries are the preferred nucleation sites for void and hillock formation. In particular, grain boundary triple points, i.e., junctions involving three grain boundaries, having one path in and two paths out lead to localized depletion of material (Fig. 3). Of particular concern are, therefore, areas which contain grain size gradients (Fig. 4). If in a given unit area more grains are contained in one part than in the other then more ions leave the fine grained part of this area via grain boundaries than are entering it. As a consequence material depletion is observed. (Similarly, hillocks might be formed in regions where more grain boundaries enter a unit area than are leaving it.) A seemingly logical solution of the electrotransport problem would thus be to utilize metallizations which contain only a small number of grain boundaries, that is, to increase the grain size to the limit of a single crystal [10,11]. This experiment involving a single crystalline aluminum film has been conducted in the laboratory with the expected result, that is, no failure after several years of dc stressing [12].

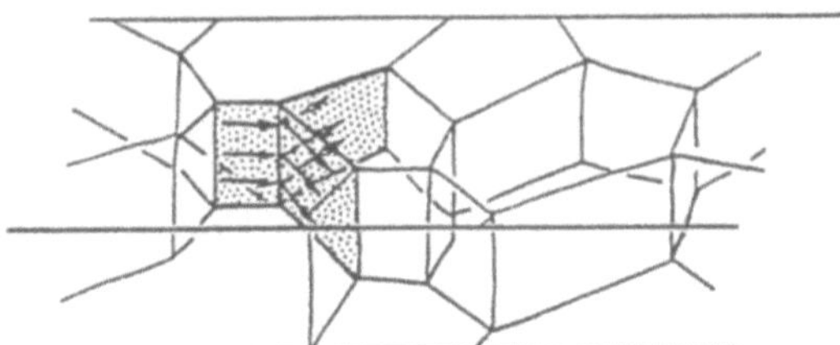

Fig. 3 Schematic representation of a grain structure emphasizing a grain boundary triple point.

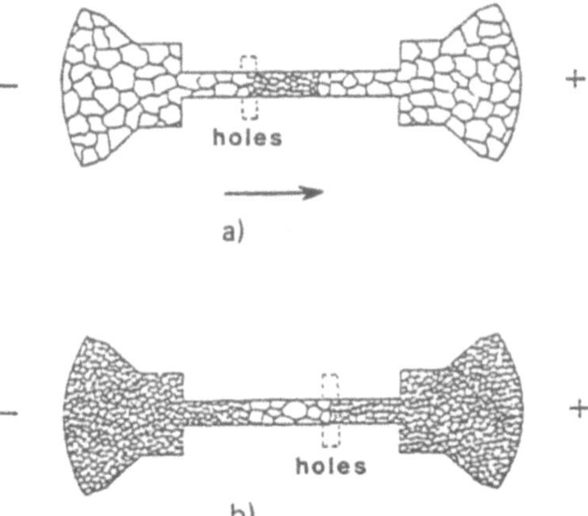

Fig. 4 Schematic representation of grain size gradients. The films are deposited by two consecutive depositions involving different substrate temperatures. A heated substrate yields a larger grain size.

Another method for reducing hole and hillock formation in thin films is to orient the grain boundaries, that is, the paths along which the ion motion would occur, perpendicularly to the current flow, similarly as the knots in a bamboo stick. Such a *bamboo structure* can indeed be achieved for the special case where the film dimensions are of that kind that only one grain is accommodated in the two transverse directions of the stripe [13–15]. This arrangement limits, however, the flexibility in designing the size of a thin film conductor.

The favorable laboratory results just described can usually not be directly utilized in actual device fabrication without disturbing other desirable properties of a chip, which makes their application generally not very feasible for industrial use. Nevertheless, it is important to understand the role of grain boundaries in the failure process involving electro- and thermomigration.

Interestingly enough, some years ago a different method to hamper the ion flow via grain boundaries was accidentally discovered. It was observed that small amounts of added elements which are nearly insoluble in aluminum (such as copper) would segregate on the grain boundaries, thus reducing there the available vacant sites [16–18]. It is assumed that the copper atoms in the grain boundaries would block the forced ion diffusion to a certain extent until eventually even the copper atoms have been removed by the electron wind force. Amounts of up to 4% Cu in aluminum are used by many manufacturers which increase the lifetime of the stripe by about one

254

order of magnitude. Other companies use 1–2% Si in aluminum which has the added advantage that it prevents silicon pick-up from the underlying device.

2.2 Temperature Gradients

Thin film stripes which are passed by a current of high current density (J) are subjected to considerable Joule heating. As a rule, the ends of the stripes are at a somewhat lower temperature because of the enhanced heat conduction to the substrate at this site. On the other hand, the main part of the conducting stripe is laid down on less heat conducting SiO_2. Because of the resulting temperature gradient, more ions leave an assumed unit volume at the cathode end of a thin film stripe than are entering it (Fig. 5). This results in a depletion of metal ions and eventually to void formation near the cathode as described already in the Introduction. Similarly, more ions are expected to enter a unit volume close to the anode than are leaving it, causing accumulation of material, that is, hillock formation near the anode. The largest divergence in ion velocity, v, occurs at a place where the temperature gradient ($\partial T/\partial x$) is maximal according to:

$$\text{div } v = \frac{D_0\, Z^*e\, \rho\, JQ}{k^2 T^3}\left(\frac{\partial T}{\partial x}\right)\exp\left(-\frac{Q}{kT}\right)\,. \tag{1}$$

(Z^*e is the effective charge on the moving atoms, Q is the activation energy of electromigration, D_0 is the pre-exponential diffusion constant, k is the Boltzmann constant, and ρ is the electrical resistivity.) A large divergence in ion velocity leads to voids or hillocks depending on the sign of $\partial T/\partial x$.

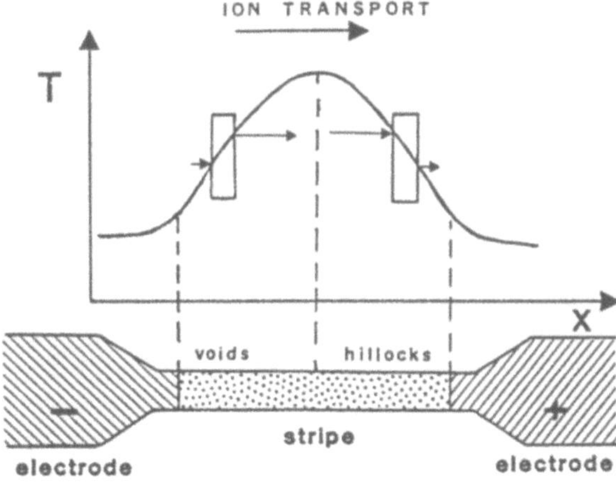

Fig. 5 Schematic representation of a temperature gradient along a thin film metallization which is caused by Joule heating and different heat dissipation near the center and on the electrodes. Compare to Figs. 1 and 2.

Besides this gross temperature gradient along the stripe, smaller temperature gradients may additionally occur perpendicular to the current flow because of the somewhat lower temperatures of the borders of the stripes. Temperature gradients may also be caused by local non-adherence of the film to the substrate (*hot spots*) because of underlying trapped gases or small particles, or by local reductions in stripe width. As it is true for grain size gradients, temperature gradients cannot be completely avoided in actual devices and are thus possible sources for void formation by thermotransport or by a combination of electromigration and thermomigration.

2.3 Other Gradients

Gradients in cross-sectional areas of a thin film metallization are sources for temperature gradients and current density gradients. They should and can be avoided in device design. Similarly, sharp bends of the stripe lines have been shown to cause divergencies in current flow and thus void and hillock formation.

2.4 Activation Energy for Electromigration

One of the most important parameters which determines the lifetime of a thin film metallization subjected to *dc* stressing is the activation energy of the migration mechanism. It can be obtained for example by measuring the time dependent resistance increase at the initial stage of an electromigration experiment [19], that is, before actual voids and hillocks are formed and visually observed (Fig. 6). It has been found that the activation energy for electromigration may have up to three distinct values depending in which

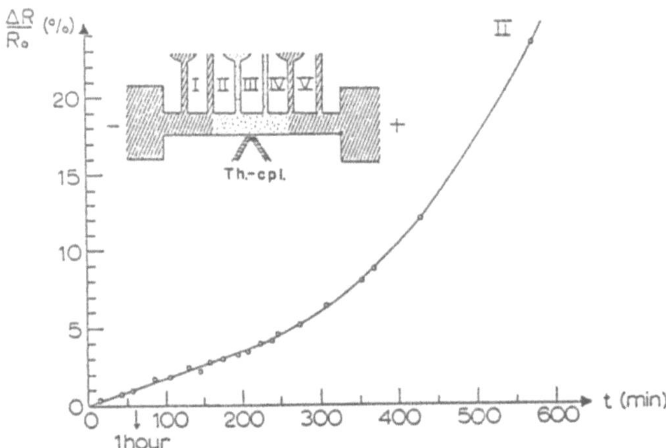

Fig. 6   Resistance increase during *dc* stressing of an aluminum film. The insert depicts the sample configuration which contains potential leads between which the change in resistance is measured. The cross-hatched areas represent thicker electrodes. ($J = 5.9 \times 10^5$ A/cm^2; $d = 1.52 \times 10^{-5}$ cm; $T = 332°$C.)

temperature range the measurements have been conducted. Their associated values are similar to those found for self-diffusion in bulk materials and correspond to diffusion through the lattice, the grain boundaries, or along the surface. Fig. 7 depicts an Arrhenius-type diagram in which the (current density normalized) marker velocity v_m is plotted versus the reciprocal device temperature. The activation energy, Q, can then be obtained from the slope in Fig. 7 by applying the equation:

$$\ln \frac{v_m}{J} = \ln \left[\frac{D_0\, Z^* e\, \rho}{kT} \right] - \frac{Q}{k} \cdot \frac{1}{T} \,.\tag{2}$$

The data in Fig. 7 yield **an** activation energy of 0.52 eV for aluminum thin films stressed at temperatures between about 200 and 330°C. This value is similar to that for aluminum self-diffusion in the grain boundaries. (At higher temperatures an activation energy of 1.16 eV can be deduced from Fig. 7 [21,22].)

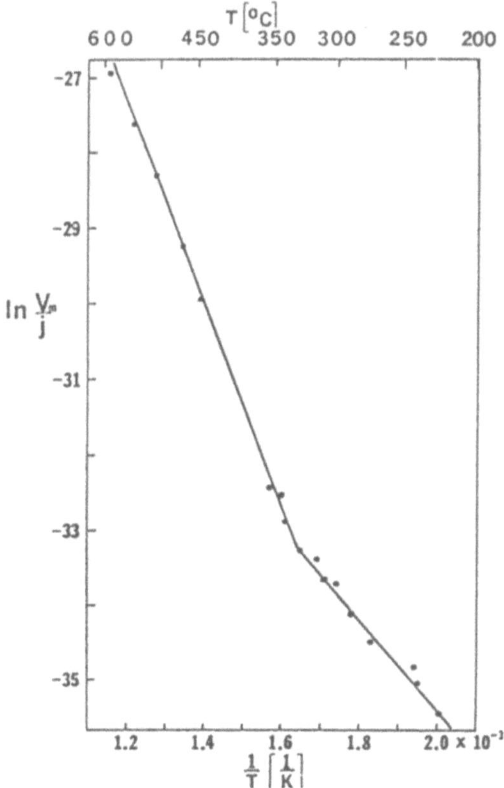

Fig. 7 Arrhenius diagram obtained for electromigration of aluminum films by the method depicted in Fig. 6. The five high temperature values (square marks) are taken from reference 20 and have been measured on bulk aluminum.

Of particular interest is the activation energy at the temperature of device operation, that is, between 100 and 200°C because it is linked directly to the failure time. The *median time to failure* (that is, the time at which 50% of a large number of identical metallizations have failed) is given by the empirical equation [23]:

$$t_{50} = A \frac{1}{J^n} \exp\left(\frac{Q}{kT}\right) \tag{3}$$

(where A is a constant which depends on the structure and on the electrical and diffusion properties of the material, and n is a numerical value which is approximately 1.5 [24]). Since t_{50} increases exponentially with the activation energy, a small increase in Q has a major impact on the device lifetime. For example, changing the metallization from pure aluminum to gold (the latter of which has a Q of 0.9 eV [25]) or to copper would increase the lifetime t_{50} by several orders of magnitude. (Specifically, for gold at 400 K the increase in lifetime would be by a factor of 6.1 × 10^4, and at 500 K by a factor of 6.8 × 10^3). This large increase in lifetime is, however, not immediately achieved in practical application. The reason for this can be found in secondary failure mechanisms which become effective once electromigration is no longer the principal cause of failure. This will be discussed below.

An even higher activation energy is found for tungsten or molybdenum. However, the lower electrical conductivity of the transition metals seems to limit their application for thin film conductors.

The question then arises why gold or copper, which have a high conductivity and a high electromigration resistance, are not more widely used by industry for thin film interconnections. One of the reasons is that these metals, once traces of them have diffused into the electronic device may provide trapping centers which reduce the number of current carriers. Secondly, the bonding of copper and gold to silicon oxide is not as good as for aluminum. Third, it has been observed that when electromigration is no longer the principal failure mode, other failure mechanisms might become important. One of them is thermal grooving of the grain boundaries [26]. These problems can be, however, prevented by inserting a thin barrier layer consisting of a proper metal (such as indium or tin) between metallization and device. This will be discussed in the following section.

3 Multiple Layer Metallizations

Gold films, when heated at moderate temperatures, eventually develop small *pin holes* which start to form at points where grain edges meet the free surface [26]. During continuing annealing, the holes increase in size, link-up with other holes and eventually cause a discontinuity of the thin film and thus failure (Fig. 8a). This thermal grooving of the grain boundaries is the response to an imbalance of surface tension which has the tendency to pull the grain boundary traces on the free surface toward the substrate.

Of special interest to industry is the effect which various barrier layers (which are sandwiched between gold film and substrate) have on the grain boundary grooving. As mentioned above, such intermediate layers are important because they increase the adhesion of the gold films to the underlying substrate and prevent interdiffusion of gold into the microelectronic device. Figs. 8 b–f depict scanning electron micrographs of gold films deposited on a variety of underlays which have been heat treated near 500°C for 1 hour in air. We notice that traces of sodium aggravate the formation of holes in gold films to the extent that only islands of gold are eventually left. Copper (and nickel) underlays also accelerate the hole formation [27]. (It has been found that the holes in gold are partially filled by copper oxide and that a copper oxide layer covers the entire gold film on the free surface.)

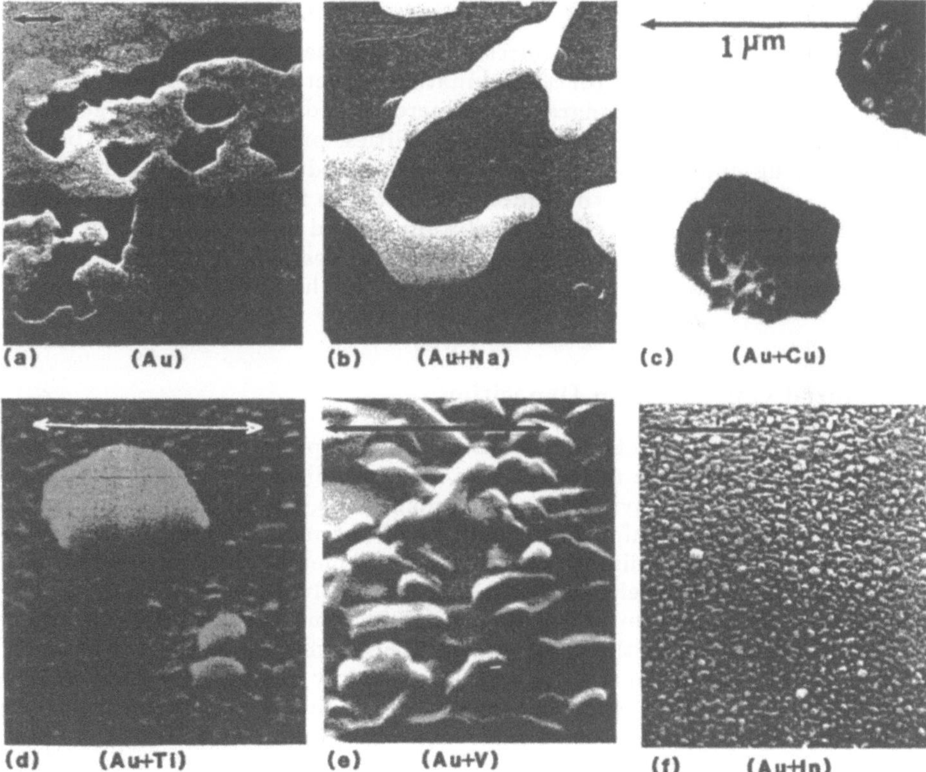

Fig. 8 Scanning electron micrographs for a pure gold film and gold films with various thin metal underlayers sandwiched between gold and pure quartz glass. The films have been annealed for one hour in air at 500°C (c,d,e) or 560°C (a,b,f). All marker lengths indicate 1 µm.

On the other hand, titanium and vanadium underlayers cause hillock formation consisting of gold which grow on the gold film [28]. Interestingly enough, indium (and to a certain extent also tin) underlays have a beneficial effect on gold films during heating [29,30]. These underlays (about 50–100 Å thick) have a stabilizing influence on the gold grains. If In/Au composite films are heated in air, In_2O_3 is formed on the free surface and also to a smaller extent within the gold film presumably near the grain boundaries [30]. The lack of grain boundary grooving may then be explained by assuming that the indium oxide pins the grain boundaries to their positions, thus inhibiting grain boundary motion and grain growth. In addition, indium oxide which forms on the free surface caps the gold film similarly as known from passivation layers (such as SiO_2) on aluminum films.

It should be mentioned in passing that alternate layers of aluminum and transition metals tend to increase the lifetime of metallizations. Once the aluminum is deteriorated by electromigration, the conduction can still take place through an alternate path, for example, through the molybdenum layer. This naturally increases the resistance of the film.

In summary, the problem of failures in thin film metallizations can be diminished by selecting metals (or a combination of metals) which cause the activation energy for electromigration to increase and which combine good adhesion properties, diffusion barrier capabilities, prevention of grain boundary grooving and, of course, high electrical conductivity. Gold or copper conductors on indium or tin underlayers seem to fulfill these requirements. There is, however, an alternative solution to gold and copper which allows us to stay with aluminum metallizations, if so desired, by applying a novel deposition technique. This will be described in the next section.

4 Ionized Cluster Beam Deposition

It appears that the electronics industry tends to prefer the continuing usage of aluminum as a base metal for thin film metallizations. Thus, techniques which are capable of laying down aluminum films, having an increased resistance against electromigration, could be of potential interest if the deposition process does not disturb other useful parameters of the microelectronic device. One such novel deposition method involves the ionization and acceleration of aluminum atoms and possibly some clusters of these atoms while they are still in the vapor phase. The aluminum films thus obtained have been shown to double the activation energy compared to that of pure aluminum [31]. The exact mechanism which causes this increase in Q is not completely understood at this time. Still, because of lack of a better descriptive name, the technique will be called here *Ionized Cluster Beam (ICB) deposition*, in accord with the literature [32].

The films are laid down in a vacuum of the 10^{-6} Torr range on thermally grown SiO_2 on a (100) p-type silicon substrate. Before deposition, aluminum vapor leaves under high pressure a carbon crucible which has a nozzle of

260

characteristically 2 mm diameter. The resulting particle beam is ionized
by impact of electrons employing an ionization voltage of 400 V and an
ionization current of 100 mA. The ionized particles are then accelerated
towards the substrate by a potential of between 1 and 7 kV (Fig. 9). A 15-
minute deposition results in a film thickness of about 120 nm. Because of
the hot crucible, the substrate may be slightly but unintentionally radiation-
heated.

Electromigration experiments have been performed on ICB deposited thin
films applying the same sample shape and the same resistance technique as
described above for conventionally-laid-down aluminum (Fig. 6). From the
slope in the Arrhenius diagram, an activation energy of about 1.1 eV can
be deduced which compares to $Q = 0.52$ eV for conventionally deposited,
pure, unpassivated aluminum films (Fig. 10). Thus, the lifetime of the ICB
films has a potential of being three or four orders of magnitude larger than
that of presently used copper-doped, passivated aluminum films. Further
studies have to substantiate this result.

The microstructure of aluminum films laid down by the ICB technique
show peculiar rounded grains and often display Moirée fringes near the
grain boundaries which are known to be characteristic for small angle grain
boundaries [32]. Those areas possess a low *vacancy density* and thus are
believed to decrease the ion mobility under the influence of an electric field.
Electrotransport in ICB films is thus probably dominated by migration
through the volume. The higher activation energy, which is close to that
for volume self-diffusion and volume electromigration (Fig. 7), supports this
hypothesis.

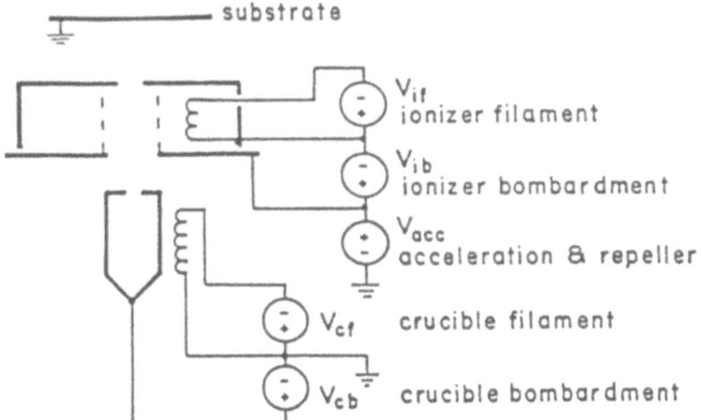

Fig. 9 Schematic representation of the Ionized Cluster Beam deposition technique designed by Yamada and Takagi [32].

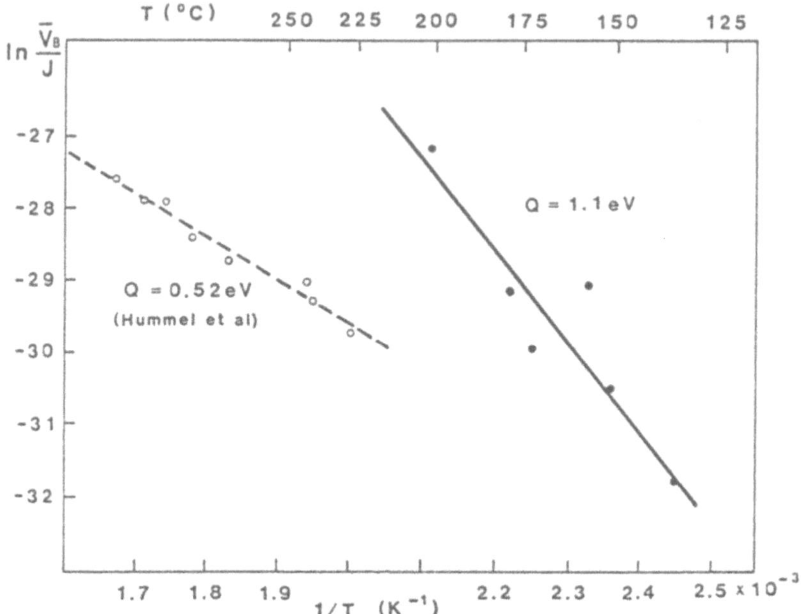

Fig. 10 Arrhenius-type diagram of unpassivated dc-stressed aluminum films deposited by the ionized cluster beam technique on 7000 Å thick SiO_2. Included are data obtained by conventional vapor deposition (see Fig. 7) for comparison.

5 Electromigration under Pulsed Conditions

In practice, the metallizations described above are rarely under the influence of a constant high current. Instead, current pulses of short duration are often employed. A duty cycle, d_c, can then be defined which is the ratio of the time during which the current flows (t_{on}) divided by the total time of operation (t_{tot}).

$$d_c = \frac{t_{on}}{t_{tot}} \; . \tag{4}$$

The median time to failure has been observed to increase under pulsed conditions [24, 33–41]. The lifetime of the device is larger the shorter the on-time, that is, the smaller the duty cycle:

$$t_{50} = A \frac{1}{d_c^m \, J^n} \exp\left(\frac{Q}{kT}\right) \; . \tag{5}$$

Some experiments seem to agree reasonably well with $m = 2$ for which there is, however, no theoretical justification. It is, instead, more reasonable to omit an exponent for d_c and modify Eq. (5) by considering the following factors:

262

(1)   Under pulsed conditions, the device temperature is somewhat lower than under continuous operation because of the intermittent cooling periods (Fig. 11).

(2)   It may be anticipated that between current pulses, some of the electro-migration induced vacancies stream back, obeying conventional laws of diffusion. This partial self-healing of the electromigration damage during off-times requires a second term in Eq. (5) which in its modified version reads [42]:

$$t_{50} = A \frac{1}{d_c J^n} \exp \left(\frac{Q}{k[(T_0 \exp(\beta d_c)]} \right) + B D_T \frac{1 - d_c}{f} \qquad (6)$$

(where β is a constant which can be obtained from an unpulsed experiment, $f \neq 0$ is the unidirectional pulse frequency, T_0 is the ambient temperature, D_T is the diffusion constant for grain boundary migration at the device temperature, and B is an adjustable parameter).

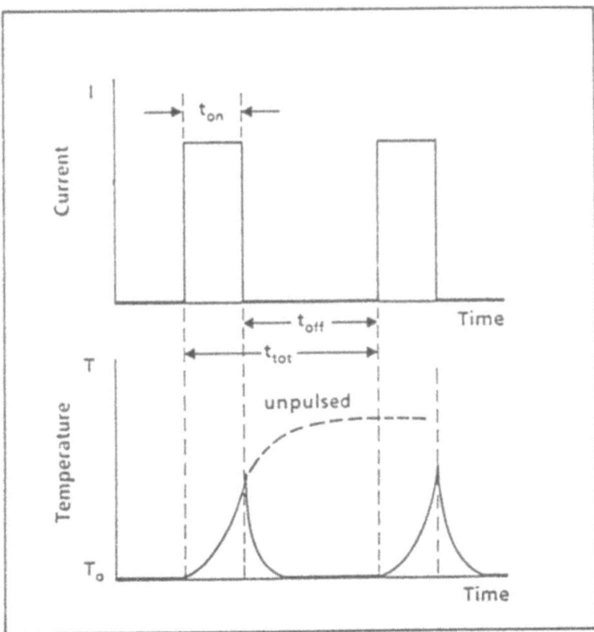

Fig. 11   Time dependence of current and temperature under pulsed conditions (schematic).

263

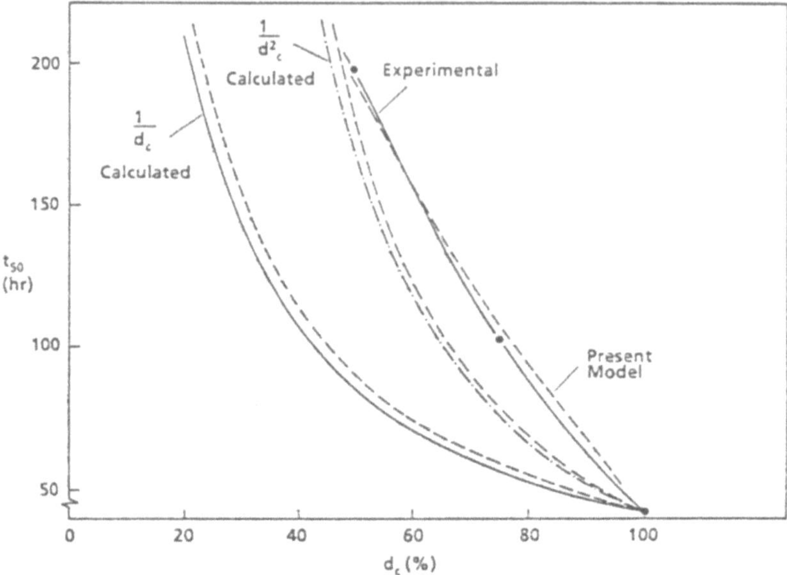

Fig. 12 Experimental and calculated median failure times as a function of duty cycle, d_c. The dash-dotted curve was calculated taking $m = 2$ in Eq. (5) ($1/d_c^2$ dependence). The dashed line next to it represents a temperature correction (Fig. 11). The solid line to the left represents $m = 1(1/d_c$ dependence). The curve marked "present model" was calculated applying Eq. (6).

Fig. 12 depicts pulsed electromigration results for f =10 kHz. The agreement between experiment and model calculation is good and is certainly better than that obtained by applying the simpler Eq. (5) and m =2. Eq. (6) predicts that for very high frequencies the median failure time is reduced, approaching eventually the $1/d_c$ curve.

Conclusions

Electromigration, thermomigration, and thermally induced grain boundary grooving are still the predominant failure modes in thin film metallizations for microelectronic devices. Much has been learned about the mechanisms which cause these failures. In particular, all types of gradients, such as temperature gradients, grain size gradients, or gradients of structural parameters (width, etc.), should be minimized during device manufacturing. Alternatives to conventionally deposited aluminum films can be obtained by utilizing metallizations consisting of gold or copper in conjunction with appropriate barrier or cladding materials, or by using multilayer metallizations. Deposition of aluminum by the ICB technique or related methods provides yet another alternative.

Acknowledgement

I am indebted to Dr. R.T. DeHoff for critical review of this manuscript. The financial support of part of this work by ARO is gratefully acknowledged.

References

[1] *I.A. Blech and H. Sello*, Physics of Failures in Electronics, Rome Air Dev. Center, USAF **5**, 496 (1966)

[2] *H. Wever*, Elektro- und Thermotransport in Metallen (J. Ambrosius Barth, Leipzig 1973)

[3] *R.E. Hummel and H.B. Huntington (Eds.)*, Electro- and Thermotransport in Metals and Alloys (AIME Publ., New York 1977)

[4] *F.M. d'Heurle, P.S. Ho*, in: Thin Films — Interdiffusion and Reactions, ed. by *J.M. Poate, K.N. Tu, J.W. Mayer* (J. Wiley and Sons, New York 1978), p. 243

[5] *H.B. Huntington*, in: Encyclopedia of Chemical Technology (J. Wiley and Sons, New York 1971), Suppl. Vol., 2nd ed., p. 278

[6] *J.R. Llloyd, J. Pierce, R.A. Levy, and R.G. Frieser (Eds.)*, Proc. of the Symp. on Electromigration of Metals, New Orleans (The Electrochemical Society Inc., Pennington, NJ 1985)

[7] *R.E. Hummel*, in: Electro- and Thermotransport in Metals and Alloys (AIME Publ., New York 1977), p. 93

[8] *P.S. Ho, F.M. d'Heurle, and A. Gangulee*, in: Electro- and Thermotransport in Metals and Alloys (AIME Publ., New York 1977), p. 108

[9] *W. Seith and O. Kubaschewski*, Z. Electrochem. **41**, 551 (1935)

[10] *J.R. Black*, Proc. 6th Ann. Int. Rel. Phys. Symp. IEEE (1967), p. 148

[11] *M.J. Attardo and R. Rosenberg*, J. Appl. Phys. **41**, 2381 (1970)

[12] *A. Gangulee and F.M. d'Heurle*, Thin Solid Films **12**, 399 (1972) and **16**, 227 (1973)

[13] *G.A. Scoggan, B.N. Agarvala, P.P. Peressini, and A. Brouillard*, Proc. 13th Ann. Int. Rel. Phys. Symp. IEEE (1975), p. 151

[14] *C.B. Oliver and D.E. Bower*, Proc. 8th Ann. Int. Rel. Phys. Symp. IEEE (1970), p. 116

[15] *A.J. Learn*, Appl. Phys. Lett. **19**, 292 (1971)

[16] *I. Ames, F.M. d'Heurle, and R. Horstman*, IBM J. Res. Develop. **4**, 461 (1970)

[17] *R. Rosenberg, A.F. Mayadas, and D. Gupta*, Surface Science **31**, 566 (1972)

[18] *F.M. d'Heurle, N.G. Ainslie, A. Gangulee, and M.C. Shine*, J. Vac. Sci. Technol. **9**, 289 (1972)

[19] *R.E. Hummel, R.T. DeHoff, and H.J. Geier*, J. Phys. Chem. Solids **37**, 73 (1976)

[20] *R.V. Penny*, J. Phys. Chem. Solids **25**, 335 (1964)

[21] *Th. Heumann and H. Meiners*, Z. Metallkunde **57**, 571 (1966)

[22] *J.J. Spokas and C.P. Slichter*, Phys. Rev. **115**, 560 (1959)

[23] *J.R. Blech*, Proc. IEEE **57**, 1587 (1969)

[24] *H.A. Schafft, T.C. Grant, A.N. Saxena, and C.-Y. Kao*, Proc. 23rd Ann. Int. Rel. Phys. Symp. IEEE (1985), p. 93

[25] *R.E. Hummel and H.J. Geier*, Thin Solid Films **25**, 335 (1975)

[26] *R.E. Hummel, R.T. DeHoff, S. Matts-Goho, and W.M. Goho*, Thin Solid Films **78**, 1 (1981)

[27] *J.Y. Kim, R.E. Hummel, and R.T. DeHoff*, J. Vac. Sci. and Technol., April 1989 (in press)

[28] *J.Y. Kim and R.E. Hummel*, J. Vac. Sci. and Technol., submitted for publication

[29] *R.E. Hummel, S.M. Goho, and R.T. DeHoff*, Proc. 22nd Ann. Int. Rel. Phys. Symp. IEEE (1984), p. 234

[30] *S.Y. Lee, R.E. Hummel, and R.T. DeHoff*, Thin Solid Films **149**, 29 (1987)

[31] *R.E. Hummel and I. Yamada*, Appl. Phys. Lett. **53**, 1765 (1988) and **54**, 18 (1989)

[32] *I. Yamada and T. Takagi*, IEEE Trans. Electron Devices, ED **34**, 1018 (1987)

[33] *J.M. Towner and E.P. van de Ven*, Proc. 21st Ann. Int. Rel. Phys. Symp. IEEE (1983), p. 36

[34] *J.R. Davis*, Proc. IEEE **123**, 1209 (1976)

[35] *J.M. Schoen*, J. Appl. Phys. **51**, 508 (1980)

[36] *R.J. Miller*, Proc. 16th Ann. Int. Rel. Phys. Symp. IEEE (1978), p. 241

[37] *A.T. English, K.L. Lai, and P.A. Turner*, Appl. Phys. Lett. **21**, 397 (1972)

[38] *L. Brooke*, Proc. 25th Ann. Int. Rel. Phys. Symp. IEEE (1987), p. 136

[39] *J.W. McPherson and P.B. Ghate*, Proc. Symp. Electromigration of Metals, ed. by *J.R. Lloyd, R.A. Levy, J. Pierce, and R.G. Frieser* (Electrochem. Soc. 1985), p. 64

[40] *J.S. Arzigian*, Proc. 21st Ann. Int. Rel. Phys. Symp. IEEE (1983), p. 32

[41] *R.A. Sigsbee*, Proc. 11th Ann. Int. Rel. Phys. Symp. IEEE (1973), p. 301

[42] *R.E. Hummel and H.H. Hoang*, J. Appl. Phys. **65**, 1929 (1989)

Quantum Dot Resonant Tunneling Spectroscopy

Mark A. Reed, John N. Randall, James H. Luscombe, William R. Frensley, Raj J. Aggarwal*, Richard J. Matyi[†], Tom M. Moore, and Anna E. Wetsel[‡]

Central Research Laboratories, Texas Instruments Incorporated, Dallas, Texas 75265, USA

Summary: The electronic transport through 3-dimensionally confined semiconductor quantum wells (*quantum dots*) is investigated and analyzed. The spectra corresponds to resonant tunneling from laterally-confined emitter contact subbands through the discrete 3-dimensionally confined quantum dot states. Momentum non-conservation is observed in these structures. Results on coupled quantum dot states (*molecules*) will be presented.

1 Introduction

Carrier confinement to reduced dimensions has led to numerous important developments in basic semiconductor physics and device technology. Until recently, this confinement has only been realizable by an interface (such as Si/SiO_2 and GaAs/AlGaAs), and thus the confinement is in one dimension. Advances in microfabrication technology [1-3] now allow one to impose quantum confinement in additional dimensions, typically done by constricting or confining in lateral dimensions an existing 2D carrier system. Though the technology to create lateral confining potentials is considerably less advanced than the degree of control that exists in the vertical (epitaxial, or interface) dimension, remarkable progress has been made in elucidating the relevant physics of these systems.

Investigations into two-dimensionally confined *quantum wires* [4] were first attempted in metallic systems. However, subband spacings in metals are considerably smaller than in semiconductors for the same dimensional size. The realization of semiconductor quantum wires [5,6] allowed the investigation of well-defined lateral subbands, and the creation of *electron waveguides* [7]. These structures are excellent laboratories to study fundamentals of

---

* Department of Electrical Engineering and Computer Science, Massachusetts Institute of Technology, Cambridge, MA.

† Department of Metallurgical and Mineral Engineering, University of Wisconsin at Madison, WI.

‡ Department of Physics, Harvard University, Cambridge, MA.

electronic transport [8,9], and allow the investigation of the Landauer formalism in ballistic quantized point contacts [10,11], non-locality [12], and waveguide mode-mixing. Unfortunately, electronic transport that reveals quantum size effects in these structures can only be done near equilibrium (i.e., low voltages and temperatures) and thus these measurements do not shed light on more common non-equilibrium semiconductor situations.

Recently, three-dimensionally confined *quantum dots* have been realized [13]. These structures are analogous to semiconductor atoms, with energy levels tunable by the confining potentials. However, quantum dot structures pose an experimental paradox distinct from the 2D and 1D structures mentioned above. Establishing transport through the single electronic states of a quantum dot implies the states cannot be totally isolated; i.e., the confining potential must be slightly leaky, and thus the states are *quasi-bound*. Additionally, contact to higher-dimensional carrier reservoirs are non-trivial from an experimental and analysis viewpoint.

We have adopted a configuration where the quasi-bound momentum component (and thus the resultant transport direction) is epitaxially defined in the form of a resonant tunneling structure, and additional confinement is fabrication-imposed. This configuration is distinct from all the above referenced quantum wire cases where the (unbound) current flow is along the interface; here it is through the interface. Because of this constraint, the behavior of a system operated far from equilibrium can be examined. This configuration is also significantly different from the case of the tunneling observed through discrete defect states [14,15], since the structural experimental parameters are, at best, difficult to control in the defect tunneling configuration whereas they are trivial to vary by epitaxy for the confined resonant tunneling structure configuration.

An alternative approach which circumvents the contact problem is to probe the density of states in isolated systems by capacitance measurements [16]. This near-equilibrium approach has the advantages of isolation, but lacks information of transport in and coupling to single electron states, since the 0D electron states are measured solely electrostatically. The advantage of studying a single dot spectrum is also sacrificed in capacitance measurements because of signal-to-noise limitations, and thus size fluctuations become an issue.

We present here a study of resonant tunneling through various quantum dot systems, and the bandstructure modeling necessary to understand the experimental electronic transport spectra.

2 Quantum Dot Fabrication and Tunneling

Our approach used to produce quantum dot nanostructures suitable for electronic transport studies is to laterally confine resonant tunneling heterostructures. This approach embeds a quasi-bound quantum dot between two quantum wire contacts.

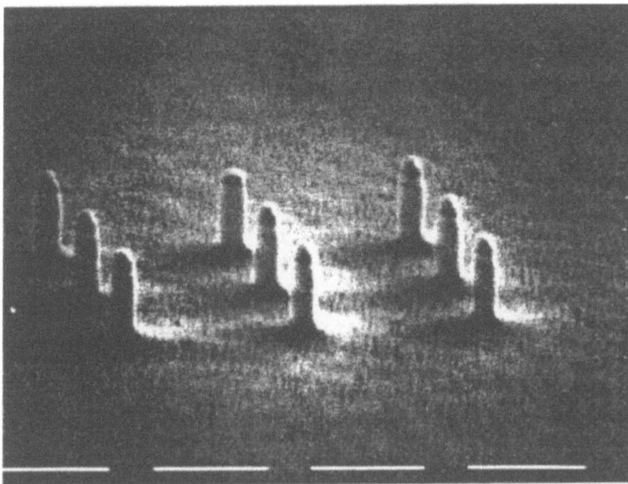

Fig. 1 Scanning electron micrograph of an array of anisotropically etched columns containing a quantum dot. The horizontal marker is 0.5 micrometer; the diameter of each of the columns is approximately 1000Å. The dark region on top of the column is the electron-beam defined ohmic contact/etch mask.

An ensemble of AuGe/Ni/Au ohmic metallization dots (single or multiple dot regions) are defined by electron-beam lithography on the surface of the grown resonant tunneling structure. Creation of dots less than 500Å is possible, though we will show that the appropriate range for the typical epitaxial structure and process used is in the range 1000Å- 2500Å in diameter. A bi-layer polymethylmethacrylate (PMMA) resist and lift-off method is used. The metal dot ohmic contact serves as a self-aligned etch mask for highly anisotropic reactive ion etching (RIE) using BCl_3 as an etch gas.

The resonant tunneling structure is etched through to the n^+ GaAs bottom contact, defining columns in the epitaxial structure. A SEM of a collection of these etched structures is seen in Figure 1.

To make contact to the tops of the columns, a planarizing and insulating polyimide is spun on the sample, then etched back by O_2 RIE to expose the metal contacts on the top of the columns. A gold contact pad was then evaporated over the top of the column(s). The bottom conductive substrate provides electrical continuity. Multiple columns can be connected in parallel for diagnostic purposes; however, it should be stressed that all data discussed hereupon is for a single isolated column.

Figure 2 schematically illustrates the lateral (radial) potential of a column containing a quantum dot and the spectrum of 3-dimensionally confined electron states under zero and applied bias. A spectrum of discrete states would be expected to give rise to a series of resonances in transmitted cur-

269

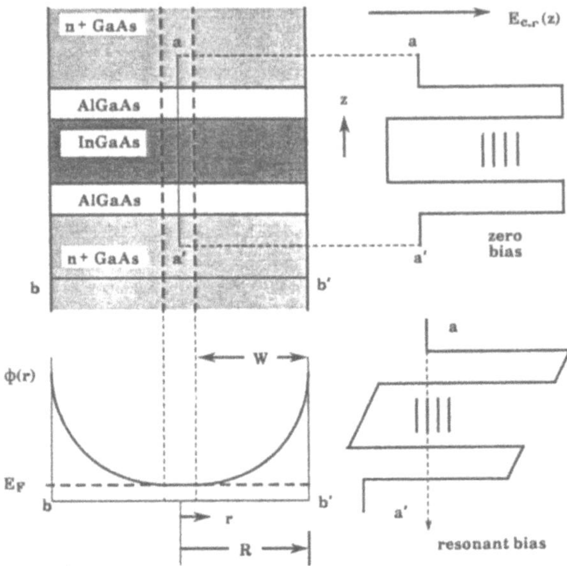

Fig. 2 Schematic illustration of the vertical (a-a') and lateral (b-b') potential of a column containing a quantum dot, under zero and applied bias. $\Phi(r)$ is the (radial) potential, R is the physical radius of the column, r is the radial coordinate, W is the depletion depth, Φ_T is the height of the potential determined by the Fermi level (E_F) pinning, and $E_{c,\Gamma}$ is the Γ-point conduction band energy.

rent as each state is biased through the source contact. To observe lateral quantization of quantum well state(s), the physical size of the structure must be sufficiently small that quantization of the lateral momenta produces energy splittings $> 3k_BT$. Concurrently, the lateral dimensions of the structure must be large enough such that pinch-off of the column by the depletion layers formed on the side walls of the GaAs column does not occur.

Due to the Fermi level pinning of the exposed GaAs surface, the conduction band bends upward (with respect to the Fermi level in the contacts), and where it intersects the Fermi level determines in real space the edge of the central conduction path core. When the lateral dimension is reduced to twice the depletion depth or less, the lateral potential becomes parabolic though conduction through the central conduction path core is pinched off.

Observation of tunneling through the discrete states of a quantum dot necessitates fabrication within a narrow design criterion.

Figure 3 shows the current-voltage-temperature characteristics of a quantum dot resonant tunneling structure successfully fabricated within these constraints. The structure lithographically is 1000Å in diameter and epitaxially is a n+ GaAs contact/AlGaAs barrier/InGaAs quantum well structure.

270

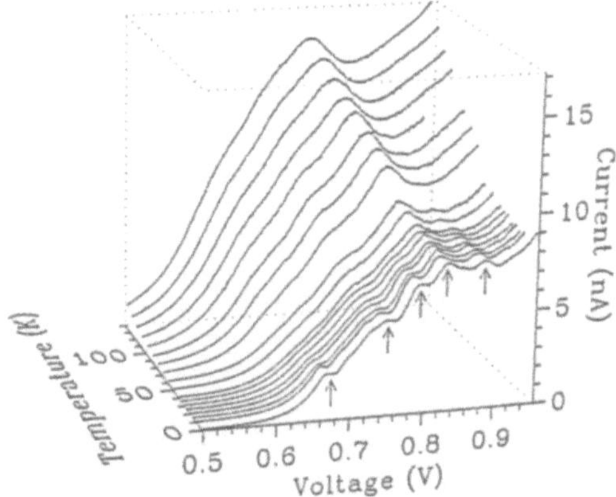

Fig. 3 Current-voltage characteristics of a single quantum dot nanostructure as a function of temperature, indicating resonant tunneling through the discrete states of the quantum dot. The arrows indicate voltage peak positions of the discrete state tunneling for the T = 1.0K curve.

At high temperature, the characteristic negative differential resistance of a double barrier resonant tunneling structure is observed. As the temperature is lowered, two effects occur. First, the overall impedence increases presumably due to the elimination of a thermally activated excess leakage current. Second, a series of peaks appears above and below the main negative differential resistance (NDR) peak. In the range of device bias 0.75V-0.90V, the peaks appear equally spaced with a splitting of approximately 50 mV. Another peak, presumably due to the ground state of the harmonic oscillator potential, occurs 80 mV below the equally spaced series.

The existence of the fine structure in the tunneling characteristics of this, and other, laterally confined resonant tunneling structures indicates the formation of laterally confined electronic states. However, a full indexing of the spectrum is needed to verify that the structure in the electrical characteristics is the discrete levels. To do this, a full 3D screening model of the quantum dot system is necessary.

3 1D Resonant Tunneling Spectroscopy

Prior to understanding the detailed spectroscopy of a full 3-dimensionally confined system, let us first model and compare with experiment 1D modeling of resonant tunneling structures. Though resonant tunneling is qualitatively well understood, until now there lacks detailed *quantitative* spectroscopy of tunneling structures. We will use the epitaxial structure of the quantum dot presented previously to verify the modeling procedure and to illustrate the effect of the 3-D localization potential.

271

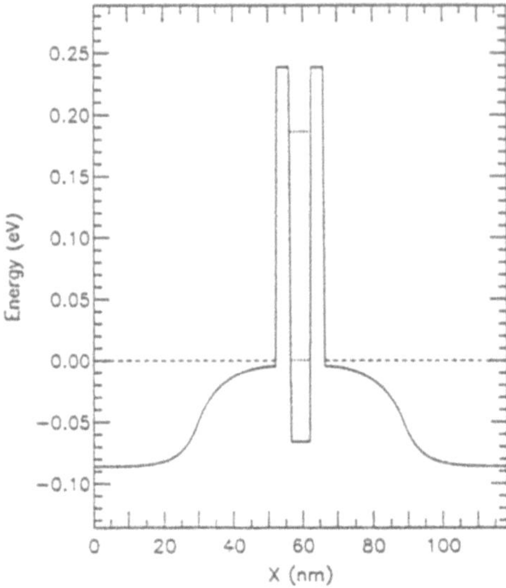

Fig. 4 Self-consistent band diagram using Poisson's equation for the electrostatic potential of the quantum dot 1D epitaxial structure. The epitaxial dimension is denoted by x, in nanometers. The electrons in the contacts are treated in a finite-temperature Thomas-Fermi approximation. The simulation does not include current flow. The structure is a 40Å $Al_{.3}Ga_{.7}As$ barrier / 60Å $In_{.07}Ga_{.93}As$ quantum well / $Al_{.3}Ga_{.7}As$ barrier structure with contacts doped to $2x10^{18}$ cm^{-3}, at 77K. The Fermi level is denoted by a dashed line and the energies of the bound states are denoted by dotted lines.

Figure 4 shows a realistic conduction band profile of the quantum dot epitaxial structure at equilibrium. The model from which this Figure was obtained finds the self-consistent solution of Poisson's equation for the electrostatic potential. The electrons in the contacts are treated in a finite-temperature Thomas-Fermi approximation (i.e., these electrons are assumed to be in local equilibrium with the Fermi levels established by their respective contacts.) A result of this calculation, illustrated in the Figure, is that the band profile near the quantum well is significantly perturbed by the contact potential of the n^+-undoped junction. This shifts the resonant state upward (with respect to the n^+ GaAs Fermi level) from that expected from a naïve flat-band picture.

Figure 5 shows the experimental current-voltage characteristics of a typical large area mesa device of this epitaxial structure at 77K. The low bias peak, shown in Figure 5(a), is the resonance due to tunneling through the ground state of the quantum well. Notice that the In content in the quantum well sufficiently lowers this quantum well state such that it lies <u>below</u> the Fermi level, but above the conduction band edge of the contact. The unique position of this state causes the observed large zero-bias conductivity of the

272

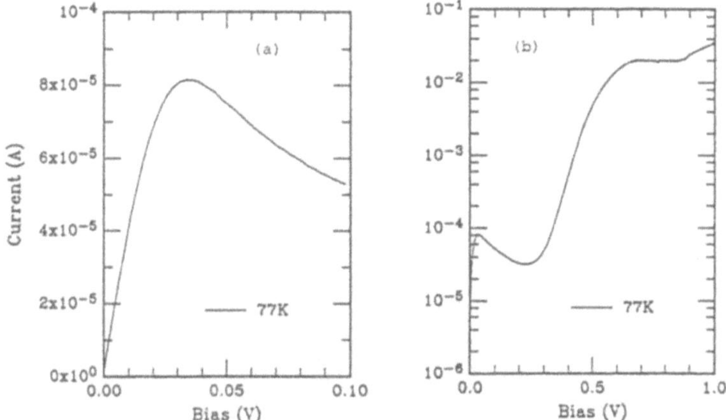

Fig. 5 Current-voltage characteristics of the quantum dot epitaxial structure large mesa device, at T=77K. (a) Expanded low bias region showing resonant tunneling through the ground state of the quantum well. Note the large zero bias conductance. (b) Full scale, showing the $n = 1$ ($V = 0.03$V) and $n = 2$ ($V \sim 0.7$V) resonances.

structure and the low (30 mV) resonance, verified in the modeling as the voltage bias where the quantum well state falls below the emitter conduction band edge.

Nominal structural parameters were not sufficiently accurate to predict the correct resonant positions. It has been demonstrated [17] that this method of modeling the resonant voltage peak positions yields greater accuracy for structural parameters than most present characterization techniques. The nominal structural parameters were accurate to within their error bars; to fit the experimental resonant peak positions (Figure 5), it was necessary to assume a 60Å undoped $In_{.07}Ga_{.93}As$ quantum well instead of the nominal 50Å undoped $In_{.08}Ga_{.92}As$ quantum well. The high zero bias conductivity and low resonance imposes two constraints on the model that allows for relatively precise determination of the structural parameters. The only parameter which is not tightly constrained is the tunnel barrier height; however, the flat conductance region at approximately 0.7V due to the $n = 2$ resonance implies that the $n = 2$ state is becoming unbound at this bias due to the lowering of the collector-side barrier below the emitter Fermi level. Figure 6 shows this effect, which implies a Al content in the tunnel barriers of 30% instead of the nominal 25%.

4 3D Resonant Tunneling Spectroscopy

Using these techniques, we can now attempt to understand the detailed spectroscopy of a full 3-dimensionally confined system. We have modeled

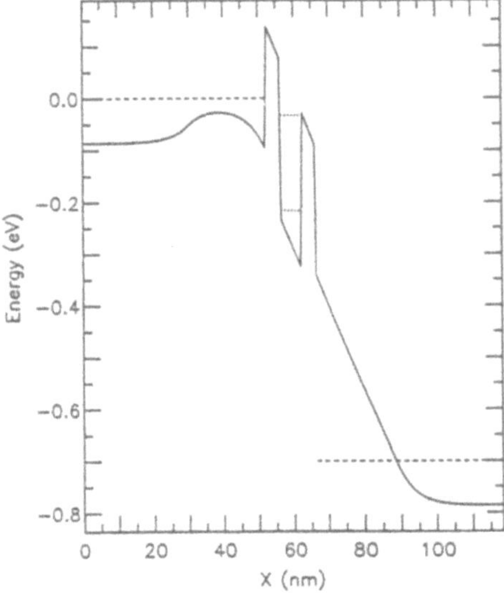

Fig. 6 Self-consistent band diagram of the quantum dot 1D epitaxial structure at a bias of 0.7V and T=77K. The structure is a 40Å $Al_{.3}Ga_{.7}As$ barrier / 60Å $In_{.07}Ga_{.93}As$ quantum well / $Al_{.3}Ga_{.7}As$ barrier structure with contacts doped to $2x10^{18}$ cm^{-3}. The Fermi level is denoted by a dashed line and the energies of the bound (or virtual) states are denoted by dotted lines.

the full screening potential of the quantum dot system taking into account the effects of lateral confinement. Cylindrical symmetry is assumed. The model self-consistently obtains the electrostatic potential in a zero-current theory from Poisson's equation utilizing a Thomas-Fermi approximation for the electron density. The solution of the electrostatic problem then provides the potential responsible for lateral quantization of electron states, which we obtain from the radial Schrödinger equation in cylindrical coordinates. The radial bound states in the contacts provide the minima of the emitter and collector subbands. Likewise the discrete quantum well levels, which in the absence of lateral confinement would otherwise form a two-dimensional subband, are obtained from a solution of the radial Schrödinger equation. We shall first consider only the zero angular momentum ($l = 0$) states.

The boundary conditions necessary for a solution to the quantum dot screening potential are considerably more complicated than for the 1D problem. At the center of the post ($r = 0$), a simple Neumann condition of zero electric field was imposed. More involved is the question of the proper Dirichlet boundary condition to employ for the contact regions of this laterally-confined system. It is not enough to set the boundary potential in the degenerately-doped contacts to achieve charge neutrality, as one would have in a one-dimensional simulation or for bulk systems where surface effects are irrelevant. The restricted lateral extent of the quantum dot system, with the Fermi level pinning at the exposed outer lateral surface, implies a solution to the Poisson equation in the radial direction which is not a simple constant. Thus, to obtain a boundary condition in the contact regions for the full quantum dot system, we first do a 1D self consistent calculation for the radial direction, using the Laplacian for cylindrical coordinates. The boundary conditions for this calculation are again a zero field condition at the origin and another Neumann condition at the external radius set by an amount of surface charge necessary to support the value of the Fermi level pinning for $r = R$. To match up with the calculation for the full problem, it is assumed that there is negligible variation of the potential in the vertical direction in the vicinity of where the contact boundary conditions are to be imposed. The calculation is quite sensitive to the boundary condition specified at the outer lateral surface. We have employed a Neumann condition where the slope is determined by the surface charge. We allow this quantity to vary in the vertical direction. Our model assumes, to a first approximation, a constant density of surface states per unit area, independent of the material composition or doping level. We assume however that these states are occupied according to a Fermi-Dirac distribution, with the value of the Fermi level pinning acting as a local chemical potential. This rudimentary model of the surface charge distribution effectively *pins* the computed potential at the external lateral surface to the desired Fermi level pinning value. The calculation itself adjusts the occupation of surface states to self-consistently achieve a constant surface potential in the vertical direction (for zero bias) independent of material or doping level variations.

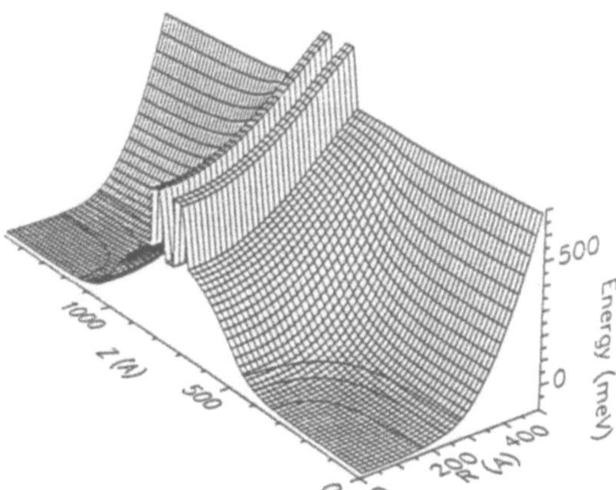

Fig. 7 Self-consistent 3D band diagram of the previously detailed single quantum dot structure (Figure 4), at equilibrium. The electron potential energy surface is plotted as a function of radius (R) and epitaxial (z) dimensions. The contours in the contact regions are the occupied laterally-defined subbands. For clarity, the quantum dot energy levels are not drawn.

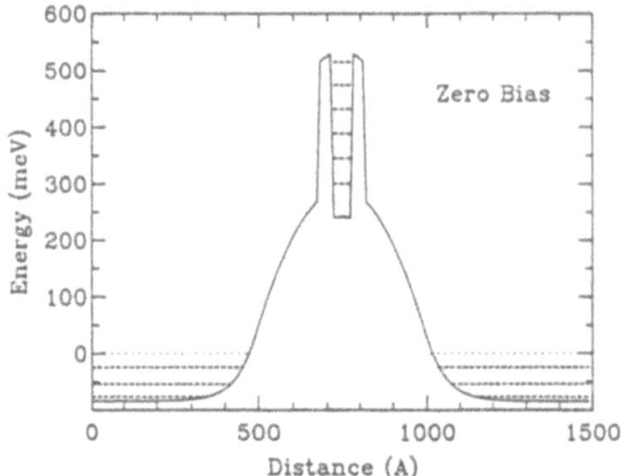

Fig. 8 Radial slice ($r = 0$) of the self-consistent 3D band diagram of the previously detailed single quantum dot structure (Figure 7) at equilibrium. The large internal depletion barrier due to the lateral confinement and the epitaxial doping profile is evident.

The equilibrium solution to the 3D screening problem using the quantum dot epitaxial structure and the measured physical radius of the column is displayed in Figure 7. The electron potential energy surface is plotted as a function of radius (R) and epitaxial (z) dimensions. The radial extent is 0-500Å and the vertical length is approximately 2000Å, centered about the double barriers. The energy scale is defined relative to the Fermi energy, thus the potential at the external radius equals 0.7V. The contours in the contact regions are the occupied laterally-defined subbands that lie below the Fermi level. For this specific case of radial dimension and contact doping level, three contact subbands are occupied. The subband energies are determined by solving the radial Schrödinger equation. For clarity, the quantum dot energy levels are not drawn in Figure 7.

The equilibrium solution demonstrates strong depletion in the region of the quantum dot due to the radial depletion, exacerbated by the z-dependent doping profile. This is demonstrated in Figure 8, which is just the $r = 0$ contour of Figure 7. The contact subbands are denoted by dashed lines below the dotted Fermi level ($E_F \equiv 0$). It is clear that the lateral depletion has a dominant effect on lifting the double barrier structure significantly above the level previously determined only by the z-doping profile. The quantum dot states determined by solving the radial Schroedinger equation are shown as dashed lines between the barriers. These are the quantum dot states arising from the previous quantum well ground state ($n_z = 1$); the excited state ($n_z = 2$) quantum dot states are virtual. Previous misidentification of the quantum dot resonances with the $n_z = 2$ resonance [13] was due to a fortuitous coincidence of the dot resonant voltages (~ 0.7-0.8V; see Figure 4) and the size of the voltage needed to overcome the internal quantum dot depletion barrier ($\sim 2\times(300$ meV$)$).

276

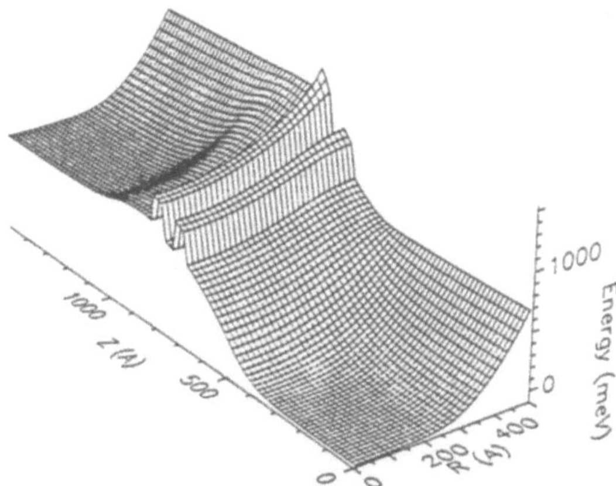

Fig. 9 Self-consistent 3D band diagram of the previously detailed single quantum dot structure, at $V = 0.835V$. For clarity, the contact subbands and quantum dot energy levels are not drawn.

It has been suggested [18] that the observed quantum dot spectrum can be explained as resonances when the quantum dot states are biased through the emitter subband states with increasing device bias. To determine if this mechanism quantitatively explains the spectrum, we solve the 3D self-consistent screening quantum dot model at applied bias, to determine the variation of the emitter and quantum dot energy levels with applied voltage. Figure 9 shows a typical electron potential energy surface at finite applied voltage. For clarity, the subbands and levels are not shown, though they are determined in exactly the same way as previously detailed and shown in Figure 8. The bias voltage positions of the quantum dot resonances are then determined by generating a family of surfaces similar to Figure 9 and determining the eigenvalues.

Figure 10 shows the crossings of the emitter subband levels (n') with the quantum dot levels (n) as a function of applied bias. The parameters of the quantum dot model were the same as detailed above except for the width of the undoped spacer layer. The 3D model at present could only accomodate *box-like* (sharp) doping profiles, instead of the graded doping profile of the epitaxial structure. It was found that the absolute voltage values of the crossings were very sensitive to spacer thickness in the box-like profile model, but that the relative spectral spacing was invariant over a wide range($\sim$ .1V); i.e., a change of spacer thickness *translates* the spectrum along the voltage axis. The spacer thickness is thus used as a fitting parameter; indeed, not only are the structural characterization or nominal growth parameters insufficiently accurate to yield this information, but the statistical fluctuations of dopants on this scale becomes important. For the spatial region included in the 3D model, there are less than 800 dopants.

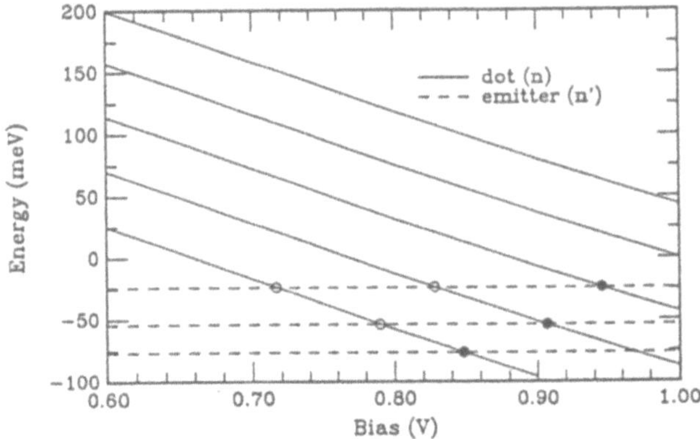

Fig. 10 Emitter subband levels (n') and the quantum dot levels (n) as a function of applied bias. The circles denote the crossings; solid for momentum-conserving $(n = n')$ transitions and open for momentum-nonconserving $(n \neq n')$ transitions.

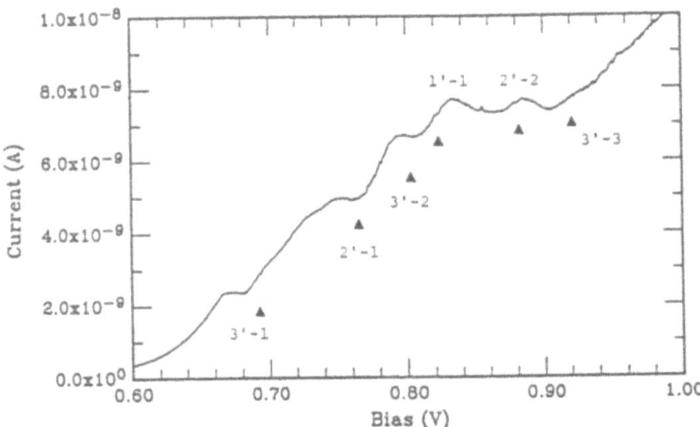

Fig. 11 Current-voltage characteristic at T=1.0K of the previously detailed single quantum dot structure, with predicted resonant peak positions and initial and final state index numbers $(n' - n)$.

Figure 11 shows the crossings of the emitter subbands with the quantum dot states, transposed onto the 1.0K current-voltage characteristic of the quantum dot, with a spacer width of 177Å and with the initial and final state index numbers labeled $(n' - n)$. There is general agreement between the experimental and predicted peak voltage positions, especially the anomalously large splitting of the first resonance. This can be seen as a consequence of the subband-level crossing mechanism, when more than one lateral subband

278

is below the Fermi level. The experimental peaks differ from the experimental peak positions by at most 15 mV, which corresponds to approximately 5 meV in energy. This is in good agreement, considering the approximations of zero-current , homogeneous dopant distributions, and perfect radial symmetry. It should also be noted that the experimental measurement is current, which implies that an integration over the density of emitter states should be done for a strict comparison. It is possible that peaks may be shifted in voltage or even *washed out* when this is correctly done; however, the qualitative and quantitative agreement of the peak positions suggests this may not be a significant effect.

An additional corroboration of this spectroscopy and the peak indexing is found in the temperature dependence of the quantum dot peaks. Figure 12 shows the spectrum of the quantum dot at T=1.0K and 50K. When the temperature is raised to 50K, the three lowest voltage peaks disappear and the spectrum is dominated by the single $1' - 1$ transition. This is expected, since when the subband spacing is less than $3k_BT$, thermal smearing will destroy the well defined subband structure. As can be seen in Figure 3, this occurs at approximately 7 meV, or 80K, in reasonable agreement with the observed temperature dependence. In this high temperature limit, the structure emulates an unconfined 1D resonant tunneling diode, with the resonance determined by the $1' - 1$ crossing. This procedure is an easy, general method to index the $1' - 1$ transition.

Finally, the predicted $3' - 3$ transition appears to be absent in the spectrum, except for a very weak structure at 0.92V-0.93V. However, this is not unexpected since the collector barrier becomes sufficiently low that the state becomes virtual, similar to that discussed earlier (see Figure 6). This has an important implication - verification that the observed resonances are due

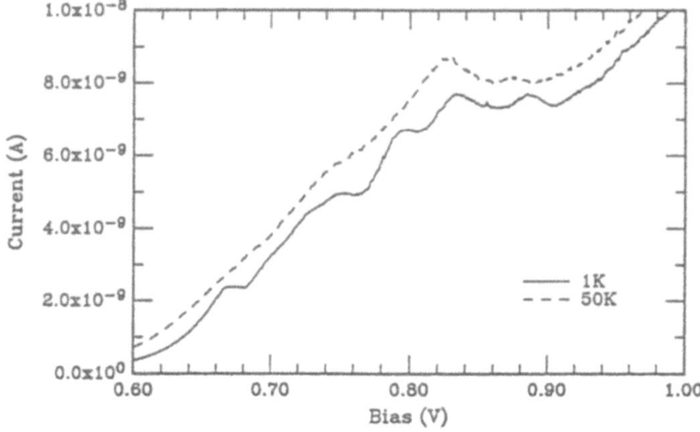

Fig. 12 Current-voltage characteristics of the previously detailed single quantum dot structure, at T=1.0K and 50K.

to states localized in the quantum dot and not due to the density of states
in the collector contact.

The preceeding calculations are for angular momenta (l) equal to zero.
Higher angular momentum states can be calculated, and effectively split
the spectra into $l \times$(number of $n' - n$ crossings). Such extra structure does
not seem evident in our experimental data, though sharper peak shapes
are very desirable. Preliminary magnetic field studies up to 9.0T show no
obvious Zeeman split discrete level peaks, which should be observable by
2.0T if higher angular momenta states were occupied. This implies not only
a $\Delta l = 0$ selection rule, but a restriction to $l = 0$ as well.

As a result of indexing the transitions, we can determine selection rules
for the transitions. The observation of the momentum-nonconserving transitions ($n' \neq n$) show that n is <u>not</u> a conserved quantity in this quantum
dot system. This is due to the *hourglass* topography of the electron energy surface determined partly by the z-dependent doping profile. This
absence of a n, n' (radial) selection rule, as well as a $\Delta l = 0$ selection rule,
is natural from the radially changing, cylindrically symmetric geometry. It
should be emphasized that selection rules derive from the symmetries of the
system, and that the breakdown of intuitive selection rules for these nanostructured semiconductor atomic analogies arise from the difference between
the fabrication-imposed potential and a r^{-1} potential. Indeed, the selection
rules are, to a certain extent, variable by the experimenter. Fabrication improvements may allow us to create structures that critically explore and test
these observations. However, it is desirable to achieve narrower resonance
linewidths to perform detailed spectroscopy of these systems.

5 Coupled Quantum Dot Tunneling

The linewidth limition of quantum wire to quantum dot tunneling is that
the emitter subband distribution can effectively broaden the resonance (in
voltage), making spectroscopy with such structures difficult. A solution
to that problem is to *inject* from another quantum dot, so that the input
distribution is as (theoretically) narrow as the state being used as a spectrometer. In this case, the intrinsic quantum dot tunneling linewidth may
be measurable. Additionally, we may be able to measure overlap integrals
between the discrete electronic states of quantum dots placed a tunneling
distance apart; i.e., *molecule* states.

The fabrication of such structures is relatively straightforward. The starting epitaxial material is a double quantum well, triple barrier structure
designed to have approximately coincident resonances in one bias direction.
The structure is superficially similar to that of Nakagawa *et al.* [19] except
that the coupling barrier is made thinner than the quantum well-to-contact
tunnel barriers; 60Å outside $Al_{.4}Ga_{.6}As$ barriers, and 50Å and 65Å GaAs
quantum wells separated by a 35Å $Al_{.4}Ga_{.6}As$ barrier. Lateral fabrication of

280

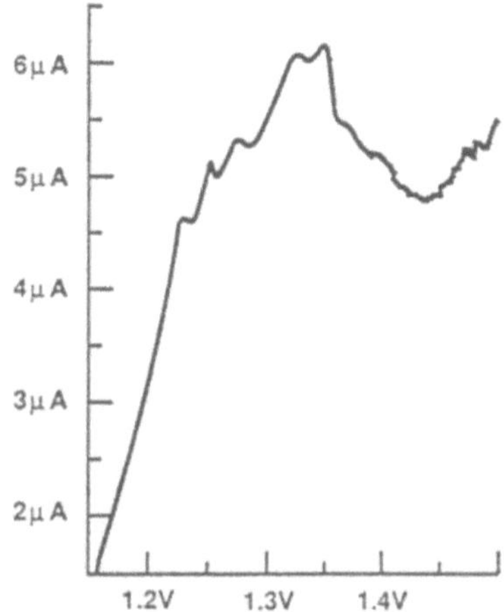

Fig. 13 Current-voltage characteristics of a coupled quantum dot (molecule) structure. The epitaxial structure is detailed in the text, and the lateral size is approximately 1000Å. T=100K.

approximately 1000Å dots was the same as before. The resultant structure has two quantum dots connected in series between quantum wire contacts.

Figure 13 shows a spectrum from one of these molecule structures. The structure is obviously complex due to the overlapping of the discrete states in the coupled dots and the emitter levels. A full modeling and indexing is beyond the scope of the present work, and will be presented elsewhere. However, it should be noted that the peaks are significantly sharper in the coupled dot spectra than the single dot spectra. Also notice that the thermal broadening of the emitter level no longer determines the temperature dependence of this structure, due to the effective filtering by the quantum dot adjacent to the quantum wire emitter.

These structures exhibit sufficiently sharp resolution over the single dot spectra due to the emitter filtering such that high resolution magnetospectroscopy can be done. Figure 14 shows the current-voltage characteristics of a similar device as that shown in Figure 13, as a function of magnetic field parallel to the current direction up to 9.0T. Other than for the change in amplitude of one peak in the spectrum, no observable Zeeman splitting is again observed, implying that the spectra is fully explainable by the $l = 0$ angular momenta states only. This has yet to be explained, though the spectra from coupled dot systems and a full modeling thereof should address the problem.

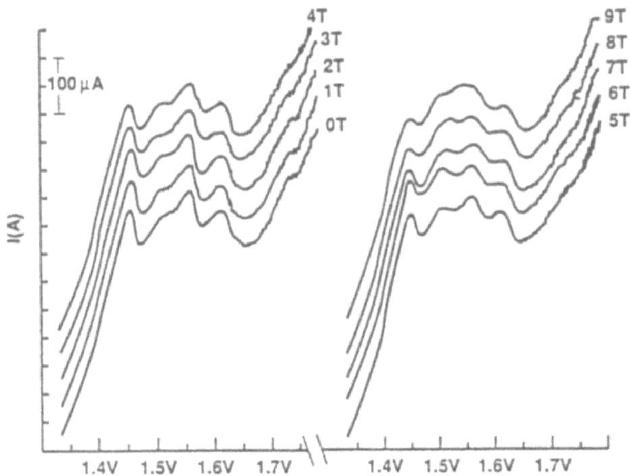

Fig. 14 Current-voltage characteristics of a coupled quantum dot structure as a function
of magnetic field (parallel to the current direction). The epitaxial structure is detailed in
the text, and the lateral size is approximately 1000Å. T=4.2K.

6 Summary

Quantum dot structures provide a unique laboratory for the exploration
of quantum transport through nanostructured semiconductors. The *atomic
states* can be varied with structural variables, allowing for the exploration
of the fundamentals of quantum-confined electronic states, transitions, and
selection rules involving those states. Coupling of *molecule* states has been
investigated, and these structures should provide an intriguing tool for the
investigation of quantum-localized electronic states and quantum effect devices.

Acknowledgement

We wish to thank R. T. Bate for constant encouragement and support, and R. K. Aldert,
E. D. Pijan, D. A. Schultz, P. F. Stickney, and J. R. Thomason for technical assistance.
This work was sponsored by the Office of Naval Research, the Army Research Office, and
the Air Force Wright Avionics Laboratory.

References

[1] A. N. Broer, W. W. Molzen, J. J. Cuomo, and N. D. Wittels, Appl. Phys. Lett. **29**, 596 (1976)

[2] R. E. Howard, P. F. Liao, W. J. Skocpol, L. D. Jackel, and H. G. Craighead, Science **221**, 117 (1983)

[3] H. G. Craighead, J. Appl. Phys. **55**, 4430 (1984)

[4] R. A. Webb, S. Washburn, C. P. Umbach, and R. B. Laibowitz, Phys. Rev. Lett. **54**, 2696 (1985)

[5] H. van Houten, B. J. van Wees, M. G. J. Heijman, and J. P. Andre, Appl. Phys. Lett. **49**, 1781 (1986)

[6] K. F. Berggren, T. J. Thornton, D. J. Newson, and M. Pepper, Phys. Rev. Lett. **57**, 1769 (1986)

[7] G. Timp, A. M. Chang, P. Mankiewich, R. Behringer, J. E. Cunningham, T. Y. Chang, and R. E. Howard, Phys. Rev. Lett. **59**, 732 (1987)

[8] R. Landauer, IBM J. Res. Dev. **1**, 223 (1957)

[9] M. Büttiker, Phys. Rev. Lett. **57**, 1761 (1986).

[10] B. J. van Wees, H. van Houten, C. W. J. Beenakker, J. G. Williamson, L. P. Kouwenhoven, D. van der Marel, and C. T. Foxon, Phys. Rev. Lett. **60**, 848 (1988)

[11] D. A. Wharam, T. J. Thornton, R. Newbury, M. Pepper, H. Ahmed, J. E. F. Frost, D. G. Hasko, D. C. Peacock, D. A. Ritchie, and G. A. C. Jones, J. Phys. C **21**, L209 (1988)

[12] G. Timp, H. U. Baranger, P. deVegvar, J. E. Cunningham, R. E. Howard, R. Beringer, and P. M. Mankiewich, Phys. Rev. Lett. **60**, 2081 (1988)

[13] M. A. Reed, J. N. Randall, R. J. Aggarwal, R. J. Matyi, T. M. Moore, and A. E. Wetsel, Phys. Rev. Lett. **60**, 535 (1988)

[14] A. B. Fowler, G. L. Timp, J. J. Wainer, and R. A. Webb, Phys. Rev. Lett. **57**, 138 (1986)

[15] T. E. Kopley, P. L. McEuen, and R. G. Wheeler, Phys. Rev. Lett. **61**, 1654 (1988)

[16] T. P. Smith III, K. Y. Lee, C. M. Knoedler, J. M. Hong, and D. P. Kern, Phys. Rev. **B38**, 2172 (1988)

[17] M. A. Reed, W. R. Frensley, W. M. Duncan, R. J. Matyi, A. C. Seabaugh, and H.-L. Tsai, Appl. Phys. Lett. **54**, 1256 (1989)

[18] G. W. Bryant, Phys. Rev. **B39**, 3145 (1989)

[19] T. Nakagawa, T. Fujita, Y. Matsumoto, T. Kojima, and K. Ohta, Appl. Phys. Lett. **51**, 445 (1987)

DC and Far Infrared Experiments on Deep Mesa Etched Single and Multi-Layered Quantum Wires

Detlef Heitmann, Thorsten Demel, Peter Grambow, and Klaus Ploog

Max-Planck-Institut für Festkörperforschung, Heisenbergstraße 1, D-7000 Stuttgart 80, Federal Republic of Germany

Summary: We discuss one-dimensional electronic systems (1DES) which have been realized by ultrafine mesa etching of modulation doped $GaAs/AlGaAs$ heterostructures and multi quantum well systems. The electronic properties of these single and multi-layered quantum wires can be characterized by dc magnetotransport measurements. Quantum wires with 400 to $150nm$ wide electron channels exhibit typical energy separations of 1 to $3meV$ for the 1D subbands. The 1D subband separation is not directly observed in the far infrared (FIR) response, rather, resonances at significantly higher frequencies are observed, indicating a dominant influence of collective effects. This gives the FIR resonances the character of plasmon modes which are localized by the microstructure. The plasmon character is in particular manifested by the observation of layer-coupled plasmon modes with n resonances in n-layered quantum wires.

1 Introduction

The progress of semiconductor technology and crystal growth techniques over the last decade, which made it possible to fabricate high mobility metal-insulator-semiconductor (MIS) systems and novel layered hetero and super-lattice structures, has initiated a broad range of fundamental research and applications in many different fields. These structures have unique physical properties which arise from the two-dimensional (2D) behaviour and, in su-perlattices, from the coupling between the 2D layers. [1] Very recently there is an increasing interest in the realization and investigation of electronic sy-stems with even lower dimensionality, i.e., 1DES and 0DES (e.g. Refs. 2-17). Here the dimensionality is referred to the energy spectrum, i.e., "1DES" or "quantum wire" means that the one-particle energy spectrum consists of a set of 1D subbands

$$E^{ij}(k_y) = \frac{\hbar^2 k_y^2}{2m^*} + E_x^i + E_x^j \quad .$$

These systems are mostly realized by starting from layered 2D systems. The energy levels E_x^j arise from the original 2D confinement due to a gate voltage in a MIS structure or due to the potential at the interface between $AlGaAs$ and $GaAs$ in a heterostructure. (The z-direction is perpendicular to the original 2D plane.) This "z-quantization" is usually very strong (typically 20meV) and only the lowest level, E_x^0, is occupied.[1] The discrete 1D subband energies E_x^i (i=0,1,2,...) arise from a lateral confinement, which is assumed to act in x-direction. The electron wavevector k_y characterizes the free motion in the y-direction. These 1DES exhibit interesting novel phenomena, i.e., in

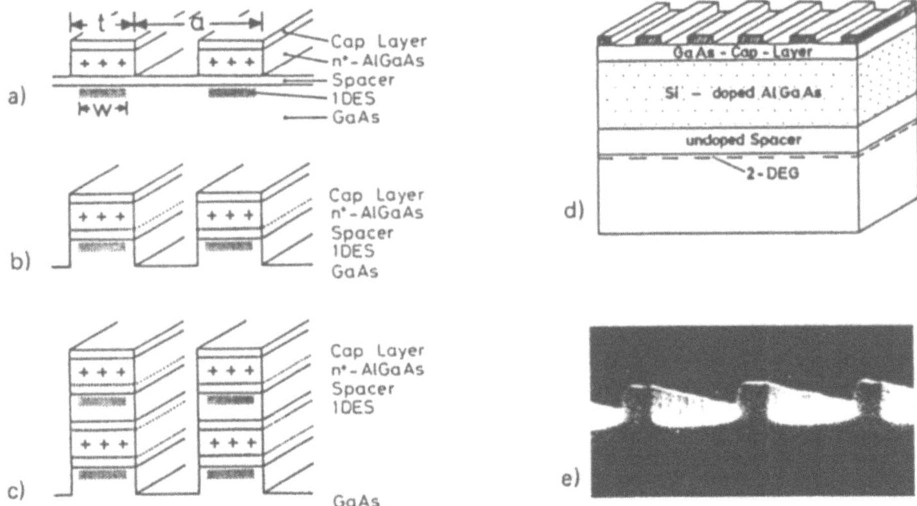

Fig. 1   Mesa etched QW wire structures: (a) shows schematically a "shallow" mesa etched structure, (b) a "deep" mesa etched single-layer structure, (c) a "deep" mesa etched double-layer structure. a and t indicate, respectively the periodicity and the geometrical width of the etched wires. (d) sketches a HEMT structure with a linear photoresist grating that serves as an etching mask (e) shows an electron micrograph of a structure (c) with $t = 400nm$ and $a = 1100nm$

the ballistic regime the transport is governed by quantum interference effects which are of great fundamental interest.[10,11] Quantum interference effects also might lead to totally new device concepts in applications.

To study 1D effects experimentally at He temperatures 1D subband separations of some meV are required. To achieve this e.g. in *GaAs* with an effective mass of $m^* = 0.067m_0$ the carriers have to be laterally confined in stripes of some 100nm. In most of the systems studied so far, the 1D confinement was achieved by split-gate configurations (e.g. Refs. 2,5,9-14), i.e., in a normally-on *AlGaAs/GaAs* HEMT structure with a front gate the original 2DES is depleted everywhere except under a narrow slit in the gate. Such systems have the advantage that the width of the channels can be tuned via the gate voltage and very narrow channel widths with correspondingly large energy separation of the 1D subbands can be achieved. In other structures small channels have been isolated by ion-beam-lithography.[6]

Here we will discuss the realization and investigation of "shallow" [3] (Fig.1a) and in particular of "deep" mesa etched structures (Fig. 1b and c) which have been prepared by etching into the *GaAs* buffer. [15,16,17] On a first glance this "geometrical" confinement seems to be a more straightforward approach. Practically it turns out to be an ambitious technological task, since the free etched surfaces trap electrons and deplete the channel. With optimized lithographic and etching techniques we have realized 1DES in "shallow" and "deep" mesa etched structures. This technique does not only allow the

286

preparation of single quantum well (SQW) wires as shown in Fig. 1b, but also
of multi quantum well (MQW) wire structures (Fig. 1c). These structures
exhibit interesting fundamental physical phenomena that will be discussed
here. They may also have adantageous technical application: It was pointed out by Sakaki[18] that for 1DES, due to reduced scattering in the 1D
k-space, the mobility is inherently increased. This is actually most strongly
pronounced if only the lowest 1D subband is occupied. Also the proposed
quantum interference devices work most efficiently under this condition. However, with only one 1D subband occupied the current density is very low.
We propose that this disadvantage can be overcome by many parallel 1D wires, not only in x-direction, but also in the z-direction, starting from MQW
structures.

2 Preparation of Shallow and Deep Mesa Etched Quantum Wires

The starting material for our structures are high mobility modulation doped $Al_xGa_{1-x}As/GaAs(x = 0.3)$ heterostructures and MQW systems, grown
by molecular beam epitaxy (MBE). The essential requirements for the fabrication of the quantum wires are, (a) to minimize all processes that cause
radiation damage or create surface states and thus act as electron traps in
the narrow wires and (b), to keep the number of process steps to a minimum,
again to minimize possible damage and to increase reproducibility. We use
an one-layer photoresist process and holographic lithography to prepare the
etch masks. The resist layer is exposed to a sinusoidal interference pattern
from the superposition of two coherent, expanded laser beams with plane
wave fronts from an Ar^+-ion-laser, working at the wavelength $\lambda = 457.9nm$.
With this holographic lithography we can fabricate periodic photoresist stripes where the periodicity, a, can be controlled via the angle of incidence,
θ, of the laser beams ($a = \lambda/2sin\theta$). A $GaAs/AlGaAs$ HEMT-structure with
a photoresist grating is shown schematically in Fig.1d. The width of the
stripes, t, can be varied in a certain range ($0.2a \leq t \leq 0.8a$) via the exposure
and developement time. With our process we can produce many parallel wires with extremely high homogeneity over large areas ($3mm$ in diameter and
more). The latter makes these structures also ideally suited for far infrared
(FIR) investigations, where large areas and many parallel wires are necessary
to investigate the systems with a sufficient signal to noise ratio.

For the mesa etching we have selected a $SiCl_4$ reactive ion etching (RIE)
process.[19] A RIE process has two components, the physical etching and
the chemical etching. Whereas the physical component etches anisotropically and gives thus steep side walls, it has less selectivity with respect to
the mask, and, crucially for our demands, creates damages. Chemical etching
has a high selectivity but etches isotropically. We have thus balanced these
two aspects, on the one hand, to have enough physical etching to achieve
an accurate transfer of the mask and steep vertical sidewalls, which is in
particular important for the multi-layered quantum wires (Fig. 1c), on the
other hand, to make the physical component small, to achieve enough selectivity for our one-layered photoresist technique, and to make any damage
as small as possible. We have optimized all process parameters according

to the criteria discussed above. We have prepared successfully multi-layered quantum wire structures with up to 5 layers.[17] Here an etching depth of about $1000nm$ was required which is larger than the width and spacing of the wires of $550nm$. An electron micrograph for a deep mesa etched structure is shown in Fig.1e.

In the following we refer to structures as shown in Figs. 1a, b and c as structures A, B and C, respectively. For the structures A and B we started from conventional modulation doped heterostructures consisting of a $GaAs$ buffer layer, a $25nm$ $Al_xGa_{1-x}As$ spacer, $50nm$ of n-doped $Al_xGa_{1-x}As$ ($N_d = 1 \cdot 10^{18}cm^{-3}, x = 0.3$) and a $10nm$ $GaAs$ cap layer. For structures C we started from modulation doped, electronically decoupled MQW systems with two periods where each period consisted of a $25nm$ undoped $Al_xGa_{1-x}As$ spacer, followed by a short-period $AlAs/GaAs$-superlattice (5 periods, each layer $22nm$ thick), a $50nm$ $GaAs$-QW, a $5nm$ $AlAs$ spacer, and a $25nm$ n-doped $Al_xGa_{1-x}As$ layer ($N_d = 1.5 \cdot 10^{18}cm^{-3}, x = 0.3$). The cap layer consisted of $25nm$ undoped $Al_xGa_{1-x}As$ ($x = 0.3$) and $2nm$ of $GaAs$. We have also prepared samples with a configuration similarly to C but consisting of five quantum wells.

3 Magnetotransport Measurements on 1DES

The characterization of an 1DES, i.e., the determination of the potential, energy levels, number of electrons and width of the channels is a complex task. The best description can be achieved by numerical self-consistent band-structure calculations, as, e.g., performed by Laux et al. [20] who modelled split-gate devices. However not always all necessary input parameters, e.g., pinning of the Fermi levels in certain regimes of the microstructure, are known. An estimate of the channel width can be extracted from weak localization experiments.[21] In most studies so far 1DES are characterized by the magnetic depopulation of the 1D subbands which occurs if the 1DES is exposed to a perpendicular magnetic field B.[2] The underlying physics can be explained without loss of generality if one assumes a parabolic confinement potential[2] $V(x) = \frac{1}{2}m^*\Omega_0^2x^2$. In this case the Schroedinger equation can be solved analytically. The magnetic field induces an additional potential $V_B(x) = \frac{1}{2}m^*\omega_c^2(x - x_0)^2$, where $\omega_c = eB/m^*$ is the cyclotron frequency. In this model the energy levels in a magnetic field are given by

$$E^i(k_y, B) = \hbar\Omega(i + \frac{1}{2}) + \frac{\hbar^2k_y^2}{2m_y^*(B)}$$

with $\Omega^2 = \omega_c^2 + \Omega_0^2$, and $m_y^*(B) = m^*\Omega^2/\Omega_0^2$. Since the 1D density of states, $D_{1D}(E, B)$, increases with increasing B the 1D subbands become successively depopulated giving rise to oscillations of the Fermi energy. In a transport measurement this leads to Shubnikov-de Haas (SdH) type of oscillations. We recall that in a 2DES the number of occupied Landau levels increases with decreasing B, leading, ideally, to an infinite number of SdH oscillations periodic in $1/B$.[1] In a 1DES however, only a finite number of 1D subbands is occupied at B=0, giving rise to a finite number of SdH oscillations and deviations from the $1/B$ period, if the hybrid-1D-subband-Landau levels are depopulated with increasing B.

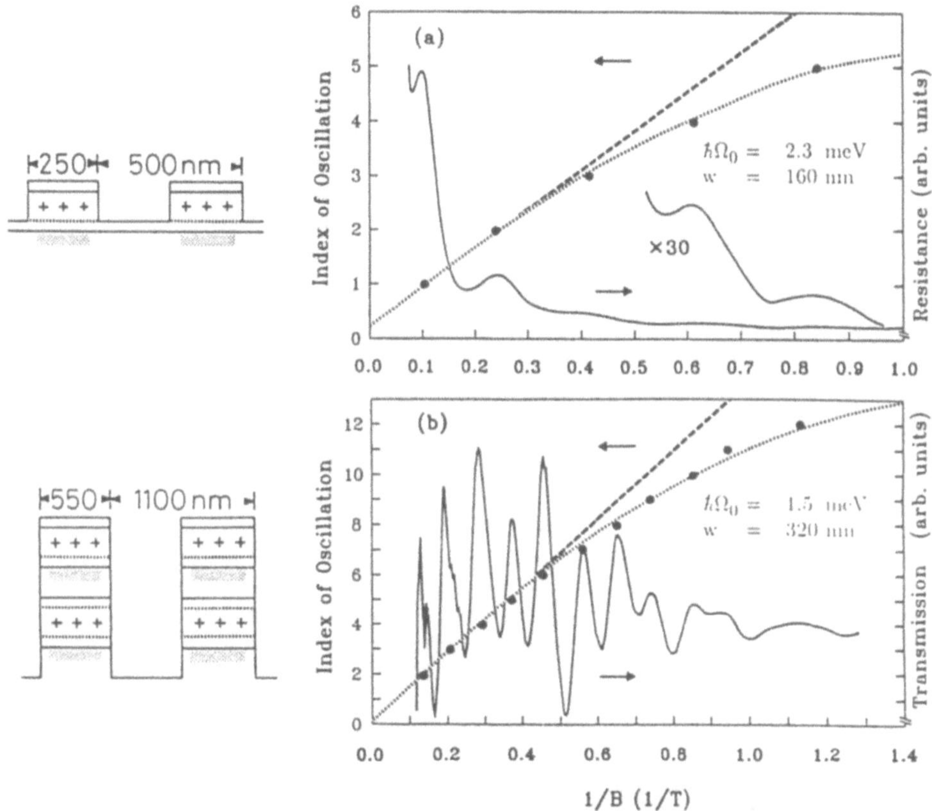

Fig. 2 Magnetotransport measurements (full lines, right scale) on "shallow" mesa etched (a) and in a "deep" mesa etched two-layered (b) structures plotted versus 1/B. A fan chart for the positions of the maxima in the magneto resistivity (full circles, left scale) exhibits deviations from a linear 1/B dependence (dashed lines) and indicates an one-dimensional energy structure in our samples. The dotted lines show the depopulation of 1D subbands within a harmonic oscillator model, calculated for the indicated values of the confining potentials $\hbar\Omega_0$ and the channel widths w. The structures and the geometrical dimensions are shown on the left.

To demonstrate the formation of 1D subbands in our structures we performed magnetotransport measurements at low temperatures (T=2.2K) in perpendicular magnetic fields B. For this purpose we defined on some of the SQW wire structures an active area of $2.5 \times 2.5 mm^2$ by chemical mesa etching. Ohmic contacts of a Au/Ge alloy were aligned perpendicular to the grating in order to measure the dc transport parallel to the stripes. On other samples a quasi-dc response was obtained by measuring the transmission of microwaves (30-40 GHz) through the sample. Since for this measurement no contacts were needed, it was especially useful for samples with MQW wires, because alloyed contacts would have short-circuited the channels of different layers.

In Fig. 2a we show measurements on a shallow mesa etched sample (type A). The period of our structures was $a = 500nm$, the remaining width of the n-doped $AlGaAs$ was $t = 250nm$. Via Ohmic contacts the two terminal resistance was measured as a function of the magnetic field, using a constant current of $1\mu A$. The original data are plotted versus $1/B$. The dc conductivity shows well pronounced SdH-type oscillations. The important point in Fig. 2a is that the period of the oscillations is not constant in $1/B$, but shows distinct deviations at large values of $1/B$ and corresponding small B. This clearly indicates the 1D character of our structure as was discussed above for the harmonic oscillator model. We calculated the depopulation of the 1D subbands by a magnetic field within this model using Ω_0 and the total 1D carrier density N_{1D} as fitting parameters. The experimental fan chart in Fig. 2a was best described for $\hbar\Omega_0 = 2.3meV$ and $N_{1D} = 4.5 \cdot 10^6 cm^{-1}$. For these values, six 1D subbands were occupied at B=0. Defining the width w of the electron channel by the amplitude at the Fermi energy: $E_F(B = 0) = V(\frac{w}{2}) = \frac{1}{2}m^*\Omega_0^2(\frac{w}{2})^2$, we found $w = 160nm$ which was smaller than the geometrical width $t = 250nm$. Therefore we defined a "lateral edge depletion region" on either side of length $w_{dl} = \frac{1}{2}(t - w) \approx 45nm$.

Let us now discuss the "deep" mesa etched structures. We prepared samples with different dimensions a and t. Our experience was that for profiles with t about $400nm$, as shown in Fig. 1e, or smaller, no free electrons were left in the channels. This was detected both from zero dc conductivity and also from a missing far-infrared (FIR) response. The explanation was that all the carriers were trapped in etching-induced surface states. However, for structures with $t = 550nm$ and periodicities $a = 1100nm$ we found both, dc conductivity and FIR response. As an example we show in Fig. 2b the quasi-dc conductivity of two-layered quantum wire structures (type C), measured in microwave transmission as described above. SdH oscillations can be clearly resolved and show again distinct deviations from a linear $1/B$ behaviour at small B. An analysis within the harmonic oscillator potential model gives for the 1D confinement $\hbar\Omega_0 = 1.5meV$, for $N_{1D} = 15 \cdot 10^6 cm^{-1}$ and 16 occupied 1D subbands. From the channel width $w = 320nm$ we deduced that there is a lateral depletion length $w_{dl} = 100nm$ on eather side of the wire. Similar values were measured on a number of samples which all had the same doping and were prepared with the same etching process as described above. We expect that w_{dl} should depend particularly on the doping concentration and the etching process. A value of $w_{dl} \approx 500nm$ was reported for chemical mesa etched stripes whose widths ranged from $800nm$ to $2400nm$.[21] Thus with our technology we achieve drastically smaller edge depletion regions which is the crucial point to realize 1DES.

All parameters here are extracted within the harmonic oscillator model. Although more exact values for the 1D subband energies of course need a more sophisticated description of the confining potential, the 1D electronic character is clearly resolved by the observation of the SdH oscillations in our structures and their deviation from the linear $1/B$ behaviour. Actually this model should give a good approximation to the potential since the confinement arises from the electric fields of the distant ionized donors. The calculation for the split-gate configuration also give, at least for a small number of occupied subbands, in good approximation parabolic potentials.[20]

4 Far Infrared Spectroscopy on Quantum Wire Structures.

With typical subband separations of some meV FIR intersubband resonance spectroscopy seems to be the most direct way of determining the 1D band-structure. We will show in the following that the optical response exhibits a very complex behaviour where the subband separation can be derived only very indirectly. Here in particular the measurements on the multi-layered quantum wire structures are very helpful to understand the nature of the excitations.

The FIR measurements were performed in a superconducting magnet cryostat, which was connected via a waveguide system to a Fourier transform spectrometer. The transmission $T(B)$ through the sample was measured at a temperature of $2.2K$ at fixed magnetic fields B, oriented normally to the surface of the sample. The spectra were normalized to a spectrum $T(B_0)$, where B_0 was chosen in such a way that the reference spectrum $T(B_0)$ was flat in the frequency region of interest. The resolution of the spectrometer was set to $0.5cm^{-1}$. The FIR measurements were performed in the same set up and under identical conditions as discussed for the characterization of the samples above.

As an example we show in Fig.3 experimental spectra for the two-layered quantum wire structure. At B=0 two resonances at $\omega_{r1} = 64cm^{-1}$ and $\omega_{r2} = 34cm^{-1}$ are observed if the incident electric field is polarized perpendicular to the wires. The high energy resonance has a larger amplitude. Both resonances shift with increasing B to higher frequencies. With increasing B resonances are observed also for parallel polarization with exactly the same resonance frequencies as for perpendicular polarization. A very similar

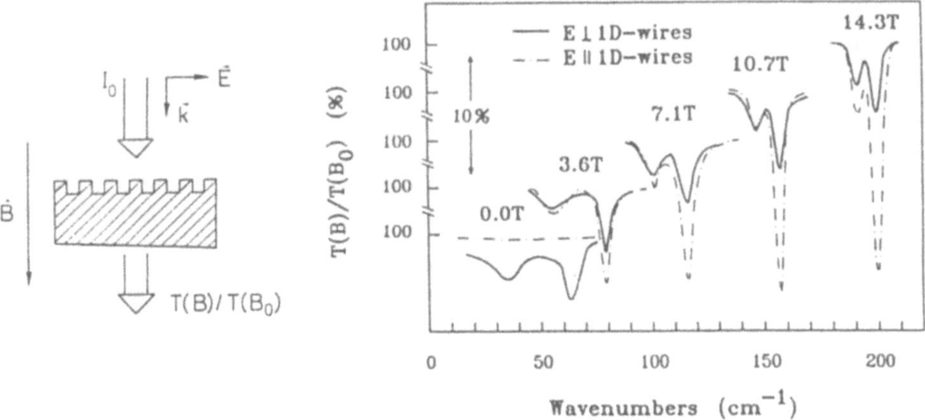

Fig. 3 Experimental FIR spectra, measured on a double-layered quantum wire structure at indicated magnetic fields B. Full lines and dash-dotted lines denote, respectively, polarization of the incident FIR radiation with the electric field vector perpendicular and parallel to the wires. On the left the experimental arrangement for FIR transmission spectroscopy is sketched.

behaviour is also found for the one-layered quantum wire system (type B) except that only one resonance is observed. (For B=0 at $\omega_r = 30cm^{-1}$.) The experimental resonance positions for the two-layered quantum wire system are plotted in Fig.4a on a linear scale, in Fig.4b on a quadratic scale, ω_{ri}^2 versus B^2. From the latter graph we find that the two resonances ω_{r1} and ω_{r2} obey the relation $\omega_{ri}^2(B) = \omega_{ri}^2(B = 0) + \omega_c^2$. The same dependence was found also for the single resonance in the one-layered system (Fig.4c). For the five-layered structure one observes five resonances where each shows the same quadratic dependence on B (Fig.4d).

The most striking result is that these resonance frequencies ω_{ri} in the FIR spectra are significantly higher in energy than one would expect from the 1D subband separation $\hbar\Omega_0$ which was determined from the dc magneto-transport measurements, i.e. at $B = 0$, for the one-layered quantum wire system: $\hbar\Omega_0 = 1meV$, $\hbar\omega_r = 4meV$, for the two-layered system: $\hbar\Omega_0 = 1.5meV$, $\hbar\omega_{r1} = 8meV$, $\hbar\omega_{r2} = 4meV$. However, it is very well known that one has to be careful in the interpretation of optical spectra. E.g. for 2DES the FIR intersubband resonance ω_r is shifted with respect to the one-particle transition due to the resonant screening effect of all electrons in the system.[22,1] This collective so-called depolarization effect is characterized by an effective plasma frequency ω_d. If we adopt for a moment this 2D model also for the 1DES, i.e., $\omega_r^2 = \Omega_0^2 + \omega_d^2$, we find that the intersubband transition is strongly governed by the depolarization effect. E.g. for sample B it is: $\omega_d^2 = \omega_r^2 - \Omega_0^2 = (4meV)^2 - (1meV)^2 = (3.9meV)^2$. The question arises if such a model can be applied at all under these conditions. The depolarization shift has originally been introduced as a correction to the one-particle model. We will come back to this point below.

In the following we demonstrate, that the strongly dominant collective contribution can be understood in terms of a "local" plasmon resonance, $\omega_d = \omega_{pl}$. The magnetic field dependence of the FIR resonances and in particular the occurrence of two (n) resonances for two (n)-layered quantum wire structures resembles a plasmon type of excitation. For a two (n)-layered, homogeneous 2DES it is known, that the collective excitation spectrum at small wavevectors q consists of two (n) branches.[23,24,25] A calculated plasmon dispersion for a two-layered homogeneous system is shown in Fig. 5. Whereas for widely separated electron sheets ($qd \gg 1$, $d = $ separation of the sheets) the plasmon branches are degenerate, for small distances ($qd \approx 1$) the coupling leads to a splitting of the plasmon dispersion. In this case the high energy branch represents an "in phase longitudinal oscillation" of both electron layers. The frequency of the lower energy branch is determined by the strength of the coupling of the two electron sheets. The observation of layer-coupled type of plasmon modes here is the first at all with FIR spectroscopy. So far these modes have only been detected in laterally homogeneous 2DES by Raman spectroscopy.[23,24]

However, the resonances observed in our 1DES differ significantly from excitations in a homogeneous system in the following points: (a) Besides the plasmon resonance one would expect to observe in a homogeneous system a cyclotron resonance. This resonance is completely quenched in our micro-structured samples, all observed resonances ω_{ri} are shifted with respect to ω_c. (b) When we calculate the plasmon frequency of a homogeneous system

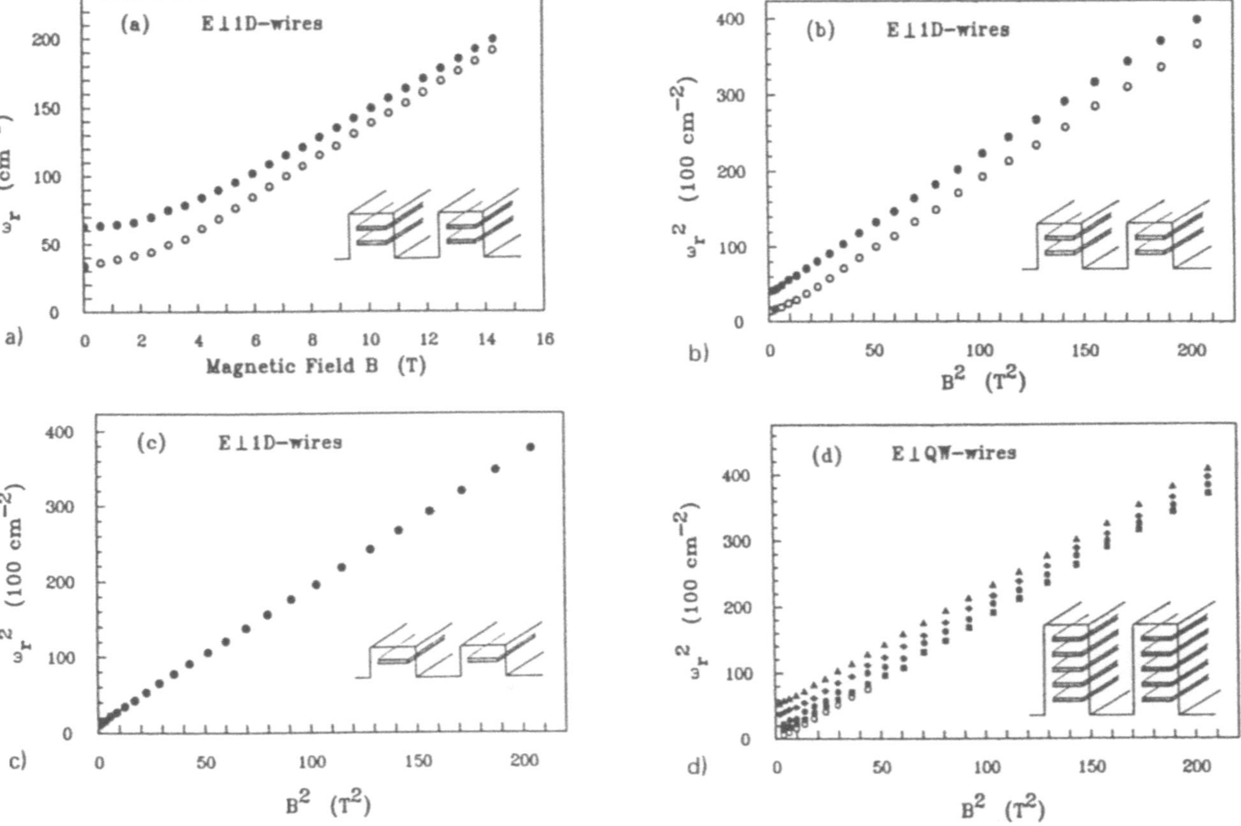

Fig. 4 Experimental resonance positions from FIR spectroscopy. Measurement on the two-layered quantum wire system are shown on a linear (a) and on a quadratic (b) scale. Measurement on a one-layered (c) and five-layered (d) structure are shown in a quadratic plot.

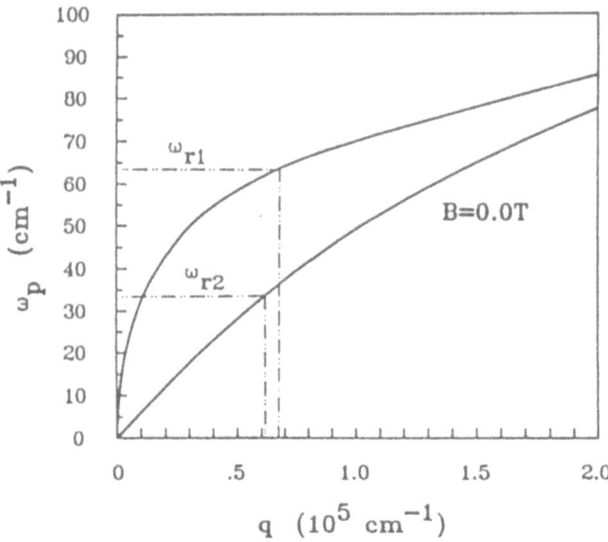

Fig. 5 Calculated plasmon dispersion for a two-layered homogeneous system at a magnetic field $B = 0$. The dash-double dotted lines indicate the experimentally observed resonance energies for $B = 0$. The intersection with the dispersion is projected onto the q-axis to the determine quantized q-value, $q = \pi/w_e$, of the local plasmons.

$\omega_p^2 = \frac{N_s e^2}{2\epsilon\epsilon_0 m^*}q$ (for a recent review see e.g. Ref.26), using the average dielectric constant $\bar{\epsilon} = 6.9$ for the microstructured region and the average 2D charge density $\bar{N}_{s2D} = N_{s1D}/a$, we find for sample A: $\omega_p = 23cm^{-1}$ and for sample B: $\omega_{p1} = 37cm^{-1}$, $\omega_{p2} = 21cm^{-1}$. Thus the experimentally observed resonances $(\omega_r = 30cm^{-1}$ and $\omega_{r1} = 64cm^{-1}$, $\omega_{r2} = 34cm^{-1})$, are significantly higher in energy compared to those of the homogeneous system.

We explain this frequency shift by "localization" of plasmons in the following sense: Let us assume a single-layer "2DES" which is additionally confined in x-direction on a width w. Then, in a very simple model, we can treat the 2D plasmon mode for the x-direction as a "plasmon in a box". The continuous 2D plasmon dispersion $\omega_p^2 = \frac{N_s e^2}{2\epsilon\epsilon_0 m^*}q$ of a homogeneous system [26] with a free wavevector q is now quantized in fixed values of $q = \pi/w_e$ and correspondingly $\omega_{p l}^2 = \frac{N_s e^2}{2\bar{\epsilon}\epsilon_0 m^*}\frac{\pi}{w_e}$. Here $\bar{\epsilon}$ is the average dielectric constant. The effective width w_e is given by $w_e = w(1 + \alpha)$, where α takes account of the phase relation if the plasmon is "reflected" at the walls of the box. This is of course a very rough model, which totally neglects Coulomb interaction with neighbouring electron stripes and leaves α, so far, undetermined. However, this model explains our experimentally observed upward shift of the resonance frequency with decreasing w. It can also be applied to the two-layer case. This is demonstrated in Fig. 5, where the calculated plasmon dispersion of a homogeneous two-layered system, using the parameters of the two-layered sample, is plotted for $B = 0$. We adopt a strictly "local" model, i.e., we take the local 2D charge density in the wire, $N_{s2D} = 6.5 \cdot 10^{11} cm^{-2}$, and the dielectric constant of the direct surroundings, $\epsilon = 12.8$, for $GaAs$ and $AlGaAs$. Projecting the experimentally observed resonance frequencies onto the dispersions we get values of q which are converted, via $w_e = \pi/q$, into the effective width. We find a value of $w_e \approx 500nm$, which is comparable to the dimensions w and t. The good agreement between the experimentally observed splitting of the two plasmon modes and our calculation confirms our interpretation of layer-coupled plasmon modes and shows that the coupling of the modes is not very different in the microstructures compared to in the homogeneous samples.

A quantitative evaluation within this strictly "local" plasmon model is not always satisfactory. E.g., for the one-layered quantum wire sample B we find that the observed resonance frequency is best described by an excitation half way between a "local" plasmon and an "extended" plasmon. The latter is a better description when coupling between the different parallel wires becomes important. Then ω_p is governed by the grating periodicity $a = 2\pi/q$ and the averaged 2D charge density $\bar{N}_{s2D} = N_{s1D}/a$ of the system. A transition from an "extended" to a localized plasmon has been observed in Si-MOS samples with a spatially periodical charge density modulation.[27,28] For a small modulation amplitude a splitting of the plasmon resonances is observed corresponding to gaps in the plasmon dispersion. These gaps occur at the Brillouin boundaries which arise from the superlattice effect of the periodical charge density modulation. In these density modulated systems plasmons with large wavevectors are increased in frequency which arises from the localization of the plasmons in the region of high electron density.[29]

A strongly dominant collective contribution that was observed here in the mesa etched quantum wire structures was also found in 1DES of $GaAs$ systems with split-gate configuration.[14] We have also prepared similar split-gate grating samples and observe an identical behaviour. Actually the geometrical resonances considered in slightly wider structures (which have not been tested to be 1DES) in Refs.30 and 31 are in many aspects very similar to the response discussed here for the one-layered quantum wire structures and the resonances found on the split-gate structures[14]. The collective contribution to the excitation in the microstructured systems can be described by various models, such as a depolarization frequency[32] or a geometrical resonance[31]. Note that for the one-layered system all models give the same dependence $\omega_p^2 = A \cdot N_s/w$. The factor A is slightly different in the different models which reflects the different potential that acts onto the electrons in different configurations. We prefer the local plasmon model, since in this model the layer-coupled plasmon effect of multi-layer quantum wires can easily be described. It also explains experiments in split-gate configuration, where a continuous transition from an extended plasmon to the local plasmon frequency is found, if via the split-gate the system is tuned from a density modulated system to a 1D system (see Ref. 14, Fig. 3). For the one-layered system our model of "plasmons in a box" is a simplified version of a calculation for a density modulated system by Lai et al.[33] In the latter calculation the coupling between different wires and the "boundary conditions" for the "reflection" of the plasmons (i.e., the determination of our α) are treated rigorously.

Let us now return to the single-particle aspects of the FIR excitation. There are several theoretical treatments (e.g. Refs. 34-36) where model 1D wave functions are assumed and the optical resonance frequencies are calculated. A general treatment is a very cumbersome task. Thus usually the assumption is made that only one or two 1D subbands contribute to the response. The result of such a calculation in the case that only the lowest subband is occupied in an isolated 1D wire is e.g. according to Ref.36

$$\omega_{res}^2 = \omega_{10}^2 + \omega_{10}e^2 N_l/(\pi\epsilon\epsilon_0).$$

Here ω_{10} is the 1D subband spacing, ϵ the dielectric constant of the material and N_l the linear density. The second term on the right side is the depolarization shift. It describes the resonant screening of the one-particle-excitation due to the collective effects of all other electrons in the system. This effect is well known from 2DES (e.g. Refs.22 or 1) and can be directly observed e.g. in grating coupler induced intersubband resonance experiments on Si-(111) MOS structures.[37] In these 2DES systems the depolarization shift is a small effect (20% frequency increase). If we calculate the depolarization shift for usual conditions in $GaAs$ quantum wires, i.e., assuming a linear density of $N_l = 10^6 cm^{-1}$ and $\omega_{10} = 2meV$, we calculate a polarization shift of $9.5meV$. This value is far beyond the limits of the approximation used in the calculation . Thus collective effects are important even at very low carrier densities and the current calculations can not be directly applied for the presently studied 1D $GaAs$ systems. The depolarization shift is decreased by interaction with other stripes in a grating structure and also by a metal gate near the wires.[35,36] A theoretical treatment of 1DES with many occupied subbands including their interactions and the transition from a "pure" one-particle resonance to a dominantly collective is still missing and seems to be a challenging task. The best coincidence between 1D subband separation (measure in dc magnetotransport) and FIR resonance (they differ only by about 30%) was found so far for quantum wires in InSb-MIS systems.[13] The reason for this is a much larger 1D subband spacing of $10meV$ which arises from the very small effective mass ($m^* = 0.014m_0$) of InSb. In addition a metal gate close to the 1D system decreases the depolarization effect.

5 Conclusions

1DES can be realized by ultrafine mesa etching techniques if by optimized lithography and etching techniques the lateral depletion region is reduced to about $100nm$. This makes it possible to prepare single and multi-layered quantum well wire systems. Typical confinement energies of $2meV$ can be determined from dc magnetotransport measurements. The FIR resonance frequencies are significantly higher than expected from the 1D subband separations, indicating a strong collective contribution that can be described in a "local" plasmon model. This model in particular explains the experimentally observed shift of the resonance frequencies and the splitting of the layer-coupled local plasmon modes.

References

[1] *T. Ando, A.B. Fowler, and F. Stern*, Rev. Mod. Phys. **54**, 437(1982)

[2] *K.-F. Berggren, T.J. Thornton, D.J. Newson, and M. Pepper* , Phys. Rev. Lett. **57**, 1769 (1986)

[3] *H. van Houten, B.J. van Wees, M.G.J. Heijman, and J.P. André, D. Andrews, and G.J. Davies* , Appl. Phys. Lett. **49**, 1781 (1986)

[4] *J. Cibert, P. M. Petroff, G. J. Dolan, S. J. Pearton, A. C. Gossard, and J. H. English* , Appl. Phys. Lett. **49**, 1275 (1986)

[5] *T. P. Smith, III, H. Arnot, J. M. Hong, C. M. Knoedler, S. E. Laux, and H. Schmid* , Phys. Rev. Lett. **59**, 2802 (1987)

[6] *M.L. Roukes, A. Scherer, S.J. Allen, Jr., H.G. Craighead, R.M. Ruthen, E.D. Beebe, and J.P. Harbison*, Phys. Rev. Lett. **59**, 3011 (1987)

[7] *H. van Houten, B.J. van Wees, J.E. Mooij, G. Roos, and K.-F. Berggren*, Superlattices and Microstructures **3**, 497 (1987)

[8] *G. Timp, A.M. Chang, P. Mankiewich, R. Behringer, J.E. Cunningham, T.Y. Chang, and R.E. Howard* , Phys. Rev. Lett. **59**, 732 (1987)

[9] *W. Hansen, M. Horst, J.P. Kotthaus, U. Merkt, Ch. Sikorski, and K. Ploog* , Phys. Rev. Lett. **58**, 2586 (1987)

[10] *B.J. van Wees, H. van Houten, C.W.J. Beenakker, J.G. Williamson, L.P. Kouwenhoven, D. van der Marel, and C.T. Foxon* , Phys. Rev. Lett. **60**, 848 (1988)

[11] *D.A. Wharam, T.J. Thornton, R. Newbury, M. Pepper, J.E.F. Frost, D.G. Hasko, D.C. Peacock, D.A. Ritchie, and G.A.C. Jones*, J. Phys. C21, L209 (1988)

[12] *K.-F. Berggren, G. Roos, and H. van Houten* , Phys. Rev. **B37**, 10118 (1988)

[13] *J. Alsmeier, Ch. Sikorski, and U. Merkt* , Phys. Rev. **B37**, 4314 (1988)

[14] *F. Brinkop, W. Hansen, J.P. Kotthaus, and K. Ploog* , Phys. Rev. **B37**, 6547 (1988)

[15] *T. Demel, D. Heitmann, P. Grambow, and K. Ploog* , Appl. Phys. Lett. **53**, 2176 (1988)

[16] *T. Demel, D. Heitmann, P. Grambow, and K. Ploog* , Phys. Rev. **B38**, 12732 (1988)

[17] *T. Demel, D. Heitmann, P. Grambow, and K. Ploog* Superlattices and Microstructures **5**,287(1989)

[18] *H. Sakaki*, Jap. J. Appl. Phys. **19**, L735 (1980)

[19] *M.B. Stern and P.F. Liao*, J. Vac. Sci. Technol. **B1**, 1053 (1983)

[20] *S. E. Laux, D. J. Frank, and F. Stern*, Surface Sci.**196**,101(1988)

[21] *K.K. Choi, D.C. Tsui, and K. Alavi* , Appl. Phys. Lett. **50**, 110 (1987)

[22] *W. P. Chen, Y. J. Chen, and E. Burstein*, Surface Sci.**58**,263(1976)

[23] *G. Fasol, N. Mestres, H.P. Hughes, A. Fischer, and K. Ploog* , Phys. Rev. Lett. **56**, 2517 (1986)

[24] *A. Pinczuk, M.G. Lamont, and A.C. Gossard* , Phys. Rev. Lett. **56**, 2092 (1986)

[25] *J.K. Jain and P.B. Allen* , Phys. Rev. Lett. **54**, 2437 (1985)

[26] *D. Heitmann*, Surface Sci.**170**,332(1986)

[27] *U. Mackens, D. Heitmann, L. Prager, J.P. Kotthaus, and W. Beinvogl* , Phys. Rev. Lett. **53**, 1485 (1984)

[28] *M.V. Krasheninnikov and A.V. Chaplik*, Sov. Phys. Semicond. **15**, 19 (1981)

[29] *D. Heitmann and U. Mackens*, Superlattices and Microstructures **4**,503(1988)

[30] *S.J. Allen, Jr., H.L. Störmer, and J.C. Hwang* , Phys. Rev. **B28**, 4875 (1983)

[31] *S.J. Allen, F. DeRosa, G.J. Dolan, and C.W. Tu*, Proc. 17th Int. Conf. Phys. Semicon., San Francisco (1984), edts. *J.D. Chadi and W.A. Harrison*, p. 313

[32] *W. Hansen, J.P. Kotthaus, A. Chaplik, and K. Ploog*, in: High Magnetic Fields in Semiconductor Physics, Proc. of the Int. Conf. Würzburg 1986, Springer Series in Solid-State Sciences 71, ed. by *G. Landwehr* (Springer, Berlin 1987), p.266

[33] *W.Y. Lai, A. Kobayashi, and S. Das Sarma* , Phys. Rev. **B34**, 7380 (1986)

[34] *S. Das Sarma and W.Y. Lai* , Phys. Rev. **B32**, 1401 (1985)

[35] *W. Que and G. Kirczenow* , Phys. Rev. **B37**, 7153 (1988)

[36] *A. V. Chaplik*, submitted to Superlattices and Microstructures

[37] *D. Heitmann and U. Mackens* Phys. Rev.**B33**, 8269 (1986)

Coherent electron focusing

C. W. J. Beenakker

Philips Research Laboratories, 5600 JA Eindhoven, The Netherlands

H. van Houten

Philips Laboratories, Briarcliff Manor, NY 10510, USA

B. J. van Wees

Applied Physics, Delft University of Technology, 2600 GA Delft, The Netherlands

Summary: Theory and experiment are reviewed of the classical and quantum mechanical focusing by a magnetic field of ballistic electrons injected through a point contact in a two-dimensional electron gas. Two alternative points of view are emphasized. On the one hand, the experiment is a realization of electron optics in the solid state. The three basic building blocks are a coherent and monochromatic point source/detector, an electrostatic mirror with little diffuse scattering, and a magnetic lens. On the other hand, coherent electron focusing is a resistance measurement in the quantum ballistic transport regime, which exhibits the characteristic features of this regime in a most extreme way. For example, large magnetoresistance oscillations occur (up to 95% amplitude modulation is observed), with a periodicity which is non-locally determined by the separation between current and voltage point contacts. A WKB calculation of the transmission probabilities shows that this effect is the result of the interference of coherently excited magnetic edge states at the electron gas boundary. Another example is the absence of local equilibrium: The measurements show that the point contacts can selectively populate (and detect) specific Landau levels, and that this highly non-equilibrium population is maintained over distances of microns.

1 Introduction

Electron focusing in metals was pioneered by Sharvin [1] and Tsoi [2] as a powerful tool to investigate the shape of the Fermi surface, surface scattering, and the electron-phonon interaction [3]. The experiment is the analogue in the solid state of magnetic focusing in vacuum (e.g. in a β —spectrometer). Required is a large mean free path for the carriers at the Fermi surface, to ensure *ballistic* motion as in vacuum. The mean free path (which can be as large as 1 cm in pure metallic single crystals) should be much larger than the length L on which the focusing takes place. Experimentally $L = 10^{-2} - 10^{-1}$ cm is the separation of two metallic needles (point contacts) pressed on the crystal surface, which serve to inject a divergent electron beam and detect its focusing by the magnetic field. In metals, electron focusing is essentially a *classical* phenomenon because of the small Fermi wave length λ_F (typically 0.5 nm, on the order of the inter-atomic separation).

The Fermi wave length is 100 times as large in the two-dimensional electron gas (2DEG) which is present at the interface of a GaAs-AlGaAs heterostructure. This length scale is within reach of electron-beam lithography, while remaining well below the mean free path in high-mobility material (10 μm can be realized at low temperatures in heterostructures grown by molecular-beam epitaxy). For these two reasons the *quantum ballistic* transport regime has become accessible in a 2DEG [4]. In the present paper theory and experiment are reviewed of electron focusing in this regime [5 − 8], which turns out to be strikingly different from the classical regime familiar from metals. This has motivated the new name: *coherent electron focusing*.

The geometry of the experiment (Fig. 1) is the transverse focusing geometry of Tsoi [2], and consists of two point contacts on the same boundary in a perpendicular magnetic field B. [In metals one can also use the geometry of Sharvin [1], with opposite point contacts in a longitudinal field. This is not possible in two dimensions.] Because the electron gas is confined to the interior of the heterostructure, one can not just use a metal needle to fabricate a point contact to a 2DEG. Instead, the point contacts are created electrostatically by depositing an electrode of a suitable shape on top of the heterostructure [9]. On applying a negative voltage to the split-gate electrode shown in Fig. 1 the electron gas underneath the gate structure is depleted, creating two 2DEG regions (i and c) electrically isolated from the rest of the 2DEG — apart from the two narrow and short constrictions (point contacts)

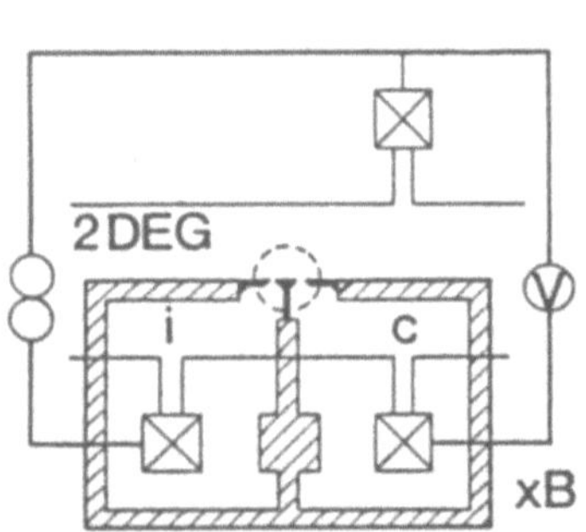

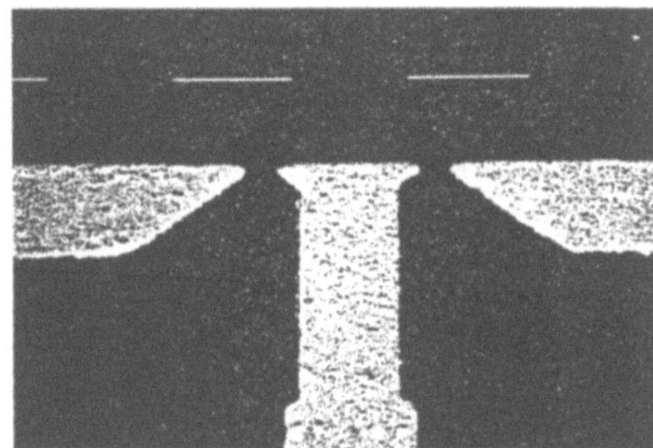

Fig.1 Schematic layout (left) of the double point contact device for the electron focusing experiments (in a three-terminal measurement configuration). The crossed squares are ohmic contacts to the 2DEG. The split-gate (shaded) separates injector (i) and collector (c) areas from the bulk 2DEG. The fine details of the gate structure inside the dashed circle are shown in a scanning electron micrograph (right). The bar denotes a length of 1 μm. In this device the point contact separation is $L = 1.5$ μm. A device with $L = 3.0$ μm was also studied. [From Ref 8.]

300

under the 250 nm wide openings in the split-gate. The devices studied had point contact separations L of 1.5 and 3.0 μm, both values being below the mean free path of 9 μm estimated from the mobility. Because the depletion potential extends laterally beyond the gate pattern for high (negative) gate voltages, one can force the constrictions to become progressively narrower (at the same time reducing the electron gas density in the constrictions) — until they are fully pinched off. By this technique it is possible to create point contacts of *variable* width W, something which is not realizable in a metal. Note that W is comparable in magnitude to λ_F (which was 40 nm in the devices studied). These are *quantum point contacts*, as evidenced by their conductance which was discovered to be approximately quantized in units of $2e^2/h$ [10,11].

Electron focusing can be seen as a transmission experiment in electron optics. The classical regime then corresponds to geometrical optics, the quantum regime to wave optics. The optical analogy is useful, both to understand the experiments and to inspire new ones [12]. An alternative point of view is that coherent electron focusing is a prototype of a non-local resistance measurement in the quantum ballistic transport regime, such as studied extensively in narrow-channel geometries [13]. Longitudinal resistances which are negative, not $\pm B$ symmetric, and dependent on the properties of the current and voltage contacts as well as on their separation; periodic and aperiodic magnetoresistance oscillations; absence of local equilibrium — these are all characteristic features of this transport regime which appear in a most extreme and bare form in the electron focusing geometry. One reason for the simplification offered by this geometry is that the current and voltage contacts, being point contacts, are not nearly as invasive as the wide leads in a Hall bar geometry [14]. Another reason is that the electrons interact with only one boundary (instead of two in a narrow channel).

The outline of this paper is as follows. In Sec. 2 the experimental results on electron focusing [5,8] are described as a transmission experiment in a 2DEG. A theoretical description [6,8] is given in Sec. 3, in terms of mode interference in the wave guide formed by the magnetic field at the 2DEG boundary. In Sec. 4 we discuss the quantum Hall effect in the electron focusing geometry [7,8] as a non-local resistance measurement. The theoretical framework used to relate these two alternative descriptions is the Landauer-Büttiker formalism [15,16], which treats a resistance measurement as a transmission experiment. We conclude in Sec. 5.

2 Mirror, lens, and point source

Fig. 2 illustrates electron focusing in two dimensions as it follows from classical mechanics. The arrangement combines three basic elements: mirror, magnetic lens, and point source/detector. The point source (i) injects electrons with the Fermi energy $E_F \equiv m v_F^2/2$ ballistically into the

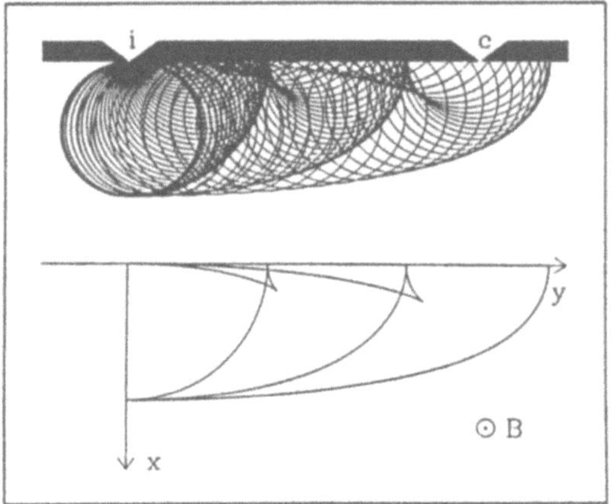

Fig.2
Top: Skipping orbits along the 2DEG boundary. The trajectories are drawn up to the third specular reflection. Bottom: Plot of the caustics, which are the collection of focal points of the trajectories. [From Ref 8.]

2DEG. The injected electrons all have the same Fermi velocity v_F, but in different directions. Electrons are detected if they reach the adjacent collector (c), after one or more specular reflections at the boundary connecting i and c. These *skipping orbits* are composed of translated circular arcs of identical radius $l_{cycl} \equiv \hbar k_F / eB$, which is the cyclotron radius in a perpendicular magnetic field B ($k_F \equiv mv_F / \hbar$ is the Fermi wave vector). The focusing action of the magnetic field is evident in Fig. 2 (top) from the black lines of high density of trajectories. These lines are known in optics as *caustics*, and are plotted separately in Fig. 2 (bottom). The caustics intersect the 2DEG boundary at multiples of the cyclotron diameter from the injector. As the magnetic field is increased, a series of these focal points shifts past the collector. The electron flux incident on the collector thus reaches a maximum whenever its separation L from the injector is an integer multiple of $2l_{cycl}$. This occurs when $B = pB_{focus}$, $p = 1,2,...$, with

$$B_{focus} = 2\hbar k_F / eL .\qquad(1)$$

For a given injected current I_i the voltage V_c on the collector is proportional to the incident flux. The classical picture thus predicts a series of equidistant peaks in the collector voltage as a function of magnetic field.
In Fig. 3 (top) we show such a classical focusing spectrum, calculated for parameters corresponding to the experiment discussed below ($L = 3.0\,\mu$m, $k_F = 0.15\,$nm^{-1}). The spectrum consists of equidistant focusing peaks of approximately equal magnitude superimposed on the Hall resistance (dashed line). The p —th peak is due to electrons injected perpendicularly to the boundary which have made $p - 1$ specular reflections between injector and collector. Such a classical focusing spectrum is commonly observed in metals [17], albeit with a decreasing

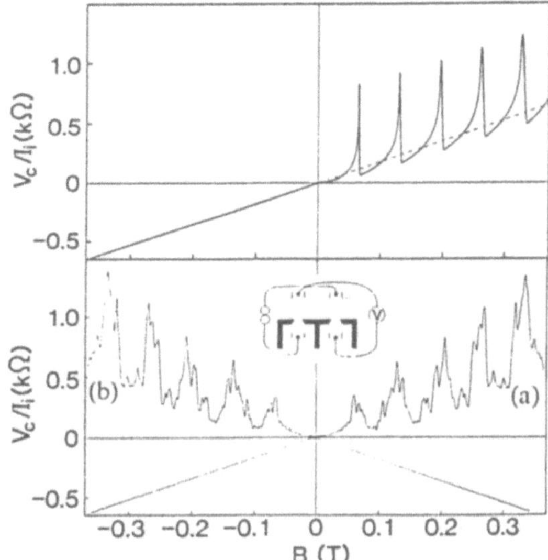

Fig.3
Bottom: Experimental electron focusing spectrum ($T = 50$ mK, $L = 3.0$ μm) in the generalized Hall resistance configuration depicted in the inset. The two traces a and b are measured with interchanged current and voltage leads, and demonstrate the injector-collector reciprocity as well as the reproducibility of the fine structure. Top: Calculated classical focusing spectrum corresponding to the experimental trace a (50 nm wide point contacts were assumed). The dashed line is the extrapolation of the classical Hall resistance seen in reverse fields. [From Ref 8.]

height of subsequent peaks because of partially diffuse scattering at the metal surface. Note that the peaks occur in one field direction only; In reverse fields the focal points are at the wrong side of the injector for detection, and the normal Hall resistance is obtained. The experimental result for a 2DEG is shown in the bottom half of Fig. 3 (trace a ; trace b is discussed below). A series of five focusing peaks is evident at the expected positions. This observation by itself has two important implications:

- A point contact acts as a monochromatic point source of ballistic electrons with a well-defined energy;
- The electrostatically defined 2DEG boundary is a good mirror with little diffuse scattering.

Fig. 3 is obtained in a measuring configuration (inset) in which an imaginary line connecting the voltage probes crosses that between the current source and drain. This is the configuration for a generalized Hall resistance measurement. Alternatively, one can measure a generalized longitudinal resistance, in the configuration shown in the inset of Fig. 4. One then measures the focusing peaks without a superimposed Hall slope. Note that the experimental longitudinal resistance (Fig. 4, bottom) becomes *negative*. This is a classical result of magnetic focusing, as demonstrated by the calculation shown in the top half of Fig. 4. Büttiker [18] has studied negative longitudinal resistances in a different (Hall bar) geometry.

On the experimental focusing peaks a fine structure is evident in Figs. 3 and 4. The fine structure is well reproducible (compare Figs. 3 and

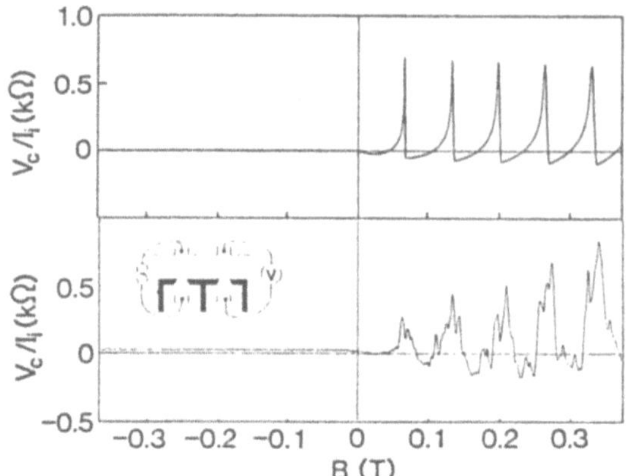

Fig.4
As Fig. 3, but in the longitudinal resistance configuration. [From Ref 8.]

4), but sample dependent. It is only resolved at low temperatures (below 1 K) and small injection voltages (the measurements shown are taken at 50 mK and a few μV AC voltage over the injector). A nice demonstration of the reproducibility of the fine structure is obtained upon interchanging current and voltage leads, so that the injector becomes the collector and vice versa. The resulting focusing spectrum shown in Fig. 3 (trace b) is almost the precise mirror image of the original one (trace a) — although this particular device had a strong asymmetry in the widths of injector and collector. The symmetry in the focusing spectra is a consequence of the fundamental reciprocity relation derived by Büttiker [16], which generalizes the familiar Onsager-Casimir symmetry relation for the *resistivity tensor* to *resistances* (see Sec. 4).
The fine structure on the focusing peaks in Figs. 3 and 4 is the first indication that electron focusing in a 2DEG is qualitatively different from the corresponding experiment in metals. At higher magnetic fields the resemblance to the classical focusing spectrum is lost, see Fig. 5. A Fourier transform of the spectrum for $B \geq 0.8$ T (inset in Fig. 5) shows that the large-amplitude high-field oscillations have a dominant periodicity of 0.1 T, which is approximately the same as the periodicity B_{focus} of the much smaller focusing peaks at low magnetic fields (B_{focus} in Fig. 5 differs from Fig. 3 because of a smaller $L = 1.5 \mu$m). This dominant periodicity is the result of quantum interference between the different trajectories in Fig. 2 which take an electron from injector to collector. [In Sec. 3 we show this in a mode picture, which in the WKB approximation is equivalent to calculating the interferences of the (complex) probability amplitude along classical trajectories. The latter ray picture is treated extensively in Ref. 8.] The theoretical analysis implies for the experiment that:

■ The injector acts as a *coherent* point source with the coherence maintained over a distance of several microns to the collector.

304

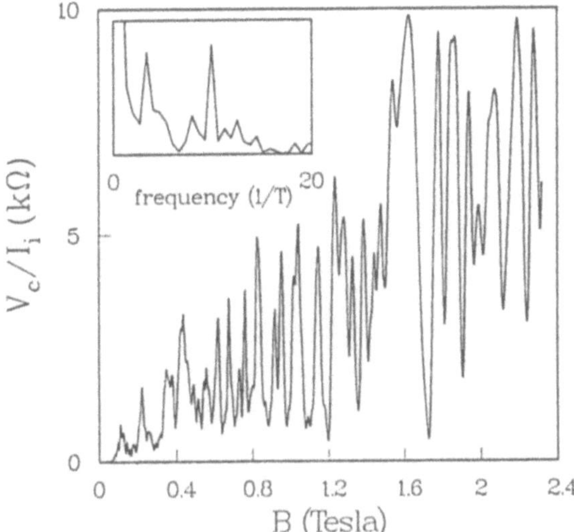

Fig.5
Experimental electron focusing spectrum over a larger field range and for very narrow point contacts (estimated width 20 − 40 nm; $T = 50$ mK, $L = 1.5\,\mu$m). The inset gives the Fourier transform for $B \geq 0.8$ T. The high-field oscillations have the same dominant periodicity as the low-field focusing peaks − but with a much larger amplitude. [From Ref 8.]

3 Edge states and skipping orbits

Magnetic edge states [19,20] are transverse modes of a wave guide of width $\sim l_{cycl}$ formed by the magnetic field at the 2DEG boundary. The edge states at the Fermi level are labelled by a quantum number $n = 1,2 \ldots N$, with $N = k_F\, l_{cycl}/2$ the total number of propagating modes or edge *channels* (for simplicity we ignore here the discreteness of N). An injector of width below λ_F excites a coherent superposition of these propagating modes (plus evanescent modes, which using the ray treatment of Ref. 8 are found to give only a small contribution for the large $k_F L$ considered, and will be neglected here). The wave function Ψ is of the form

$$\Psi(x, y) = \sum_{n=1}^{N} a_n f_n (x)\, e^{ik_n y} . \qquad (2)$$

Here k_n is the wave number for propagation of mode n in the y −direction (along the 2DEG boundary, see Fig. 2 for our choice of axes), $f_n(x)$ is the transverse amplitude profile of mode n, and a_n its excitation factor. For $k_F L \gg 1$ the phase factors $\exp(ik_n L)$ vary rapidly as a function of n. Constructive interference of modes at the collector then requires that $k_n L$ differs by multiples of 2π for a series of n. To find out what this condition implies for the magnetic field, we determine k_n in WKB approximation (which should be sufficiently accurate for this purpose).

Consider again the classical skipping orbits (Fig. 2). The position (x, y) of the electron on the circle with center coordinates (X, Y) can be expressed in terms of its velocity $\mathbf{v}$ by

$$x = X + v_y/\omega_c , \quad y = Y - v_x/\omega_c , \tag{3}$$

with $\omega_c = eB/m$ the cyclotron frequency. Note that the separation X of the center from the boundary is constant on a skipping orbit, only the center coordinate Y parallel to the boundary changes at each specular reflection. The canonical momentum of the electron is $\mathbf{p} = m\mathbf{v} - e\mathbf{A}$. In the Landau gauge $\mathbf{A} = (0, Bx, 0)$ we have

$$p_x = mv_x , \quad p_y = -eB\, X . \tag{4}$$

The wave number k corresponds classically to the canonical momentum component $p_y = \hbar k$, so that in view of Eq. (4) we have the correspondence $k = - (eB/\hbar) X$. Since the motion projected on the x-axis is periodic, one can apply the Bohr-Sommerfeld quantization rule [21]

$$\frac{1}{\hbar} \oint p_x\, \mathrm{d}x + \gamma = 2\pi n . \tag{5}$$

The integral is over one period of the motion, n is an integer, and γ is the sum of the phase shifts acquired at the two turning points of the motion. The phase shift upon reflection at the boundary is π (for an infinite barrier potential, to ensure that incident and reflected waves cancel); The other turning point is a caustic of the skipping orbits with constant X, leading to a phase shift of $-\pi/2$ [22]. This totals to $\gamma = \pi/2$. Using also Eqs. (3) and (4) we may thus write Eq. (5) in the form

$$\frac{eB}{\hbar} \oint (Y - y)\, \mathrm{d}x = 2\pi \left(n - \frac{1}{4} \right) , \quad n = 1, 2, \dots N. \tag{6}$$

This quantization rule has the simple geometrical interpretation [20] that the flux enclosed by one arc of the skipping orbit and the boundary equals $(n - 1/4)$ times the flux quantum h/e (see insets in Fig. 6).
Eq. (6) determines, for a given magnetic field, the energy $E = mv^2/2$ as a function of the quantum number n and the wave number $k = - (eB/\hbar) X$. To carry out the integration in Eq. (6) we express y in terms of x by means of Eq. (3). The resulting energy spectrum $E_n(k)$ is given by

$$\frac{2E}{\hbar\omega_c} \left(\arccos \xi - \xi (1 - \xi^2)^{1/2} \right) = 2\pi \left(n - \frac{1}{4} \right) , \quad \xi \equiv \hbar k\, (2mE)^{-1/2} , \tag{7}$$

and is plotted in Fig. 6 (solid curves). Also plotted in Fig. 6 is the exact solution of the Schrödinger equation (dashed curves, taken from Ref.

306

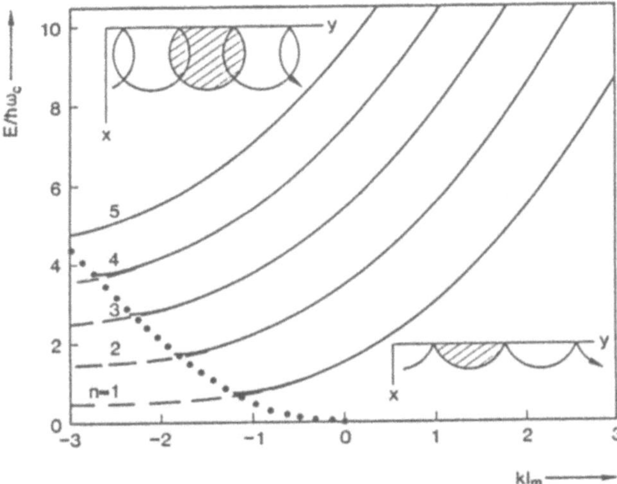

Fig.6 Energy spectrum $E_n(k)$ of magnetic edge states at an infinite barrier potential boundary. Note that $kl_m = -X/l_m$, with X the separation of the orbital center from the boundary and $l_m \equiv (\hbar/eB)^{1/2}$ the magnetic length. The insets show classical skipping orbits for positive and negative k. In the semi-classical approximation the magnetic flux through the shaded areas is quantized. The result from Eq. (7) (solid curves) is indistinguishable from the exact solution (dashed curves, from Ref. 23), unless k is within $1/l_m$ of the transition from skipping to cyclotron orbits (dotted curve). [From Ref. 8.]

23). The (semi-classical) WKB approximation (7) is indistinguishable on this scale from the exact solution, except just before the transition from skipping orbits to bulk cyclotron orbits at $X = mv/eB$ (dotted curve in Fig. 6). The quantized wave numbers k_n at the Fermi energy satisfy $E_n(k_n) = E_F$, so that k_n is determined by Eq. (7) with the substitutions $E \equiv E_F$, $\zeta \equiv k_n/k_F$. As shown in Fig. 7 the resulting dependence on n of the phase $k_n L$ is close to linear in a broad interval,

$$k_n L = \text{constant} - 2\pi n B/B_{\text{focus}} + k_F L \times \text{order} (1 - 2n/N)^3 . \tag{8}$$

It follows from this expansion that if B/B_{focus} is an integer, a fraction of order $(1/k_F L)^{1/3}$ of the N edge channels interfere constructively at the collector. Because of the 1/3 power, this is a substantial fraction even for the large $k_F L \sim 10^2$ of the experiment. The relevant states have quantum number n in an interval centered around $N/2$, corresponding classically to skipping orbits which reach the boundary at approximately right angles. The edge states outside the domain of linear n—dependence of the phase give rise to fine structure without a simple periodicity.

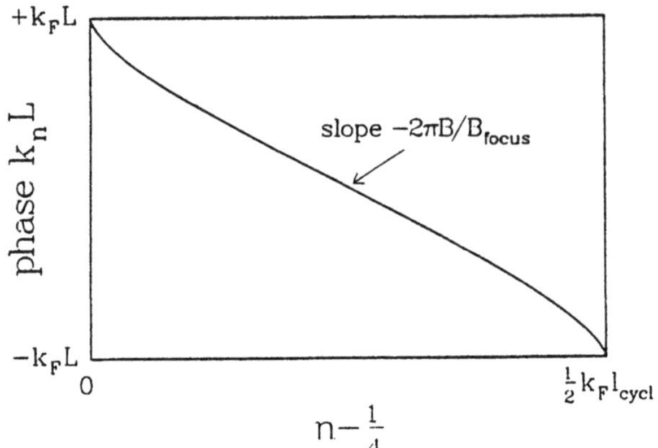

Fig.7
Phase $k_n L$ of the edge channels at the collector, calculated from Eq. (7). Note the domain of approximately linear n-dependence of the phase, responsible for the oscillations with B_{focus}-periodicity. [From Ref. 8.]

To determine the amplitude of the oscillations in the collector voltage, we need to know the excitation factors of the modes by the injector and the transmission amplitude through the collector. In Ref. 8 we calculated these quantities using a point-dipole injector and a transmission amplitude proportional to the derivative $\partial\Psi/\partial x$ of the unperturbed wave function at the collector — thereby neglecting the finite width of the injector and collector point contacts. The result obtained there can be written in the form

$$\frac{V_{\text{c}}}{I_{\text{i}}} = \frac{h}{2e^2} \left| \frac{1}{N} \sum_{n=1}^{N} e^{ik_n L} \right|^2 . \tag{9}$$

In Fig. 8 we have plotted the focusing spectrum from Eq. (9), corresponding to the experimental Fig. 5. The inset shows the Fourier transform for $B \geq 0.8\,\text{T}$. There is no detailed one-to-one correspondence between the experimental and theoretical spectra. No such correspondence was to be expected in view of the sensitivity of the experimental spectrum to small variations in gate voltage (which defines the point contacts and the 2DEG boundary). Those features of the experimental spectrum which are insensitive to the precise measurement conditions are, however, well reproduced by the calculation: We recognize in Fig. 8 the low-field focusing peaks and the large-amplitude high-field oscillations with the same periodicity. [The reason that the periodicity B_{focus} in Fig. 8 is somewhat larger than in Fig. 5 is most likely the experimental uncertainty in the effective point contact separation of the order of the split-gate opening (250 nm).] The high-field oscillations range from about 0 to $10\text{k}\Omega$ in both theory and experiment.

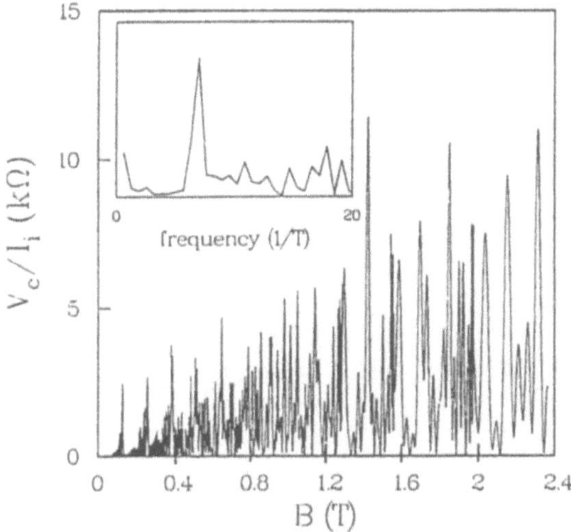

Fig.8 Focusing spectrum calculated from Eq. (9), for parameters corresponding to the experimental Fig. 5. The inset shows the Fourier transform for $B \geq 0.8$ T. Infinitesimally small point contact widths are assumed in the calculation.

This maximum amplitude is not far below the theoretical upper bound of $h/2e^2 \approx 13$ kΩ, which follows from Eq. (9) if we assume that *all* the modes interfere constructively. This indicates that a *maximal phase coherence* is realized in the experiment, and implies that:

■ The experimental injector and collector point contacts resemble the idealized point source/detector in the calculation;

■ Scattering events other than specular scattering on the boundary can be largely ignored (since any other inelastic *as well as elastic* scattering events would scramble the phases and reduce the oscillations with B_{focus} — periodicity).

It follows from Eq. (9) that if interference of the modes is ignored, the normal quantum Hall resistance $h/2Ne^2$ is obtained. This is *not* a general result, but depends specifically on the properties of the injector and collector point contacts — as we will discuss in the following section.

4 Quantum point contacts as Landau level selectors

Mode interference becomes unimportant if the magnetic field is sufficiently strong, and the point contacts are sufficiently wide, that the electrostatic potential in the point contact region does not cause scattering between the modes. The requirement for such *adiabatic* transport is that the potential varies slowly on the scale of l_{cycl} (in the quantum Hall effect regime where $N \sim 1$ and $E_F \sim \hbar\omega_c$, the cyclotron radius is the magnetic length $l_m \equiv (\hbar/eB)^{1/2}$). In this field regime the form of the

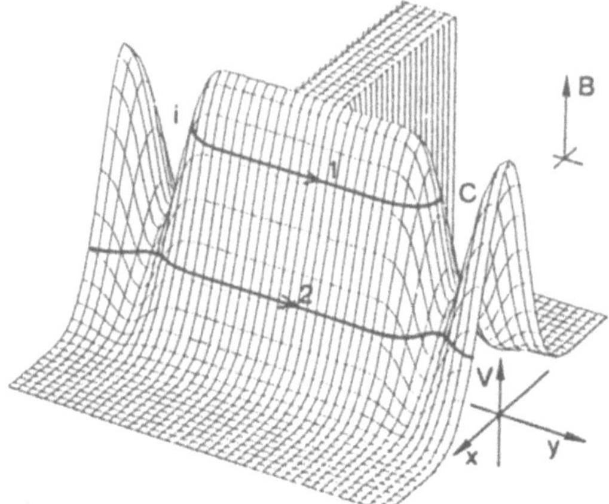

Fig.9
Schematic potential landscape, showing the 2DEG boundary and the saddle-shaped injector and collector point contacts. In a strong magnetic field the edge channels are extended along equipotentials (Eq. (10)), as indicated here for $n = 1,2$ (the arrows point in the direction of motion). In this case a Hall conductance of $(2e^2/h) N$ with $N = 1$ would be measured by the point contacts — in spite of the presence of 2 occupied Landau levels in the bulk 2DEG.

electrostatic potential $V(x, y)$ defining the point contacts becomes important, and the point injector/detector model used in the previous section — while adequate at lower magnetic fields — is insufficient.
Schematically, $V(x, y)$ is represented in Fig. 9. Fringing fields from the split-gate create a potential barrier in the point contacts, so that V has a saddle form as shown. The heights of the barriers E_i, E_c in the injector and collector are separately adjustable by means of the voltages on the split-gates, and can be determined from the conductances of the individual point contacts [24]. The width of the point contacts does not play a role, because it is larger than l_{cycl}. The adiabatic transport is along equipotentials as indicated in Fig. 9 (arrows point in the direction of motion, determined by the potential gradient). The energy of the equipotential is the *guiding center energy* E_G, which is given for edge channel n by

$$E_G = E_F - (n - \frac{1}{2}) \hbar\omega_c \tag{10}$$

(Zeeman spin-splitting of the energy levels should be included at large magnetic fields, but is ignored here for simplicity). The edge channels can only be transmitted through a point contact if E_G exceeds the potential barrier height (disregarding tunneling through the barrier). The injector thus injects $N_i \approx (E_F - E_i)/\hbar\omega_c$ edge channels into the 2DEG, while the collector is capable of detecting $N_c \approx (E_F - E_c)/\hbar\omega_c$ channels. Along the boundary of the 2DEG, however, a larger number of $N \approx E_F/\hbar\omega_c$ edge channels, equal to the number of bulk Landau levels in the 2DEG, are available for the current transport. The selective population, and detection, of Landau levels leads to deviations from the normal Hall resistance.
These considerations can be put on a theoretical basis by applying the general Landauer-Büttiker formalism [15,16], which relates resistances

to transmission probabilities into current and voltage probes. Consider the geometry in Fig. 1 of a three-terminal conductor with point contacts in two of the probes. The probes are connected by perfect leads to reservoirs which have a constant electro-chemical potential. We denote by μ_i and μ_c the chemical potentials of the two reservoirs connected, respectively, to the injector and collector point contact, and by μ_d the chemical potential of the third reservoir (the current drain). Following Büttiker [16], we can relate the currents I_α ($\alpha = $ i,c,d) in the three leads to these chemical potentials via the transmission probabilities $T_{\alpha \to \beta}$ (from reservoir α to reservoir β) and reflection probabilities R_α (from reservoir α back to the same reservoir),

$$\frac{h}{2e} I_\alpha = (N_\alpha - R_\alpha)\mu_\alpha - \sum_{\beta \neq \alpha} T_{\beta \to \alpha} \mu_\beta , \tag{11}$$

N_α being the number of occupied modes in the lead α. We now impose the condition that the collector draws no net current, which implies $I_c = 0$ and $I_d = -I_i$, and choose our zero of energy such that $\mu_d = 0$. One then finds from Eq. (11) the two equations

$$\mu_c = \frac{T_{i \to c}}{N_c - R_c} \mu_i , \quad \frac{h}{2e} I_i = (N_i - R_i) \mu_i - T_{c \to i} \mu_c , \tag{12}$$

and obtains for the ratio of collector voltage $V_c = \mu_c/e$ (measured relative to the voltage of the current drain) to injected current I_i the result

$$\frac{V_c}{I_i} = \frac{2e^2}{h} \frac{T_{i \to c}}{G_i G_c - \delta} . \tag{13}$$

Here $\delta \equiv (2e^2/h)^2 T_{i \to c} T_{c \to i}$, and $G_i \equiv (2e^2/h)(N_i - R_i)$ and $G_c \equiv (2e^2/h)(N_c - R_c)$ denote the conductances of the injector and collector point contact, respectively. The injector-collector reciprocity in electron focusing, demonstrated in Fig. 3, is manifest in Eq. (13), since G_i and G_c are even in B and [16] $T_{i \to c}(B) = T_{c \to i}(-B)$.
In the electron focusing geometry the term δ in Eq. (13) can be neglected, since $T_{c \to i} \approx 0$. An additional simplification is possible in the adiabatic transport regime. We consider the case that the barrier in one of the two point contacts is sufficiently higher than in the other, to ensure that electrons which are transmitted over the highest barrier will have a negligible probability of being reflected at the lowest barrier. Then $T_{i \to c}$ is dominated by the transmission probability over the highest barrier, $T_{i \to c} \approx \min(N_i - R_i, N_c - R_c)$. Substitution into Eq. (11) gives the remarkable result that the *Hall conductance* $G_{11} \equiv I_i/V_c$ measured in the electron focusing geometry can be expressed entirely in terms of the *contact conductances* G_i and G_c,

$$G_{11} \approx \max(G_i, G_c) . \tag{14}$$

311

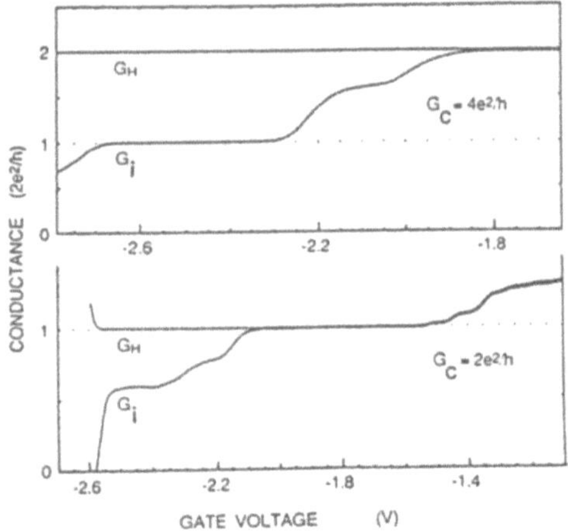

Fig.10.
Experimental correlation between the conductances G_i, G_c of injector and collector, and the Hall conductance $G_{11} \equiv I_i / V_c$, shown to demonstrate the validity of Eq. (14) ($T = 1.3$ K, $L = 1.5\,\mu$m). The magnetic field was kept fixed (top: $B = 2.5$ T, bottom: $B = 3.8$ T). By increasing the gate voltage on one half of the split-gate defining the injector, G_i was varied at constant G_c. [From Ref 7.]

Eq. (14) tells us that quantized values of G_{11} occur not at $(2e^2/h)\,N$, as one would expect from the N Landau levels in the 2DEG — but at the smaller value of $(2e^2/h)\max(N_i, N_c)$. Moreover, there is no quantized Hall conductance unless the largest of the two contact conductances is quantized. As shown in Fig. 10, this is indeed observed experimentally. Notice in particular how any deviation from quantization in $\max(G_i, G_c)$ is faithfully reproduced in G_{11}. The implication of this experiment is that:

- Point contacts can be used to *selectively* populate and detect Landau levels at the 2DEG boundary;
- Adiabatic transport (i.e. transport in the absence of inter-Landau level scattering) has been realized over a distance of $1.5\,\mu$m along the 2DEG boundary.

As discussed by Büttiker [25], the fundamental origin for deviations from the normal quantum Hall effect is the absence of local equilibrium among the edge channels. Selective population is indeed an extreme example of a non-equilibrium population. Recent related experiments [26,27] have demonstrated that a non-equilibrium population of edge channels can be maintained on even longer length scales, possibly as large as several hundred microns. It remains a theoretical challenge to explain these surprisingly long relaxation lengths.

5 Conclusion

In Sec. 1 we emphasized that the *length scales* relevant for the electron focusing experiment are very different in a metal and in a 2DEG. Both the ratios λ_F/L and λ_F/W are much larger in a 2DEG, typically by factors of 10^4 and 10^2, respectively. As we showed in Secs. 2 and 3, coherent electron focusing is possible in a 2DEG because of this rela-

312

tively large value of the Fermi wave length. The adiabatic transport discussed in Sec. 4 is also made possible by the large λ_F, since now $l_{cycl} = \hbar/eB\lambda_F$ can become comparable to W at magnetic fields of a few Tesla. To achieve the same in a metal would require fields over 100 T. The difference in *energy scale* between a metal and a 2DEG manifests itself in the dependence of the focusing spectrum on the voltage drop over the injector. In metals, electrons are injected at energies above E_F which are generally much less than $E_F \sim 5\,\text{eV}$ [28]. In contrast, $E_F \sim 10\,\text{meV}$ in a 2DEG, and DC-biasing the small AC injection voltage used in the electron focusing experiment should lead to a noticeable shift in the focusing peaks, in analogy with a β −spectrometer. In the simplest model one would have (cf. Eq. (1)) $B_{focus} \propto (E_F + eV_{DC})^{1/2}$, so that for a DC bias $V_{DC} = 1\,\text{mV}$ one would expect a 5% shift in the focusing periodicity − provided the hot electrons remain ballistic. This is indeed observed [29], although deviations from this simple behavior are found for larger DC biases (possibly related to the non-linear current-voltage characteristics of the point contacts themselves[30]). The observation of hot-electron transport over several microns is remarkable, and unexpected from related work in different systems [31].

The main result of the theoretical analysis of coherent electron focusing in Sec. 3 is the demonstration of high-field oscillations with B_{focus} −periodicity, but much larger amplitude than the low-field focusing peaks. This is also the feature of the experimental focusing spectra which is insensitive to small changes in gate voltage and which is found in both the devices studied. The theory can be improved in several ways. This will affect the detailed form of the spectra, but probably not the fundamental periodicity. Since the exact wave functions of the edge states are known (Weber functions), one could go beyond the WKB approximation. This will become important at large magnetic fields, when the relevant edge states have small quantum numbers. In this regime one would also have to take into account a possible B − dependence of E_F relative to the conduction band bottom (due to pinning of the Fermi energy at the Landau levels). It would be interesting to find out to what extent this bulk effect is reduced at the 2DEG boundary by the presence of edge states to fill the gap between the Landau levels.

Another direction of improvement is towards a more realistic modelling of the injector and collector point contacts. Since the maximum amplitude of the theoretical and experimental oscillations is about the same (compare Figs. 5 and 8), the loss of spatial coherence due to the finite point contact size does not seem to be particularly important in this experiment (infinitesimal point contact width was assumed in the calculation). On the other hand, the experimental focusing spectrum does not contain as much rapid oscillations as the calculation would predict.

Energy averaging due to a finite temperature is not the reason for this difference (temperatures on the order of 10 K are necessary to smear out the rapid oscillations). We surmise that the rapid oscillations are reduced by the *collimation* effect proposed originally [32] to explain the non-additivity of the resistance of two opposite point contacts in series [33] (and more recently invoked [34] to explain the quenching of the Hall resistance in a narrow-channel geometry [13]). Both the flaring of a point contact to form a horn, and the presence of a potential barrier in the point contact region tend to collimate the injected electron beam [32], so that electrons are predominantly injected at right angles to the boundary. The quantum mechanical correspondence discussed in Sec. 3 then implies that such a point contact excites (and detects) predominantly the edge channels with quantum number n close to $N/2$, at the expense of channels with smaller or larger n. Since the former edge channels are responsible for the oscillations with B_{focus} —periodicity, while the latter give rise to rapid aperiodic oscillations (see Fig. 7 and the accompanying discussion), the collimation effect provides one mechanism for the absence of rapid oscillations in the experimental focusing spectrum.

Acknowledgement

The authors have greatly benefitted from their collaboration with M.E.I. Brockaart, C.T. Foxon, C.J.P.M. Harmans, J.J. Harris, L.P. Kouwenhoven, P.H.M. van Loosdrecht, D. van der Marel, J.E. Mooij, M.F.H. Schuurmans, J.A. Pals, E.M.M. Willems, and J.G. Williamson.

References

[1] *Yu.V. Sharvin*, Zh.Eksp.Teor.Fiz. **48**, 984 (1965) [Sov.Phys.JETP **21**, 655 (1965)]

[2] *V.S. Tsoi*, Pis'ma Zh.Exp.Teor.Fiz. **19**, 114 (1974) [JETP Lett. **19**, 70 (1974)]

[3] *P.C. van Son, H. van Kempen*, and *P. Wyder*, Phys.Rev.Lett. **58**, 1567 (1987)

[4] Physics and Technology of Submicron Structures, ed. by *H. Heinrich, G. Bauer*, and *F. Kuchar* (Springer, Berlin 1988); Nanostructure Physics and Fabrication, ed. by *M. Reed* and *W.P. Kirk* (Academic Press, New York, to be published)

[5] *H. van Houten, B.J. van Wees, J.E. Mooij, C.W.J. Beenakker, J.G. Williamson*, and *C.T. Foxon*, Europhys.Lett. **5**, 721 (1988)

[6] *C.W.J. Beenakker, H. van Houten*, and *B.J. van Wees*, Europhys.Lett. **7**, 359 (1988)

[7] *B.J. van Wees, E.M.M. Willems, C.J.P.M. Harmans, C.W.J. Beenakker, H. van Houten, J.G. Williamson, C.T. Foxon*, and *J.J. Harris*, Phys.Rev.Lett. **62** 1181 (1989)

[8] *H. van Houten, C.W.J. Beenakker, J.G. Williamson, M.E.I. Broekaart, P.H.M. van Loosdrecht, B.J. van Wees, J.E. Mooij, C.T. Foxon,* and *J.J. Harris,* Phys.Rev. B (April 15, 1989)

[9] *T.J. Thornton, M. Pepper, H. Ahmed, D. Andrews,* and *G.J. Davies,* Phys.Rev.Lett. **56**, 1198 (1986); *H.Z. Zheng, H.P. Wei, D.C. Tsui,* and *G. Weimann,* Phys.Rev. **B34**, 5635 (1986)

[10] *B.J. van Wees, H. van Houten, C.W.J. Beenakker, J.G. Williamson, L.P. Kouwenhoven, D. van der Marel,* and *C.T. Foxon,* Phys.Rev.Lett. **60**, 848 (1988)

[11] *D.A. Wharam, T.J. Thornton, R. Newbury, M. Pepper, H. Ahmed, J.E.F. Frost, D.G. Hasko, D.C. Peacock, D.A. Ritchie,* and *G.A.C. Jones,* J.Phys. **C21**, L209 (1988)

[12] *H van Houten* and *C.W.J. Beenakker,* in: Analogies in Optics and Micro-electronics, ed. by *W. van Haeringen* and *D. Lenstra* (Kluwer, Deventer, to be published)

[13] *G. Timp, A.M. Chang, P. Mankiewich, R. Behringer, J.E. Cunningham, T.Y. Chang,* and *R.E. Howard,* Phys.Rev.Lett. **59**, 732 (1987); *M.L. Roukes, A. Scherer, S.J. Allen, Jr., H.G. Craighead, R.M. Ruthen, E.D. Beebe,* and *J.P. Harbison,* Phys.Rev.Lett. **59**, 3011 (1987); *C.J.B. Ford, T.J. Thornton, R. Newbury, M. Pepper, H. Ahmed, D.C. Peacock, D.A. Ritchie, J.E.F. Frost,* and *G.A.C. Jones,* Phys.Rev. **B38**, 8518 (1988); *G. Timp,* in: Mesoscopic Phenomena in Solids, ed. by *P.A. Lee, R.A. Webb,* and *B.L. Al'tshuler* (Elsevier, New York, to be published)

[14] *G. Timp, H.U. Baranger, P. deVegvar, J.E. Cunningham, R.E. Howard, R. Behringer,* and *P.M. Mankiewich,* Phys.Rev.Lett. **60**, 2081 (1988); *Y. Takagaki, K. Gamo, S. Namba, S. Ishida, S. Takaoka, K. Murase, K. Ishibashi,* and *Y. Aoyagi,* Solid State Comm. **68**, 1051 (1988)

[15] *R. Landauer,* IBM J.Res.Dev. **1**, 223 (1957); **32**, 306 (1988); Z.Phys. **B68**, 217 (1987)

[16] *M. Büttiker,* Phys.Rev.Lett. **57**, 1761 (1986); IBM J.Res.Dev. **32**, 317 (1988); See also: *A.D. Stone* and *A. Szafer,* IBM J.Res.Dev. **32**, 384 (1988)

[17] *P.A.M. Benistant, G.F.A. van de Walle, H. van Kempen,* and *P. Wyder,* Phys.Rev. **B33**, 690 (1986)

[18] *M. Büttiker,* Phys.Rev. **B38**, 12724 (1988)

[19] *R.E. Prange* and *T.-W. Nee,* Phys.Rev. **168**, 779 (1968)

[20] *M.S. Khaikin,* Adv.Phys. **18**, 1 (1969)

[21] *A.M. Kosevich* and *I.M. Lifshitz,* Zh.Eksp.Teor.Fiz. **29**, 743 (1955) [Sov.Phys.JETP **2**, 646 (1956)]

[22] *L.D. Landau* and *E.M. Lifshitz,* The Classical Theory of Fields (Pergamon, Oxford 1987) § 54

[23] *A.H. MacDonald* and *P. Streda,* Phys.Rev. **B29**, 1616 (1984)

[24] *B.J. van Wees, L.P. Kouwenhoven, H. van Houten, C.W.J. Beenakker, J.E. Mooij, C.T. Foxon,* and *J.J. Harris,* Phys.Rev. **B38**, 3625 (1988)

[25] *M. Büttiker,* Phys.Rev. **B38**, 9375 (1988)

[26] *S. Komiyama, H. Hirai, S. Sasa,* and *S. Hiyamizu* (preprint)

[27] *B.J. van Wees, E.M.M. Willems, L.P. Kouwenhoven, C.J.P.M. Harmans, J.G. Williamson, C.T. Foxon,* and *J.J Harris* Phys.Rev. B (April 15, 1989)

[28] *P.C. van Son, H. van Kempen,* and *P. Wyder,* J.Phys. **F17**, 1471 (1987)

[29] *J.G. Williamson et al.* (unpublished)

[30] *L.P. Kouwenhoven, B.J. van Wees, C.J.P.M. Harmans, J.G. Williamson, H. van Houten, C.W.J. Beenakker, C.T. Foxon,* and *J.J. Harris,* Phys.Rev. **B** (April 15, 1989)

[31] *A. Palevski, M. Heiblum, C.P. Umbach, C.M. Knoedler, A.N. Broers,* and *R.H. Koch* (preprint)

[32] *C.W.J. Beenakker* and *H. van Houten,* Phys.Rev. **B** (to be published)

[33] *D.A. Wharam, M. Pepper, H. Ahmed, J.E.F. Frost, D.G. Hasko, D.C. Peacock, D.A. Ritchie,* and *G.A.C. Jones,* J.Phys. **C21,** L887 (1988)

[34] *H.U. Baranger* and *A.D. Stone* (preprint); *A.M. Chang* and *T.Y. Chang* (preprint); *C.J.B. Ford, S. Washburn, M. Büttiker, C.M. Knoedler,* and *J.M. Hong* (preprint)

316

The Size-Induced Metal-Insulator Transition and Related Electron Interference Phenomena in Modern Microelectronics

Peter Marquardt and Guenther Nimtz

II. Physikalisches Institut der Universität zu Köln, D-5000 Köln 41, Federal Republic of Germany

Summary: The localization of conduction electrons in defect-induced or size-induced confinements is a universal effect of potential barriers on the wave nature of electrons. Both elastic and inelastic scattering disturb the coherence of electron waves. When the distance between potential barriers approaches the coherence length, classical conduction breaks down. Instead, an electron standing wave pattern is built up analogous to that of electromagnetic waves in a heterogeneous waveguide. The decay of the quasi-static Drude conductivity with crystal volume in well-separated mesoscopic conductors represents an experimental manifestation of electron interference effects. This size-induced metal-insulator transition (SIMIT) is in agreement with the quantum model Gor'kov and Eliashberg developed some 25 years ago. The SIMIT indicates the operational size limit for microelectronic devices based on particle-like transport. Future minuscule devices will exploit the electron wave nature. The variety of these novel designs comprises heterostructures and size-induced potential barriers. Here we introduce a new class of devices whose energy gap is simply taylored by structuring their cross section.

1 Electrons: Waves for Non-destructive Interference Studies

In photon experiments the number of quanta is not conserved, since photons are annihilated in the process of detection. Owing to their non-zero rest mass m and charge e, electrons can be probed without annihilation, e.g. by a conduction measurement. The de Broglie wavelength λ of electrons

$$\lambda = h/mv = \frac{h}{(2mE)^{1/2}} \tag{1}$$

(with $E = mv^2/2$ and $h = 2\pi\hbar$ Planck's constant) typically is 0.5 nm in metals and may be as large as 1 μm in semiconductors in consequence of electron velocity v and effective mass m. Single electron interference makes accessible a regime of wave phenomena where classical field theory fails [1]. Both being based on a probabilistic interpretation of spatial distribution of quanta [2], electron waves and electromagnetic waves have fundamental properties in common. With the advent of mesoscopic structures in solid state physics it is now possible to investigate *non-classical* interference with Fermions localized in sub-micrometer (also called mesoscopic) confinements. As all confinements are constructed of potential barriers, universal effects are observed independent of whether the confinement is defect-induced or size-induced. The wave model describes quantization, localization, tunneling, penetration into a classically forbidden region, and other related phenomena.

2 Coherence of Electron Wave Packets

Electrons are modeled by superpositions of monochromatic de Broglie waves with wavelengths continuously distributed over a small range on both sides of the mean wavelength λ given by Eq. 1. Left to themselves, electrons in free space tend to spread out becoming unlocalizable plane waves eventually. According to Bloch's theorem this happens as well in the presence of an ideal array of identical potentials. Any deviation from the ideal array, e.g. barriers of defect potentials, however, affects the phase relationship between the partial waves via elastic and inelastic scattering. While elastic scattering randomizes the angle distribution of the partial waves, inelastic processes alter the partial wave spectrum. The ubiquitous presence of potential barriers determines the phase coherence length L_c of the wave trains. The localization of conduction electrons by these interactions, including all measuring processes, is known as the collapse of the wave packet [2]. It has become customary to attribute phase breaking to inelastic scattering [3] and to estimate L_c from

$$L_c = (D\tau_i)^{1/2} \tag{2}$$

with $D = v^2\tau_e/d$ the diffusion coefficient in d dimensions and τ_i and τ_e the life times of energy and momentum eigenstates, respectively. As the role of elastic scattering is hidden in the diffusion coefficient, it is more intelligible to introduce the elastic and inelastic mean free paths $L_e = v\tau_e$ and $L_i = v\tau_i$ instead:

$$L_c = \left(\frac{1}{3}L_e L_i\right)^{1/2}, \tag{3}$$

or the time τ_c during which the wave is coherent:

$$\tau_c = \left(\frac{1}{3}\tau_e\tau_i\right)^{1/2}. \tag{4}$$

Typically, at 300 K, in a metal $\tau_e \approx 10^{-14}$ s and in a semiconductor $\tau_e \approx 10^{-12}$ s. In general, elastic scattering dominates by two orders of magnitude and thus $\tau_i \gg \tau_e$ as well as $L_i \gg L_e$ hold. Hence, L_c always exceeds the classical mean free path and classical concepts of electron propagation, including ballistic transport, have to be discarded on a mesoscopic scale. As a result of averaging over many scattering events, the conductivity of a macroscopic crystal still can be interpreted in terms of the phenomenological momentum relaxation time τ_e as a result of averaging over many scattering events. In a mesoscopic crystal, the averaging scheme breaks down. The definition of classical transport progressively loses its meaning as the crystal size is reduced. The condition $L_c \geq s$ defines the criterion for the passage from classical to quantum conductance. Apart from a geometry factor, the latter is quantized in units of e^2/h, a length-independent quantity characterizing the electron motion in the ground state of the most elementary electron resonator: the hydrogen atom.

The estimate Eq. 3 of L_c is a consequence of longitudinal interference. By virtue of their spin, electrons have rotational degrees of freedom. Hence the complete wave function including the spinor functions is not a scalar [2]. Electromagnetic waves cannot be focused to a spot smaller than half of their wavelength. Similarly, the propagation of sufficiently slow electrons is cut-off in a conductor like that of electromagnetic waves in an under-sized waveguide. This analogy elucidates recent observations of quantized transport in narrow conducting structures.

3 Interference of Electron Waves in Mesoscopic Conductors

Lately, electron interference effects were observed and investigated in quite different physical confinements. The earliest studies in this field are devoted to transport properties of disordered electronic systems [3]. Here, as mentioned above, the interference of electron waves is induced by defect potentials. The metal-insulator transitions of Anderson, Mott, and that according to the Joffe-Regel criterion belong to this class [4]. More recently, ordered electronic systems have been studied, e.g. heterostructures and doped superlattice structures [5]. Their basic dimensions being shorter than L_c, all these structures represent electron resonators. A heterostructure prepared and studied by Reed et al. [6] consists of an InGaAs electron cavity placed between two potential barriers of AlGaAs layers (see Fig. 1). The sophisticated device allows to investigate the electron transport by resonant tunneling.

In the present report, we focus on size-induced interference effects in homogeneous systems. With one spatial dimension, say s_x, becoming smaller than L_c and free electron wave vectors k_y, k_z in the other directions, we have the following 1d quantization ($N_x = N = 1, 2, 3, \ldots$) of a 2d electron gas:

$$E_N = \left[\left(\frac{N_x h}{2s_x} \right)^2 + (\hbar k_y)^2 + (\hbar k_z)^2 \right] \Big/ 2m. \tag{5}$$

For a 3d confinement, the quantum box, the eigenvalues of a zero-d electron gas are

$$E_N = h^2 \left[\left(\frac{N_x}{s_x} \right)^2 + \left(\frac{N_y}{s_y} \right)^2 + \left(\frac{N_z}{s_z} \right)^2 \right] \Big/ 8m. \tag{6}$$

The minimum ground state energy due to size-induced quantization is related to the maximum electron wavelength. In the case of a semiconductor, the size quantization results in a blue shift of the fundamental absorption

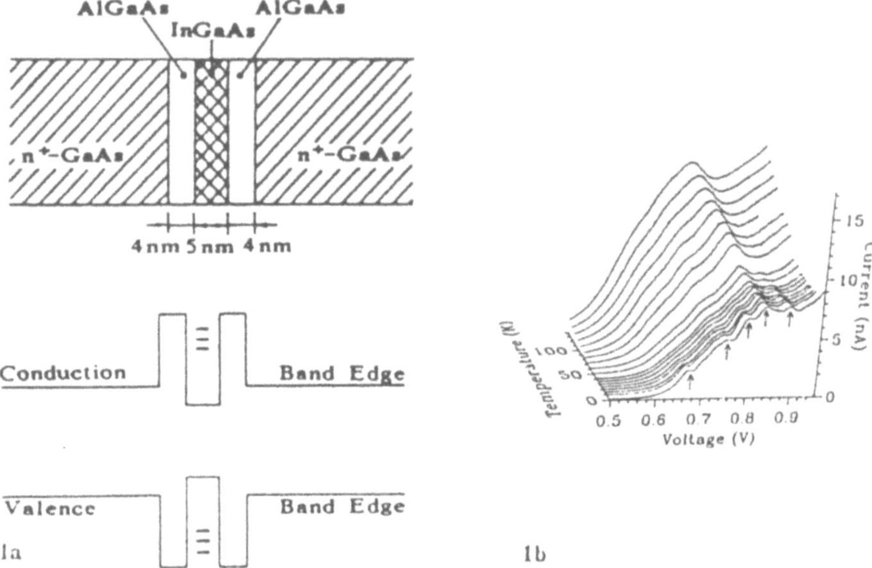

Fig. 1 (a) Sketch of a *quantum dot* and of its confining potential barriers realized by doped semiconductor heterostructures. (b) The current-voltage characteristic of this sophisticated device reflects the quantized level structure inside the quantum well [6].

edge as was shown by Ekimov et al. [7] for several mesoscopic semiconductors, e.g. CdS (see Fig. 2).

The effective mass approximation of the Fröhlich gap

$$\Delta E(s) = C/m^* s^2 \qquad (7)$$

with m^* the combined conduction and valence band mass, and C a constant, in fact fits the experimental data. The agreement between experiment and this calculation even holds for crystals consisting of as few as 125 atoms, i.e. with a nominal diameter of 2.4 nm.

The quantization effect can also be used to model the band structure quite similar to semiconductor heterostructures like that sketched in Fig. 1a. Being the simpler method, the size-tayloring has some important advantages over tayloring devices by heterostructures.

Owing to the transverse nature of spin-carrying waves, there is a transport cut-off effect which can be related to the Joffe-Regel criterion in disordered conductors. This criterion, translated into the wave picture, tells us that an electron with a wavelength twice the mean distance between neighbouring defects is localized. The transverse spread of an electron wave packet scales with its wavelength and the transport will be quenched finally when the scattering centers are separated less than $\lambda/2$ [8]. In this sense, the

320

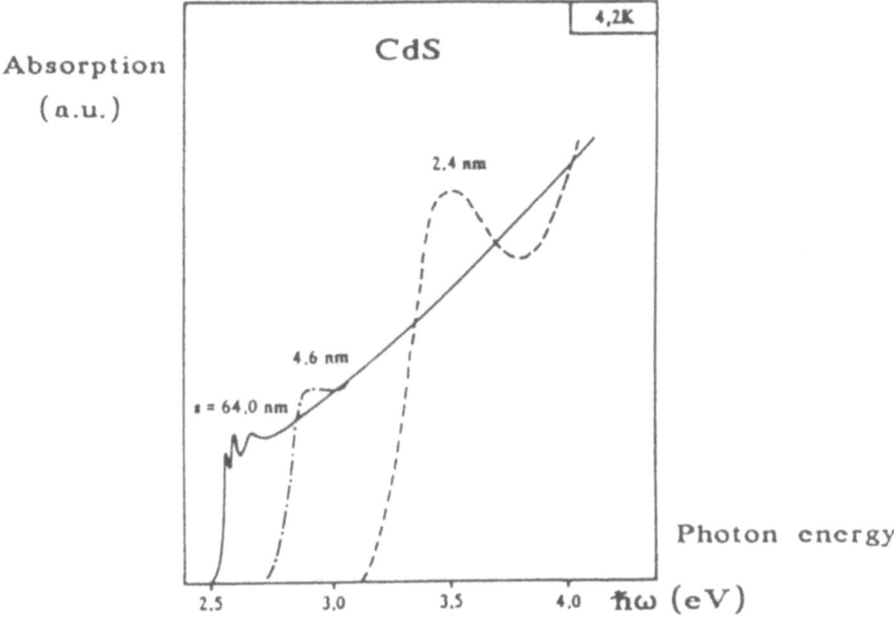

Fig. 2 The effective band gap of mesoscopic CdS crystals having size s is shifted proportional to s^{-2} in accordance with Eq. 7 for the size dependence of the Fröhlich gap [7].

scattering centers guide the electron. In a narrow conductor having width $w < L_e$, the wall potential guides the electron wave. By analogy to electromagnetic waves, electron propagation will be blocked in an undersized waveguide with cross sections less than $\lambda/2$. In the case of electromagnetic waves, this is known as the cut-off wavelength (COW) condition. Similarly, in a disordered conductor having a mean distance L_e between the scattering centers, electron transport is blocked if $\lambda > 2L_e$. The cut-off effect on electron transport by a suitably narrow geometry of an otherwise undisturbed conductor constitutes the principle of novel minuscule transistors and sensors (see Section 6).

4 The Size-Induced Metal-Insulator Transition in Mesoscopic Conductors

Mesoscopic conductors, well-separated from each other in an insulating matrix, constitute ideal model systems for size-confinement of electrons. According to the arguments given in Section 2, their conductivity is expected to decrease when $s \leq L_c$. This is equivalent to electron localization. At low temperatures, L_c may become as large as many μm in mesoscopic structures coupled to a macroscopic heat reservoir, e.g. contacts or substrates

321

[9]. What are the conditions to see effects of large values of L_c in isolated conductors at elevated temperatures? The surprising answer of contactless microwave measurements in the 10 GHz frequency range is that the conductivity is found to be size-affected up to the micrometer regime between 1 and 500 K. We concluded from this result that both elastic and inelastic scattering rates are markedly reduced due to the lack of long-wavelength phonons in isolated sub-micrometer crystals (Section 5). E.g. in a 1 μm crystal the spectrum of propagating phonons is already reduced by some 10% having in mind the condition $s \gg \lambda_{phonon}$ for phonon propagation. The gradual decay of the quasi-static conductivity σ in proportion to the crystal volumes was first observed with metal colloids which allowed a continuous in-situ variation of particle sizes by thermal coalescence [10]. The colloids were produced by direct evaporation of a low-melting metal (indium) into a stirred oil surface under high vacuum conditions. The filling factor f of the metal component obtained by this preparation technique is of the order 10^{-3} and can be enhanced up to 0.3 by subsequent centrifugation. The coalescence allows to rule out electron transfer between the metal crystals via hopping or tunneling: At constant f, the interparticle distances grow in proportion to the crystal sizes which exponentially reduce hopping and tunneling contributions to the effective conductivity of the heterogeneous medium.

In order to study the coalescence effect, the colloids were placed in the center of a waveguide of a microwave bridge. Phase shift and attenuation of the samples were recorded as a function of temperature between 100 K and 500 K. From these data, the effective dielectric function $\bar{\varepsilon}$ of the colloids was evaluated. The contribution $\varepsilon = \varepsilon_1 + i\varepsilon_2$ by the metal component was separated from $\bar{\varepsilon}$ by means of the Looyenga-Landau-Lifshitz equation [11] to obtain the Drude conductivity $\sigma_D = \varepsilon_0 \omega \varepsilon_2$. This conductivity corresponds to the static value if $\omega \tau_e \ll 1$ holds. The onset of irreversible particle coalescence while heating the samples in the waveguide is accompanied by a conspicuous increase of s. Interrupting the coalescence at different stages by heating and cooling cycles allows to cover a continuous range of particle sizes in a single experiment. Correspondingly, the measured conductivity σ of the metal particles traced these cycles as a function of particle size s (see Fig. 3 b). Since no other parameter besides s was varied in the experiment, σ depends on s only. Crystal sizes and structures were controlled by electron microscopy and X-ray diffraction. Further experiments on silver, platinum and semiconductor colloids having different (but fixed) sizes and filling factors add to the general result shown in Fig. 3 a: The conductivities $\sigma(s)$ normalized to the corresponding bulk values $\sigma(\infty)$ cover the shaded area whose width is due to natural size distributions in the colloids. The log-log presentation of σ vs s demonstrates that σ decays proportional to s^3, i.e. to the crystal volume.

The gradual size-induced metal-insulator transition is in agreement with the prediction of Gor'kov and Eliashberg (GE) [12] for the size range indicated by the solid line. The GE model is based on the discrete electron energy

322

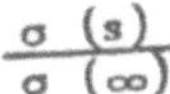

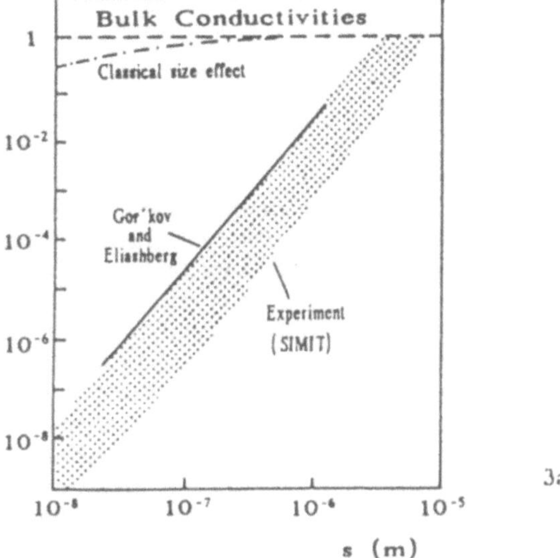

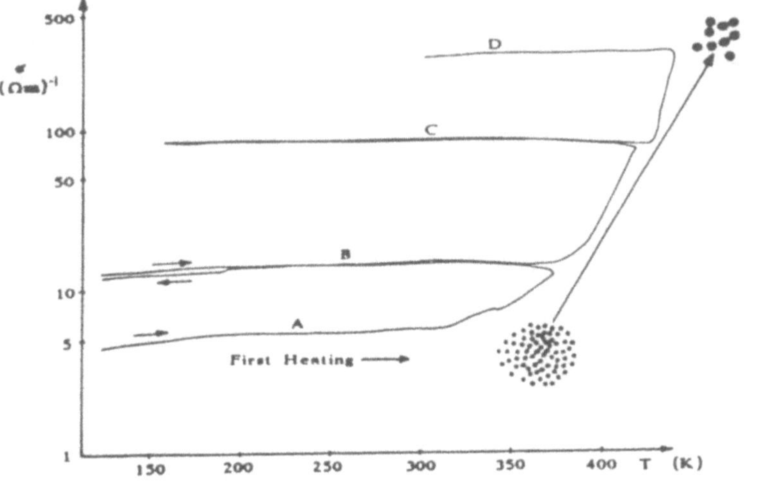

Fig. 3 (a) Conductivity $\sigma(s)$ normalized to the bulk values $\sigma(\infty)$ vs particle size s for various mesoscopic conductors investigated by a contact-free quasi-static microwave method (shaded area, see text). Note the log-log scaling. (b) Conductivity of the metal component vs temperature in an indium-oil colloid which allowed continuous variation of s by an in-situ heat treatment. Whereas σ increases in proportion to s^3, it does not significantly depend on temperature.

spectrum in a quantum box with the degeneracy of the levels (Eq. 6) lifted due to imperfections, i.e. a reduced symmetry, in the box. The mean spacing δ (Kubo gap) between subsequent levels is approximated by the inverse free electron density of states. As a rule, $\delta \approx E_F/N_0$ where E_F is the Fermi energy of the metal and N_0 the total number of conduction electrons per particle. In the limit $\hbar\omega \gg \delta$ and for dipolar response to an applied electric field of frequency ω, GE obtain the conductivity

$$\sigma_{\mathrm{GE}} \approx \frac{m^2 e^2 \omega^2 s^3}{h^3} \tag{8}$$

apart from a factor of order unity. Evaluated for microwaves (10 GHz), Eq. 8 describes the measured dependence of σ on particle volume within the experimental accuracy without fitting any parameters (solid line). In the Drude model, σ is size-reduced only when $s < L_e$, i.e. when boundary scattering prevails (broken line) which is in clear contradiction with experiment. The SIMIT stands out as a universal quantum size effect valid for all isolated mesoscopic conductors. Besides its significance for studies of electron interference in isolated confinements and for refining quantum models of conductivity, the SIMIT bears also promising prospects for a variety of technical applications due to the possibility of manipulating any desired conductivity between a metal and an insulator in the sub-micrometer range of conductors.

5 Applications and Consequences of the SIMIT

SIMIT materials, i.e. insulators doped with mesoscopic conductors, have potential applications as wide band microwave absorbers to be operated almost independent of temperature. With the proper choice of particle sizes, filling factors, and gradients of these parameters, artificial dielectrics can be taylored to realize e.g. reflexion free high frequency terminations, impedance transformers, and a better ac-matching of pyroelectric detectors. From the viewpoint of the electron wave nature, the SIMIT determines the ultimate size limit to devices based on Drude transport [9]. Semiconductor devices operated under thermal conditions at room temperature are expected to face standing electron waves, i.e. electron localization, already in confinements of some 100 nm. We emphasize that the data in Fig. 3 a are given for room temperature and that σ shows only weak temperature dependence even above melting of the indium particles, see Fig. 3 b. This seems to contradict reports about coherence lengths as large as 1 μm in mesoscopic systems only at very low temperatures (as mentioned in Section 4) [13]. However, it should be borne in mind that these systems, being coupled to macroscopic electron and phonon reservoirs, are not strictly mesoscopic. The isolated crystals investigated in the SIMIT experiments have discrete electron and phonon spectra whose long-wavelength limits experience an important blue shift. The truncation of the phonon spectrum reduces the electron-phonon scattering rate and thus gives rise to prolonged relaxation times τ_i and τ_e of the conduction electrons. One may be tempted to assume that the elastic boundary scattering rate $1/\tau_e$ is enhanced as the crystals become smaller and the surface scattering more dominant. The crystal rather behaves like an electron resonator with an increasing quality factor as its size is reduced. In an ideal resonator, the reflected wave amplitude equals the incident amplitude and the standing wave ratio is infinite. This corresponds to specular reflexion and complete absence of diffuse scattering. The phase coherent transport in an electron resonator is to be described in terms of quantum conductance instead of Drude conductivity. The SIMIT indicates the size regime where devices based on Drude transport fail. Future still smaller devices will exploit the quantum transport of electron waves. Currently, devices such as the quantum dot transistor based on *zero-dimensional conductance* are widely discussed.

6 Electron Interference Devices

The conductivity experiments on sub-micrometer crystals have revealed a size-induced metal-insulator transition. In this case of a 3d confinement, we obtain a zero-d conductor or quantum dot eventually. However, due to the transverse character of electron waves, the electron transport in mesoscopic conductors is already affected by 1d or 2d confinements. Recently, it was shown that quenching the cross section of the electron channel in a field-effect transistor, the conductance decreases due to electron interference in steps of $e^2/\pi\hbar$ [14] as seen in Fig. 4.

The minimum conductance of a 2d or 3d electron gas is $e^2/\pi\hbar$ or $e^2/3\hbar$, respectively [8]. The accuracy of the *point contact* experiment (Fig. 4) does not allow to distinguish between the factors $1/\pi$ or $1/3$, i.e. to tell the effective dimensionality of the confinement in said experiment. The steps can be related to multiples of $\lambda/2$ and the conduction finally breaks down when the channel diameter becomes smaller than $\lambda/2$. The cut-off was observed at $w \approx 22$ nm and the wavelength was ≈ 42 nm [14]. We attribute this behaviour to the electron cut-off wavelength of the conductor (COW) [8]. This COW principle can be used to construct simple interference devices as transistors, diodes, and photodetectors. Fig. 5 shows a sketch of such a COW transistor together with the size-induced band structure. The conductor is shaped like a bottleneck with the narrowest dimension w being smaller than $\lambda/2$, allowing only electrons with $\lambda < 2w$ to pass. (In the case of non-degeneracy we propose to make $w < \lambda/4$ taking into consideration Maxwell's distribution.) For electrons with $\lambda > 2w$, the conductance is zero while hot carriers can move through the channel. Carriers may be heated either by temperature T,

$$\lambda(T) = h/\left(3mk_{\mathrm{B}}T\right)^{1/2} \tag{9}$$

or by an electric field F,

$$\lambda(F) = h/eF\left(2\tau_e\tau_i\right)^{1/2} \tag{10}$$

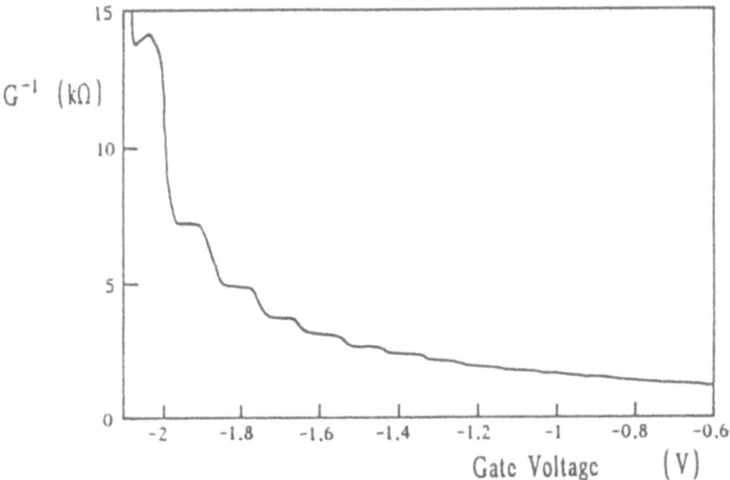

Fig. 4 Reciprocal channel conductance vs gate voltage of a split-gate field effect transistor in which a specially prepared gate determines the channel confinement [14]. The observed quantization can be explained in terms of transverse coherence of electron waves [8]. The transport is cut-off when the channel diameter becomes smaller than $\lambda/2$ (left-hand side of diagram, the gate voltage is proportional to the channel diameter).

or by absorbing electromagnetic energy $h\nu$

$$\lambda(\nu) = (h/2m\nu)^{1/2} \, . \tag{11}$$

The voltage applied in order to heat the carriers (U_2 in Fig. 5) may be directed perpendicular to the direction of propagation and to the drift voltage U_1. At room temperature, a typical electron wavelength e.g. in GaAs is ≈ 25 nm. In order to cut-off the transport, the bottleneck has a width $\omega \leq 10$ nm. As illustrated in Fig. 5, the size-induced wavelength cut-off can also be presented in terms of the electron potential which peaks at the narrow part of the structure. The narrow path, of course, has to be longer than its width in order to suppress significant tunneling since the tunneling length L_t is given by

$$L_t = h/(8m\Delta E)^{1/2} \, . \tag{12}$$

This is the same relation as that of the size-induced blue shift (Eq. 8). In this way, the size-induced potential of the device and the penetration depth of electron waves mutually condition each other. The fascinating aspect of the quantum size effect is that the band edge can be modelled by geometry instead of doping or forming heterostructures. At present, it is not yet easy to produce bottlenecks or other structures of the order of 10 nm. However, having in mind the progress in miniaturization during the last decades, the COW effect will be commercialized in the near future. To test such devices now, one may scale-up these structures and study their properties at correspondingly lower temperatures.

326

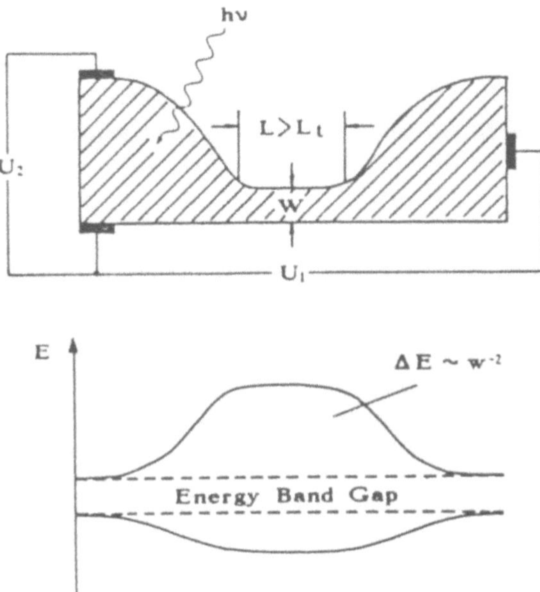

Fig. 5 Sketch of a cut-off wavelength (COW) transistor which is operated on the basis of variable electron wavelengths. The width w of the bottleneck is chosen to be smaller than $\lambda/2$ of the electrons in thermal equilibrium and smaller than the tunneling length L_t. Heating the electrons either by electromagnetic ($h\nu$), thermal ($k_B T$) or electric (eU_1) means, the transistor can be switched from the non-conducting to the conducting state. The band gap structure of the device shown below is size-taylored.

7 Summary

The progress of modern miniaturization techniques has brought forward mesoscopic structures in solid state physics which necessitate to take into account the electron wave nature in conduction phenomena. This reveals the ultimate size limit of devices devoted to classical charge transport [9]. The size-induced metal-insulator transition observed in well-separated meso-scopic conductors by contactfree quasi-static microwave and high frequency ($10^6 \ldots 10^9$ Hz, [15]) measurements is a particularly pronounced manifes-tation of interference phenomena. When the coherence length of electron wave packets becomes greater than the confinement size, classical conduc-tion passes over to quantum conduction. Having in mind the transverse spread of electron wave packets, we propose to make use of the cut-off effect by tayloring suitably shaped structures which can be operated as transis-tors, photodetectors, and thermosensors.

327

Acknowledgement

We gratefully acknowledge financial support by the Deutsche Forschungsgemeinschaft (SFB 341), the technical assistance of Daniela Pfannkuch, Susanne Mäcker, and Reiner Zorn, as well as discussions with Achim Enders, Dirk Kreimer, and Professor Peter Mittelstaedt.

References

[1] R.E. Peierls, Surprises in Theoretical Physics (Priceton University Press, Princeton, N.J. 1979)

[2] D. Bohm, Quantum Theory (Prentice Hall, Englewood Cliffs, N.J. 1951)

[3] B.L. Altshuler and P.A. Lee, Physics Today, Dec. 1988, p. 36

[4] D.J. Thouless, Physics Reports 13, 93 (1974)

[5] M.J. Kelly and R.J. Nicholas, Rep. Progr. Phys. 48, 1699 (1985)

[6] M.A. Reed, J.N. Randall, R.J. Aggarwal, R.J. Matyi, T. M. Moore, and A.E. Wetsel, Phys. Rev. Lett. 60, 535 (1988)

[7] A.J. Ekimov and A.A. Onushchenko, Sov. Phys. Semicond. 16, 775 (1982)

[8] G. Nimtz and P. Marquardt, Appl. Phys. A 47, 317 (1988)

[9] P. Marquardt and G. Nimtz, Semicond. Sci. Technol. 2, 833 (1987)

[10] G. Nimtz, P. Marquardt, and H. Gleiter, J. Crystal Growth 86, 66 (1988)

[11] P. Marquardt, Physics Lett. 123 A, 365 (1987)

[12] L.P. Gor'kov and G.M. Eliashberg, Sov. Phys. JETP 21, 940 (1965)

[13] A. Benoit, C.P. Umbach, R.B. Laibowitz, and R.A. Webb, Phys. Rev. Lett. 58, 2343 (1978)

[14] B.J. van Wees, H. van Houten, C.W.J. Beenakker, J.G. Williamson, L.P. Kouwenhoven, D. van der Marel, and C.T. Foxson, Phys. Rev. Lett. 60, 848 (1988)

[15] B. Binggeli, P. Marquardt, G. Nimtz, and R. Pelster, Proc. 19th Int. Conf. Physics of Semiconductors (Warsaw 1988)

Contents of volumes I...29

Index of Authors

332